aterials 1

# Engineering Materials 1

## An Introduction to Properties, Applications, and Design

Fourth Edition

**Michael F. Ashby**
Royal Society Research Professor Emeritus,
University of Cambridge and Former Visiting
Professor of Design at the Royal College
of Art, London

**David R. H. Jones**
President, Christ's College
Cambridge

AMSTERDAM • BOSTON • HEIDELBERG • LONDON
NEW YORK • OXFORD • PARIS • SAN DIEGO
SAN FRANCISCO • SINGAPORE • SYDNEY • TOKYO
Butterworth-Heinemann is an imprint of Elsevier

Butterworth-Heinemann is an imprint of Elsevier
The Boulevard, Langford Lane, Kidlington, Oxford, OX5 1GB UK
225 Wyman Street, Waltham, MA 02451 USA

First published 1980
Second edition 1996
Reprinted 1998 (twice), 2000, 2001, 2002, 2003
Third edition 2005
Reprinted 2006 (twice), 2007, 2008, 2009

**Notices**

Knowledge and best practice in this field are constantly changing. As new research and experience broaden our understanding, changes in research methods, professional practices, or medical treatment may become necessary.

Practitioners and researchers must always rely on their own experience and knowledge in evaluating and using any information, methods, compounds, or experiments described herein. In using such information or methods they should be mindful of their own safety and the safety of others, including parties for whom they have a professional responsibility.

To the fullest extent of the law, neither the Publisher nor the authors, contributors, or editors, assume any liability for any injury and/or damage to persons or property as a matter of products liability, negligence or otherwise, or from any use or operation of any methods, products, instructions, or ideas contained in the material herein.

**Library of Congress Cataloging-in-Publication Data**
Application submitted.

**British Library Cataloguing-in-Publication Data**
A catalogue record for this book is available from the British Library.

ISBN: 978-0-08-096665-6

For information on all Butterworth-Heinemann publications,
visit our website at *www.books.elsevier.com*

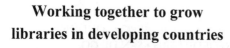

**Working together to grow
libraries in developing countries**

www.elsevier.com | www.bookaid.org | www.sabre.org

ELSEVIER          Sabre Foundation

# Contents

# Part C   Yield Strength, Tensile Strength, and Ductility

## Part H   Friction, Abrasion, and Wear

# Preface to the Fourth Edition

In preparing this fourth edition of *Engineering Materials 1*, I have taken the opportunity to make significant changes, while being careful not to alter the essential character of the book. At the most obvious level, I have added many new photographs to illustrate both the basic coursework and also the case studies—many of these have been taken during my travels around the world investigating materials engineering problems. These days, the Internet is the essential tool of knowledge and communication—to the extent that textbooks should be used alongside web-based information sources.

So, in this new edition, I have given frequent references in the text to reliable web pages and video clips—ranging from the Presidential Commission report on the space shuttle *Challenger* disaster, to locomotive wheels losing friction on Indian Railways. And whenever a geographical location is involved, such as the Sydney Harbour Bridge, I have given the coordinates (latitude and longitude), which can be plugged into the search window in Google Earth to take you right there. Not only does this give you a feel for the truly global reach of materials and engineering, it also leads you straight to the large number of derivative sources and references, such as photographs and web pages, that can help you follow up your own particular interests.

I have added Worked Examples to many of the chapters to develop or illustrate a point without interrupting the flow of the chapter. These can be what one might call "convergent"—like putting numbers into a specific data set of fracture tests to calculate the Weibull modulus (you need to be able to do this, but it is best done offline)—or "divergent," such as recognizing the fatigue design details in the traffic lights in Manhattan and thus challenging you to look around the real world and think like an engineer.

I have made some significant changes to the way in which some of the subject material is presented. So, in the chapters on fatigue, I have largely replaced the traditional stress-based analysis with the total strain approach to fatigue life. In the creep chapters, the use of creep maps is expanded to show strain-rate contours and the effect of microstructure on creep régimes. In the corrosion chapters, Pourbaix diagrams are used for the first time in order to show the regions of immunity, corrosion, and passivation, and how these depend on electrochemical potential and pH.

In addition, I have strengthened the links between the materials aspects of the subject and the "user" fields of mechanics and structures. Thus, at the ends of the relevant chapters, I have put short compendia of useful results: elastic bending, vibration, and buckling of beams after Chapter 3; plastic bending and torsion after Chapter 11; stress intensity factors for common crack geometries after Chapter 13; and data for calculating corrosion loss after Chapter 26. A simple introductory note on tensor notation for depicting stress and strain in three dimensions has also been added to Chapter 3.

Many new case studies have been added, and many existing case studies have either been replaced or revised and updated. The number of examples has been significantly expanded, and of these a large proportion contain case studies or practical examples relevant to materials design and avoidance of failure. In general, I have tried to choose topics for the case studies that are interesting, informative, and connected to today's world. So, the new case study on the *Challenger* space shuttle disaster—which derives from the earlier elastic theory (Hooke's law applied to pressurized tubes and chain sliding in rubber)—is timeless in its portrayal of how difficult it is in large corporate organizations for engineers to get their opinions listened to and acted on by senior management. The *Columbia* disaster 17 years later, involving the same organization and yet another materials problem, shows that materials engineering is about far more than just materials engineering.

Materials occupy a central place in all of engineering for without them, nothing can be made, nothing can be done. The challenge always is to integrate an intimate knowledge of the characteristics of materials with their applications in real structures, components, or devices. Then, it helps to be able to understand other areas of engineering, such as structures and mechanics, so that genuine collaborations can be built that will lead to optimum design and minimum risk. The modern airplane engine is one of the best examples, and the joints in the space shuttle booster one of the worst. In-between, there is a whole world of design, ranging from the excellent to the terrible (or not designed at all). To the materials engineer who is always curious, aware and vigilant, the world is a fascinating place.

## Acknowledgments
The authors and publishers are grateful to a number of copyright holders for permission to reproduce their photographs. Appropriate acknowledgments are made in the individual figure captions. Unless otherwise attributed, all photographs were taken by Dr. Jones.

**David Jones**

# General Introduction

## To the Student

Innovation in engineering often means the clever use of a new material—new to a particular application, but not necessarily (although sometimes) new in the sense of recently developed. Plastic paper clips and ceramic turbine blades both represent attempts to do better with polymers and ceramics what had previously been done well with metals. And engineering disasters are frequently caused by the misuse of materials. When the plastic bristles on your sweeping brush slide over the fallen leaves on your backyard, or when a fleet of aircraft is grounded because cracks have appeared in the fuselage skin, it is because the engineer who designed them used the wrong materials or did not understand the properties of those used. So, it is vital that the professional engineer should know how to select materials that best fit the demands of the design—economic and aesthetic demands, as well as demands of strength and durability. The designer must understand the properties of materials, and their limitations.

This book gives a broad introduction to these properties and limitations. It cannot make you a materials expert, but it can teach you how to make a sensible choice of material, how to avoid the mistakes that have led to difficulty or tragedy in the past, and where to turn for further, more detailed, help.

You will notice from the Contents that the chapters are arranged in *groups*, each group describing a particular class of properties: elastic modulus; fracture toughness; resistance to corrosion; and so forth. Each group of chapters starts by *defining the property*, describing how it is *measured*, and giving *data* that we use to solve problems involving design with materials. We then move on to the *basic science* that underlies each property and show how we can use this fundamental knowledge to choose materials with better properties. Each group ends with a chapter of *case studies* in which the basic understanding and the data for each property are applied to practical engineering problems involving materials.

At the end of each chapter, you will find a set of examples; each example is meant to consolidate or develop a particular point covered in the text. Try to do the examples from a particular chapter while this is still fresh in your mind. In this way, you will gain confidence that you are on top of the subject.

No engineer attempts to learn or remember tables or lists of data for material properties. But you *should* try to remember the broad orders of magnitude of these quantities. All food stores know that "a kg of apples is about 10 apples"—salesclerks still weigh them, but their knowledge prevents someone from making silly mistakes that might cost the stores money.

In the same way an engineer should know that "most elastic moduli lie between 1 and $10^3$ GN m$^{-2}$ and are around $10^2$ GN m$^{-2}$ for metals"—in any real design you need an accurate value, which you can get from suppliers' specifications; but an order of magnitude knowledge prevents you from getting the units wrong, or making other silly, possibly expensive, mistakes. To help you in this, we have added at the end of the book a list of the important definitions and formulae that you should know, or should be able to derive, and a summary of the orders of magnitude of materials properties.

## To the Lecturer

This book is a course in Engineering Materials for engineering students with no previous background in the subject. It is designed to link up with the teaching of Design, Mechanics, and Structures, and to meet the needs of engineering students for a first materials course, emphasizing design applications.

The text is deliberately concise. Each chapter is designed to cover the content of one 50-minute lecture, 30 in all, and allows time for demonstrations and graphics. The text contains sets of worked case studies that apply the material of the preceding block of lectures. There are examples for the student at the end of the chapters.

We have made every effort to keep the mathematical analysis as simple as possible while still retaining the essential physical understanding and arriving at results, which, although approximate, are useful. But we have avoided mere description: most of the case studies and examples involve analysis, and the use of data, to arrive at solutions to real or postulated problems. This level of analysis, and these data, are of the type that would be used in a preliminary study for the selection of a material or the analysis of a design (or design failure).

It is worth emphasizing to students that the next step would be a detailed analysis, using *more precise mechanics* and *data from the supplier of the material or from in-house testing*. Materials data are notoriously variable. Approximate tabulations like those that are given here, though useful, should never be used for final designs.

## Accompanying Resources

The following web-based resources are available to teachers and lecturers who adopt or recommend this text for class use. For further details and access to these resources, please go to *http://www.textbooks.elsevier.com*

### Instructor's Manual

A full Solutions Manual with worked answers to the exercises in the main text is available for downloading.

### Image Bank

An image bank of downloadable figures from the book is available for use in lecture slides and class presentations.

### Online Materials Science Tutorials

A series of online materials science tutorials accompanies *Engineering Materials 1* and *2*. These were developed by Alan Crosky, Mark Hoffman, Paul Munroe, and Belinda Allen at the University of New South Wales (UNSW) in Australia; they are based on earlier editions of the books. The group is particularly interested in the effective and innovative use of technology in teaching. They realized the potential of the material for the teaching of Materials Engineering to their students in an online environment and have developed and then used these very popular tutorials for a number of years at UNSW. The results of this work have also been published and presented extensively.

The tutorials are designed for students of materials science as well as for those studying materials as a related or elective subject—for example, mechanical and/or civil engineering students. They are ideal for use as ancillaries to formal teaching programs and also may be used as the basis for quick refresher courses for more advanced materials science students. In addition, by picking selectively from the range of tutorials available, they will make ideal subject primers for students from related faculties.

The software has been developed as a self-paced learning tool, separated into learning modules based around key materials science concepts.

### About the authors of the tutorials

*Alan Crosky* is a Professor in the School of Materials Science and Engineering, University of New South Wales. His teaching specialties include metallurgy, composites, and fractography.

*Belinda Allen* is an educational designer and adjunct lecturer in the Curriculum Research, Evaluation and Development team in the Learning and Teaching Unit, UNSW. She contributes to strategic initiatives and professional development programs for curriculum renewal, with a focus on effective integration of learning technologies.

*Mark Hoffman* is a Professor in the School of Materials Science and Engineering, UNSW. His teaching specialties include fracture, numerical modeling, mechanical behavior of materials, and engineering management.

*Paul Munroe* has a joint appointment as Professor in the School of Materials Science and Engineering and Director of the Electron Microscope Unit, UNSW. His teaching specialties are the deformation and strengthening mechanisms of materials and crystallographic and microstructural characterization.

# Engineering Materials 1

## CHAPTER 1

# Engineering Materials and Their Properties

## CONTENTS

## 1.1 INTRODUCTION

There are maybe more than 50,000 materials available to the engineer. In designing a structure or device, how is the engineer to choose from this vast menu the material that best suits the purpose? Mistakes can cause disasters. During the Second World War, one class of welded merchant ship suffered heavy losses, not by enemy attack, but by breaking in half at sea: the *fracture toughness* of the steel—and, particularly, of the welds—was too low.

More recently, three Comet aircraft were lost before it was realized that the design called for a *fatigue strength* that—given the design of the window frames— was greater than that possessed by the material. You yourself will be familiar with poorly designed appliances made of plastic: their excessive "give" is because the designer did not allow for the low *modulus* of the polymer. These bulk properties are listed in Table 1.1, along with other common classes of property that the designer must consider when choosing a material. Many of these properties will be unfamiliar to you—we will introduce them through examples in this chapter. They form the basis of this course on materials.

In this course, we also encounter the *classes of materials* shown in Table 1.2 and Figure 1.1. More engineering components are made of *metals and alloys* than of any other class of solid. But increasingly, *polymers* are replacing metals because they offer a combination of properties that are more attractive to the designer.

Engineering Materials I: An Introduction to Properties, Applications, and Design, Fourth Edition

Table 1.1 Classes of Property

| Class | Property |
|-------|----------|
| Economic and environmental | Price and availability |
| | Recyclability |
| | Sustainability |
| | Carbon footprint |
| General physical | Density |
| Mechanical | Modulus |
| | Yield and tensile strength |
| | Hardness |
| | Fracture toughness |
| | Fatigue strength |
| | Creep strength |
| | Damping |
| Thermal | Thermal conductivity |
| | Specific heat |
| | Thermal expansion coefficient |
| Electrical and magnetic | Resistivity |
| | Dielectric constant |
| | Magnetic permeability |
| Environmental interaction | Oxidation |
| | Corrosion |
| | Wear |
| Production | Ease of manufacture |
| | Joining |
| | Finishing |
| Aesthetic | Color |
| | Texture |
| | Feel |

And if you've been reading the newspaper, you will know that the new *ceramics*, at present under development worldwide, are an emerging class of engineering material that may permit more efficient heat engines, sharper knives and bearings with lower friction. The engineer can combine the best properties of these materials to make *composites* (the most familiar is fiberglass) which offer especially attractive packages of properties. And—finally—one should not ignore *natural materials*, such as wood and leather, which have properties that are— even with the innovations of today's materials scientists—difficult to beat.

In this chapter we illustrate, using a variety of examples, how the designer selects materials to provide the properties needed.

## Table 1.2 Classes of Materials

| Class | Material |
|---|---|
| Metals and alloys | Iron and steels |
| | Aluminum and alloys |
| | Copper and alloys |
| | Nickel and alloys |
| | Titanium and alloys |
| Polymers | Polyethylene (PE) |
| | Polymethylmethacrylate (acrylic and PMMA) |
| | Nylon or polyamide (PA) |
| | Polystyrene (PS) |
| | Polyurethane (PU) |
| | Polyvinylchloride (PVC) |
| | Polyethylene terephthalate (PET) |
| | Polyethylether ketone (PEEK) |
| | Epoxies (EP) |
| | Elastomers, such as natural rubber (NR) |
| Ceramics and glasses* | Alumina ($Al_2O_3$, emery, sapphire) |
| | Magnesia (MgO) |
| | Silica ($SiO_2$) glasses and silicates |
| | Silicon carbide (SiC) |
| | Silicon nitride ($Si_3N_4$) |
| | Cement and concrete |
| Composites | Fiberglass (GFRP) |
| | Carbon-fiber reinforced polymers (CFRP) |
| | Filled polymers |
| | Cermets |
| Natural materials | Wood |
| | Leather |
| | Cotton/wool/silk |
| | Bone |
| | Rock/stone/chalk |
| | Flint/sand/aggregate |

*Ceramics are crystalline, inorganic, nonmetals. Glasses are noncrystalline (or amorphous) solids. Most engineering glasses are nonmetals, but a range of metallic glasses with useful properties is now available.

## 1.2 EXAMPLES OF MATERIALS SELECTION

A typical screwdriver (Figure 1.2) has a shaft and blade made of carbon steel, a metal. Steel is chosen because its *modulus* is high. The modulus measures the resistance of the material to elastic deflection. If you made the shaft out of a polymer like polyethylene instead, it would twist far too much. A high modulus

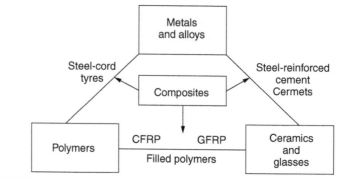

**FIGURE 1.1**

The classes of engineering materials from which articles are made.

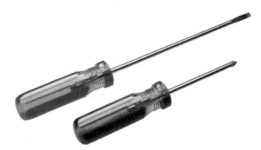

**FIGURE 1.2**

Typical screwdrivers, with steel shaft and polymer (plastic) handle. *(Courtesy of Elsevier.)*

is one criterion but not the only one. The shaft must have a high *yield strength*. If it does not, it will bend or twist permanently if you turn it hard (bad screwdrivers do). And the blade must have a high *hardness*, otherwise it will be burred-over by the head of the screw.

Finally, the material of the shaft and blade must not only do all these things, it must also resist fracture—glass, for instance, has a high modulus, yield strength, and hardness, but it would not be a good choice for this application because it is so brittle—it has a very low *fracture toughness*. That of steel is high, meaning that it gives before it breaks.

The handle of the screwdriver is made of a polymer or plastic, in this instance polymethylmethacrylate, otherwise known as PMMA, plexiglass or perspex. The handle has a much larger section than the shaft, so its twisting, and thus its modulus, is less important. You could not make it satisfactorily out of a soft rubber (another polymer) because its modulus is much too low, although a thin skin of rubber might be useful because its *friction coefficient* is high, making it easy to grip. Traditionally, of course, tool handles were made of a natural composite—wood—and, if you measure importance by the volume consumed per year, wood is still by far the most important composite available to the engineer.

Wood has been replaced by PMMA because PMMA becomes soft when hot and can be molded quickly and easily to its final shape. Its *ease of fabrication* for this application is high. It is also chosen for aesthetic reasons: its *appearance*, and feel or *texture*, are right; and its *density* is low, so that the screwdriver is not unnecessarily heavy. Finally, PMMA is cheap, and this allows the product to be made at a reasonable *price*.

A second example (Figure 1.3) takes us from low technology to the advanced materials design involved in the turbofan aeroengines that power most planes. Air is propelled past the engine by the turbofan, providing aerodynamic thrust. The air is further compressed by the compressor blades, and is then mixed with fuel and burnt in the combustion chamber. The expanding gases drive the turbine blades, which provide power to the turbofan and the compressor blades, and finally pass out of the rear of the engine, adding to the thrust.

The *turbofan blades* are made from a titanium alloy, a metal. This has a sufficiently good modulus, yield strength and fracture toughness. But the metal must also resist *fatigue* (due to rapidly fluctuating loads), *surface wear* (from striking everything from water droplets to large birds) and *corrosion* (important when taking off over the sea because salt spray enters the engine). Finally, *density* is extremely important for obvious reasons: the heavier the engine, the less the payload the plane can carry. In an effort to reduce weight even further, composite blades made of carbon-fiber reinforced polymers (CFRP) with density less than one-half of that of titanium, have been tried. But CFRP, by itself, is

**FIGURE 1.3**
Cross-section through a typical turbofan aero-engine. *(Courtesy of Rolls-Royce plc.)*

not tough enough for turbofan blades. Some tests have shown that they can be shattered by "bird strikes."

Turning to the *turbine blades* (those in the hottest part of the engine) even more material requirements must be satisfied. For economy the fuel must be burnt at the highest possible temperature. The first row of engine blades (the "HP1" blades) runs at metal temperatures of about 1000°C, requiring resistance to *creep* and *oxidation*. Nickel-based alloys of complicated chemistry and structure are used for this exceedingly stringent application; they are a pinnacle of advanced materials technology.

An example that brings in somewhat different requirements is the *spark plug* of an internal combustion engine (Figure 1.4). The *spark electrodes* must resist *thermal fatigue* (from rapidly fluctuating temperatures), *wear* (caused by spark erosion) and *oxidation and corrosion* from hot upper-cylinder gases containing nasty compounds of sulphur. Tungsten alloys are used for the electrodes because they have the desired properties.

The *insulation* around the central electrode is an example of a nonmetallic material—in this case, alumina, a ceramic. This is chosen because of its electrical insulating properties and because it also has good thermal fatigue resistance and resistance to corrosion and oxidation (it is an oxide already).

The use of nonmetallic materials has grown most rapidly in the consumer industry. Our next example, a sailing cruiser (Figure 1.5), shows just how extensively polymers and synthetic composites and fibers have replaced the traditional materials of steel, wood and cotton. A typical cruiser has a *hull* made from GFRP, manufactured as a single molding; GFRP has good *appearance* and, unlike steel or wood, does not rust or become eaten away by marine worm. The *mast* is made from aluminum alloy, which is lighter for a given strength than wood; advanced masts are now made from CFRP. The sails, formerly of the natural material cotton, are now made from the polymers nylon, Terylene or Kevlar, and, in the running rigging, cotton ropes have been replaced by polymers also. Finally, polymers like PVC are extensively used for things like fenders, buoyancy bags and boat covers.

**FIGURE 1.4**
A petrol engine spark plug, with tungsten electrodes and ceramic body. *(Courtesy of Elsevier.)*

**FIGURE 1.5**
A sailing cruiser, with composite (GFRP) hull, aluminum alloy mast and sails made from synthetic polymer fibers. *(Courtesy of Catalina Yachts, Inc.)*

Two synthetic composite materials have appeared in the items we have considered so far: GFRP and the much more expensive CFRP. The range of composites is a large and growing one (refer to Figure 1.1); during the next decade composites will compete even more with steel and aluminum in many traditional uses of these metals.

So far we have introduced the mechanical and physical properties of engineering materials, but we have yet to discuss two considerations that are often of overriding importance: *price and availability.*

Table 1.3 shows a rough breakdown of material prices. Materials for large-scale structural use—wood, concrete and structural steel—cost between US$200 and

**Table 1.3** Breakdown of Material Prices

| Class of Use | Material | Price per ton |
|---|---|---|
| Basic construction | Wood, concrete, structural steel | US$200–$500 |
| Medium and light engineering | Metals, alloys and polymers for aircraft, automobiles, appliances, etc. | US$500–$30,000 |
| Special materials | Turbine-blade alloys, advanced composites (CFRP, BFRP), etc. | US$30,000–$100,000 |
| Precious metals, etc. | Sapphire bearings, silver contacts, gold microcircuits, industrial diamond cutting and polishing tools | US$100,000–$60m |

$500 per ton. Many materials have all the other properties required of a structural material—but their use in this application is eliminated by their price.

The value that is added during light and medium-engineering work is larger, and this usually means that the economic constraint on the choice of materials is less severe—a far greater proportion of the cost of the structure is that associated with labor or with production and fabrication. Stainless steels, most aluminum alloys and most polymers cost between US$500 and $30,000 per ton. It is in this sector of the market that the competition between materials is most intense, and the greatest scope for imaginative design exists. Here polymers and composites compete directly with metals, and new structural ceramics (e.g., silicon carbide and silicon nitride) may compete with both in certain applications.

Next there are the materials developed for high-performance applications, some of which we have mentioned already: nickel alloys (for turbine blades), tungsten (for spark-plug electrodes), and special composite materials such as CFRP. The price of these materials ranges between US$30,000 and $100,000 per ton. This the régime of high materials technology, actively under research, in which major new advances are continuing to be made. Here, too, there is intense competition from new materials.

Finally, there are the so-called precious metals and gemstones, widely used in engineering: gold for microcircuits, platinum for catalysts, sapphire for bearings, diamond for cutting tools. They range in price from US$100,000 to more than US$60m per ton.

As an example of how price and availability affect the choice of material for a particular job, consider how the materials used for building bridges in Cambridge, England have changed over the centuries. As the photograph of Queens' Bridge (Figure 1.6) suggests, until 150 years or so ago wood was commonly used for bridge building. It was cheap, and high-quality timber was still available in large sections from natural forests. Stone, too, as the picture of Clare Bridge (Figure 1.7) shows, was widely used. During the eighteenth

**FIGURE 1.6**
The wooden bridge at Queens' College, Cambridge, a 1902 reconstruction of the original bridge built in 1749 to William Etheridge's design. – 52 12 07.86 N 0 06 54.12 E

**FIGURE 1.7**
Clare Bridge, built in 1640, is Cambridge's oldest surviving bridge; it is reputed to have been an escape route from the college in times of plague. – 52 12 17.98 N 0 06 50.40 E

**FIGURE 1.8**

Magdalene Bridge, built in 1823 on the site of the ancient Saxon bridge over the Cam. The present cast-iron arches carried, until recently, loads far in excess of those envisaged by the designers. Fortunately, the bridge has now undergone restoration and strengthening. – 52 12 35.46 N 0 06 59.43 E

century, the ready availability of cast iron, with its relatively low assembly costs, led to many cast-iron bridges of the type exemplified by Magdalene Bridge (Figure 1.8).

Metallurgical developments of the late nineteenth century allowed large mild-steel structures to be built (the Fort St George footbridge, Figure 1.9). Finally, the advent of reinforced concrete led to graceful and durable structures like that of the Garret Hostel Lane bridge (Figure 1.10). This evolution clearly illustrates how availability influences the choice of materials.

Nowadays, stone, steel, and reinforced concrete are often used interchangeably in structures, reflecting the relatively small *price* differences between them. The choice of which of the three materials to use is mainly dictated by the kind of structure the architect wishes to build: chunky and solid (stone), structurally efficient (steel), or slender and graceful (prestressed concrete).

So engineering design involves many considerations (Figure 1.11). The choice of a material must meet certain criteria for bulk and surface properties (e.g., strength and corrosion resistance). But it must also be easy to fabricate; it must appeal to potential consumers; and it must compete economically with

**FIGURE 1.9**
A typical twentieth-century mild-steel bridge; a convenient crossing to the Fort St George Inn!
– 52 12 44.67 N 0 07 42.09 E

**FIGURE 1.10**
The reinforced concrete footbridge in Garret Hostel Lane. An inscription carved nearby reads: "This bridge was given in 1960 by the Trusted family members of Trinity Hall. It was designed by Timothy Guy Morgan an undergraduate of Jesus College who died in that year." – 52 12 21.03 N 0 06 50.19 E

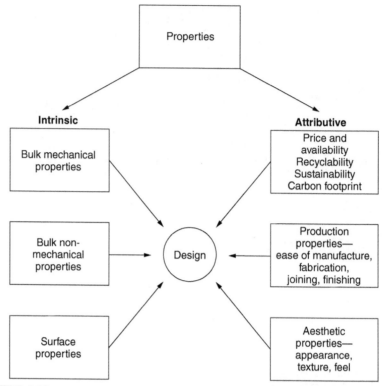

**FIGURE 1.11**

How the properties of engineering materials affect the way in which products are designed.

other alternative materials. Finally, it is becoming even more important that materials can be recycled, can be sourced in a sustainable way and can be manufactured, transported and used with the lowest possible carbon footprint.

# Price and Availability

## CHAPTER 2

# The Price and Availability of Materials

## CONTENTS

## 2.1 INTRODUCTION

In the first chapter we introduced the range of properties required of engineering materials by the design engineer, and the range of materials available to provide these properties. We ended by showing that the *price* and *availability* of materials were important and often overriding factors in selecting the materials for a particular job. In this chapter we examine these economic properties of materials in more detail.

## 2.2 DATA FOR MATERIAL PRICES

Table 2.1 ranks materials by their relative cost per unit weight. The most expensive materials—platinum, diamonds, gold—are at the top. The least expensive—wood, cement, concrete—are at the bottom. Such data are obviously important in choosing a material. Financial journals such as *The Wall Street Journal* (*www.wsj.com*) or the *Financial Times* (*www.ft.com*) give some

**Table 2.1** Approximate Relative Price per Ton (mild steel = 100)

| Material | Relative price |
|---|---|
| Platinum | 12 m |
| Diamonds, industrial | 10 m |
| Gold | 9.6 m |
| Silver | 290,000 |
| CFRP (materials 70% of cost; fabrication 30% of cost) | 20,000 |
| Cobalt/tungsten carbide cermets | 15,000 |
| Tungsten | 5000 |
| Cobalt alloys | 7000 |
| Titanium alloys | 2000 |
| Nickel alloys | 6000 |
| Polyimides | 8000 |
| Silicon carbide (fine ceramic) | 7000 |
| Magnesium alloys | 1000 |
| Nylon 66 | 1500 |
| Polycarbonate | 1000 |
| PMMA | 700 |
| Magnesia, MgO (fine ceramic) | 3000 |
| Alumina, $Al_2O_3$ (fine ceramic) | 3000 |
| Tool steel | 500 |
| GFRP (materials 60% of cost; fabrication 40% of cost) | 1000 |
| Stainless steels | 600 |
| Copper, worked (sheets, tubes, bars) | 2000 |
| Copper, ingots | 2000 |
| Aluminum alloys, worked (sheet, bars) | 650 |
| Aluminum ingots | 550 |
| Brass, worked (sheet, tubes, bars) | 2000 |
| Brass, ingots | 2000 |
| Epoxy | 1000 |
| Polyester | 500 |
| Glass | 400 |
| Foamed polymers | 1000 |
| Zinc, worked (sheet, tubes, bars) | 550 |
| Zinc, ingots | 450 |
| Lead, worked (bars, sheet, tube) | 650 |
| Lead, ingots | 550 |
| Natural rubber | 300 |
| Polypropylene | 200 |
| Polyethylene, high density | 200 |
| Polystyrene | 250 |
| Hard woods | 250 |
| Polyethylene, low density | 200 |

| Table 2.1  Cont'd | |
|---|---|
| **Material** | **Relative price** |
| Polyvinyl chloride | 300 |
| Plywood | 150 |
| Low-alloy steels | 200 |
| Mild steel, worked (angles, sheet, bars) | 100 |
| Cast iron | 90 |
| Soft woods | 50 |
| Concrete, reinforced (beams, columns, slabs) | 50 |
| Fuel oil | 190 |
| Cement | 20 |
| Coal | 20 |

Note: *At April 2011 mild steel was $500/ton*

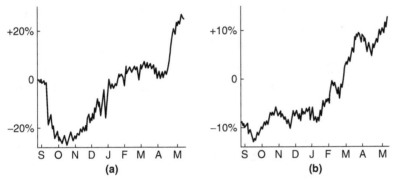

**FIGURE 2.1**
Typical fluctuations in the prices of copper (a), and rubber (b).

raw commodity prices, as does the London Metal Exchange *www.lme.com*. However, for detailed costs of finished or semi-finished materials it is best to consult the price lists of potential suppliers.

Figure 2.1 shows typical variations in the price of two materials—copper and rubber. These price fluctuations have little to do with the real scarcity or abundance of materials. They are caused by differences between the rate of supply and demand, magnified by speculation in commodity futures. The volatile nature of the commodity market can result in significant changes over a period of a few weeks. There is little that an engineer can do to foresee these changes, although the financial impact on the company can be controlled by taking out *forward contracts* to fix the price.

The long-term changes are of a different kind. They reflect, in part, the real cost (in capital investment, labor, and energy) of extracting and transporting the ore or feedstock and processing it to give the engineering material. Inflation and increased energy costs obviously drive the price up; so, too, does the necessity to extract materials, like copper, from increasingly lean ores; the leaner the ore, the more machinery and energy are required to crush the rock containing it, and to concentrate it to the level that the metal can be extracted.

In the long term, then, it is important to know which materials are basically plentiful, and which are likely to become scarce. It is also important to know the extent of our dependence on materials.

## 2.3 THE USE-PATTERN OF MATERIALS

The way in which materials are used in an industrialized nation is fairly standard. It consumes steel, concrete, and wood in construction; steel and aluminum in general engineering; copper in electrical conductors; polymers in appliances, and so forth; and roughly in the same proportions. Among metals, steel is used in the greatest quantities by far: 90% of all of the metal that is produced in the world is steel. But the nonmetals wood and concrete beat steel—they are used in even greater volume.

About 20% of the total import bill is spent on engineering materials. Table 2.2 shows how this spend is distributed. Iron and steel, and the raw materials used

| Table 2.2 Imports of Engineering Materials, Raw and Semis (percentage of total cost) | |
|---|---|
| **Material** | **Percentage** |
| Iron and steel | 27 |
| Wood and lumber | 21 |
| Copper | 13 |
| Plastics | 9.7 |
| Silver and platinum | 6.5 |
| Aluminum | 5.4 |
| Rubber | 5.1 |
| Nickel | 2.7 |
| Zinc | 2.4 |
| Lead | 2.2 |
| Tin | 1.6 |
| Pulp/paper | 1.1 |
| Glass | 0.8 |
| Tungsten | 0.3 |
| Mercury | 0.2 |
| Other | 1.0 |

to make them, account for about a quarter of it. Next are wood and lumber—widely used in light construction. More than a quarter is spent on the metals copper, silver and platinum, aluminum, and nickel. All polymers taken together, including rubber, account for little more than 10%. If we include the further metals zinc, lead, tin, tungsten, and mercury, the list accounts for 99% of all the money spent abroad on materials, and we can safely ignore the contribution of materials which do not appear on it.

## 2.4 UBIQUITOUS MATERIALS

### The composition of the earth's crust

Let us now shift attention from what we *use* to what is widely *available*. A few engineering materials are synthesized from compounds found in the earth's oceans and atmosphere: magnesium is an example. Most, however, are won by mining their ore from the earth's crust, and concentrating it sufficiently to allow the material to be extracted or synthesized from it. How plentiful and widespread are these materials on which we depend so heavily? How much copper, silver, tungsten, tin, and mercury in useful concentrations does the crust contain? All five are rare: workable deposits of them are relatively small, and are so highly localized that many governments classify them as of strategic importance, and stockpile them.

Not all materials are so thinly spread. Table 2.3 shows the relative abundance of the commoner elements in the earth's crust. The crust is 47% oxygen by weight, but—because oxygen is a big atom—it occupies 96% of the volume. Next in abundance are the elements silicon and aluminum; by far the most plentiful solid materials available to us are silicates and alumino-silicates.

A few metals appear on the list, among them iron and aluminum, both of which feature also in the list of widely used materials. The list extends as far as carbon because it is the backbone of virtually all polymers, including wood. Overall, then, oxygen and its compounds are overwhelmingly plentiful—on every side we are surrounded by oxide-ceramics, or the raw materials to make them.

Some materials are widespread, notably iron and aluminum; but even for these the local concentration is frequently small, usually too small to make it economic to extract them. In fact, the raw materials for making polymers are more readily available at present than those for most metals. There are huge deposits of carbon in the earth: on a world scale, we extract a greater tonnage of carbon every month than we extract iron in a year, but at present we simply burn it. And the second ingredient of most polymers—hydrogen—is also one of the most plentiful of elements. Some materials—iron, aluminum, silicon, the elements to make glass, and cement—are plentiful and widely available. But others (e.g., platinum, silver, tungsten) are scarce and highly localized, and—if the current pattern of use continues—may not last very long.

**Table 2.3** Abundance of Elements

| Crust | | Oceans | | Atmosphere | |
|---|---|---|---|---|---|
| **Element** | **Weight %** | **Element** | **Weight %** | **Element** | **Weight %** |
| Oxygen | 47 | Oxygen | 85 | Nitrogen | 79 |
| Silicon | 27 | Hydrogen | 10 | Oxygen | 19 |
| Aluminum | 8 | Chlorine | 2 | Argon | 2 |
| Iron | 5 | Sodium | 1 | Carbon as carbon dioxide | 0.04 |
| Calcium | 4 | Magnesium | 0.1 | | |
| Sodium | 3 | Sulphur | 0.1 | | |
| Potassium | 3 | Calcium | 0.04 | | |
| Magnesium | 2 | Potassium | 0.04 | | |
| Titanium | 0.4 | Bromine | 0.007 | | |
| Hydrogen | 0.1 | Carbon | 0.002 | | |
| Phosphorus | 0.1 | | | | |
| Manganese | 0.1 | | | | |
| Fluorine | 0.06 | | | | |
| Barium | 0.04 | | | | |
| Strontium | 0.04 | | | | |
| Sulphur | 0.03 | | | | |
| Carbon | 0.02 | | | | |

Note: *The total mass of the crust to a depth of 1 km is 3 $\times 10^{21}$ kg; the mass of the oceans is $10^{20}$ kg; that of the atmosphere is 5 $\times$ $10^{18}$ kg.*

## 2.5 EXPONENTIAL GROWTH AND CONSUMPTION DOUBLING-TIME

How do we calculate the lifetime of a resource such as platinum? Like almost all materials, platinum is being consumed at a rate that is growing exponentially with time (Figure 2.2), simply because both population and living standards grow exponentially. We analyze this in the following way. If the current rate of consumption in tons per year is $C$ then exponential growth means that

$$\frac{dC}{dt} = \frac{r}{100} C \qquad (2.1)$$

where, for the generally small growth rates we deal with here (1–5% per year), $r$ can be thought of as the%age fractional rate of growth per year. Integrating gives

$$C = C_0 \exp\left\{\frac{r(t - t_0)}{100}\right\} \qquad (2.2)$$

where $C_0$ is the consumption rate at time $t = t_0$. The *doubling-time* $t_D$ of consumption is given by setting $C/C_0 = 2$ to give

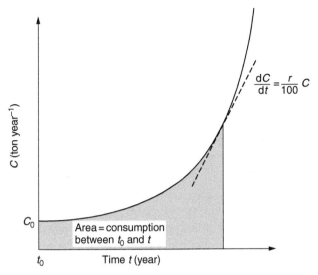

**FIGURE 2.2**
The exponentially rising consumption of materials.

$$t_D = \frac{100}{r} \log_e 2 \approx \frac{70}{r} \qquad (2.3)$$

Even with a growth late of only 2%, the doubling time is 35 years. For 4%, it is 18 years. In this context, growth rates of the order of 10%—as in China at the moment—hold out a frightening prospect if they are maintained at current levels.

## 2.6 RESOURCE AVAILABILITY

The availability of a resource depends on the degree to which it is *localized* in one or a few countries (making it susceptible to production controls or cartel action); on the *size* of the reserves, or, more accurately, the resource base (explained shortly); and on the *energy* required to mine and process it. The influence of the last two (size of reserves and energy content) can, within limits, be studied and their influence anticipated.

The calculation of resource life involves the important distinction between *reserves* and *resources*. The current reserve is the known deposits that can be extracted profitably at today's price using today's technology; it bears little relationship to the true magnitude of the resource base; in fact, the two are not even roughly proportional.

The resource base includes the current reserve. But it also includes all deposits that might become available given diligent prospecting and which, by various extrapolation techniques, can be estimated. And it includes, too, all known and unknown deposits that cannot be mined profitably now, but which—due to higher prices, better technology or improved transportation—might reasonably become available in the future (Figure 2.3). The reserve is like money in the bank—you know you have got it.

The resource base is more like your total potential earnings over your lifetime—it is much larger than the reserve, but it is less certain, and you may have to work very hard to get it. The resource base is the realistic measure of the total available material. Resources are almost always much larger than reserves, but because the geophysical data and economic projections are poor, their evaluation is subject to vast uncertainty.

Although the resource base is uncertain, it is obviously important to have some estimate of how long it can last. Rough estimates do exist for the size of the resource base, and, using these, our exponential formula gives an estimate of how long it would take us to use up *half* of the resources. The half-life is an important measure: at this stage prices would begin to rise so steeply that supply would become a severe problem. For a number of important materials these half-lives lie within your lifetime: for silver, tin, tungsten, zinc, lead, platinum, and oil (the feed stock of polymers) they lie between 40 and 70 years. Others (most notably iron, aluminum, and the raw materials from which most ceramics and glasses are made) have enormous resource bases, adequate for hundreds of years, even allowing for continued exponential growth.

The cost of energy enters here. The extraction of materials requires energy (Table 2.4). As a material becomes scarcer—copper is a good example—it must

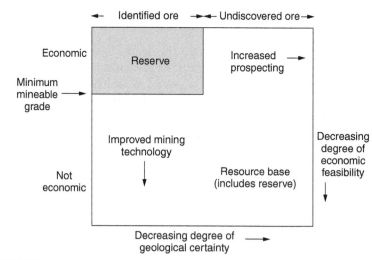

**FIGURE 2.3**
The distinction between the reserve and the resource base, illustrated by the McElvey diagram.

Table 2.4 Approximate Energy Content of Materials (GJ ton$^{-1}$)

| Material | Energy |
|----------|--------|
| Aluminum | 280 |
| Plastics | 85–180 |
| Copper | 140, rising to 300 |
| Zinc | 68 |
| Steel | 55 |
| Glass | 20 |
| Cement | 7 |
| Brick | 4 |
| Timber | 2.5–7 |
| Gravel | 0.2 |
| Oil | 44 |
| Coal | 29 |

be extracted from leaner ores. This expends more energy, per ton of copper *metal* produced, in the operations of mining, crushing, and concentrating the ore. The rising energy content of copper shown in Table 2.4 reflects the fact that the richer copper ores are, right now, being worked out.

## 2.7 THE FUTURE

How are we going to cope with shortages of engineering materials in the future? Some obvious strategies are as follows.

### Material-efficient design

Many current designs use more material than necessary, or use potentially scarce materials where the more plentiful would serve. Often, for example, it is a surface property (e.g., low friction, or high corrosion resistance) that is wanted; then a thin surface film of the rare material bonded to a cheap plentiful substrate can replace the bulk use of a scarcer material.

### Substitution

It is the property, not the material itself, that the designer wants. Sometimes a more readily available material can replace the scarce one, although this usually involves considerable outlay (new processing methods, new joining methods, etc.). Examples of substitution are the replacement of stone and wood by steel and concrete in construction; the replacement of copper by polymers in plumbing; the change from wood and metals to polymers in household goods; and from copper to aluminum in electrical wiring.

There are, however, technical limitations to substitution. Some materials are used in ways not easily filled by others. Platinum as a catalyst, liquid helium as a refrigerant, and silver on electrical contact areas cannot be replaced; they perform a unique function. Others—a replacement for tungsten lamp filaments, for example—are the subject of much development work at this moment. Finally, substitution increases the demand for the replacement material, which may also be in limited supply. The massive trend to substitute plastics for other materials puts a heavier burden on petrochemicals, at present derived from oil.

### Recycling

Recycling is not new: old building materials have been recycled for millennia; scrap metal has been recycled for centuries; both are major industries. Recycling is labor intensive, and therein lies the problem in expanding its scope.

## 2.8 CONCLUSION

Overall, the materials-resource problem is not as critical as that of energy. Some materials have an enormous base or (like wood) are renewable—and fortunately these include the major structural materials. For others, the resource base is small, but they are often used in small quantities so that the price could rise a lot without having a drastic effect on the price of the product in which they are incorporated; and for some, substitutes are available. But such adjustments can take time—up to 25 years if a new technology is needed; and they need capital too.

Rising energy costs mean that the relative costs of materials will change in the next 20 years: designers must be aware of these changes, and continually on the look-out for opportunities to use materials as efficiently as possible. But increasingly, governments are imposing compulsory targets on recycling materials from a wide range of mass-produced consumer goods (e.g., cars, electronic equipment, and white goods). Manufacturers must now design for the whole life cycle of the product: it is no longer sufficient for one's mobile phone to work well for two years and then be thrown into the trash can—it must also be designed so that it can be dismantled easily and the materials recycled into the next generation of mobile phones.

### Environmental impact

As well as simply consuming materials, the mass production of consumer goods places two burdens on the environment. The first is the volume of waste generated. Materials that are not recycled go eventually to landfill sites, which cause groundwater pollution and are physically unsustainable. Unless the%age of materials recycled increases dramatically in the near future, a significant proportion of the countryside will end up as a rubbish dump. The second is the production of the energy necessary to extract and process materials, and manufacture and distribute the goods made from them. Fossil fuels, as we have seen, are a finite resource. And burning fossil fuels releases carbon dioxide into the atmosphere, with serious implications for climate change. Governments are setting targets for carbon dioxide emissions—and also are imposing carbon taxes—the overall effect being to drive up energy costs.

## EXAMPLES

**2.1 a.** Commodity A is currently consumed at the rate $C_A$ tons per year, and commodity B at the rate $C_B$ tons per year ($C_A > C_B$). If the two consumption rates are increasing exponentially to give growths in consumption after each year of $r_A\%$ and $r_B\%$, respectively ($r_A < r_B$), derive an equation for the time, measured from the present day, before the annual consumption of B exceeds that of A.

**b.** The table shows figures for consumption and growth rates of steel, aluminum, and plastics. What are the doubling-times (in years) for consumption of these commodities?

**c.** Calculate the number of years before the consumption of (a) aluminum and (b) polymers would exceed that of steel, *if exponential growth continued.*

| Material | Current Consumption (tons year$^{-1}$) | Projected Growth Rate in Consumption (% year$^{-1}$) |
|---|---|---|
| Iron and steel | $3 \times 10^8$ | 2 |
| Aluminum | $4 \times 10^7$ | 3 |
| Polymers | $1 \times 10^8$ | 4 |

**Answers**

**a.**

$$t = \frac{100}{r_B - r_A} \ln\left(\frac{C_A}{C_B}\right)$$

**b.** Doubling-times: steel, 35 years; aluminum, 23 years; plastics, 18 years.

**c.** If exponential growth continued, aluminum would overtake steel in 201 years; polymers would overtake steel in 55 years.

**2.2** Discuss ways of conserving engineering materials, and the technical and social problems involved in implementing them.

**2.3 a.** Explain what is meant by *exponential growth* in the consumption of a material.

**b.** A material is consumed at $C_0$ ton year$^{-1}$ in 2012. Consumption in 2012 is increasing at $r\%$ year$^{-1}$. If the resource base of the material is $Q$ tons, and consumption *continues* to increase at $r\%$ year$^{-1}$, show that the resource will be half exhausted after a time, $t_{1/2}$, given by

$$t_{1/2} = \frac{100}{r} \ln\left\{ \frac{rQ}{200C_0} + 1 \right\}$$

**2.4** Discuss, giving specific examples, the factors that might cause a decrease in the rate of consumption of a potentially scarce material.

**PART**

# The Elastic Moduli

# The Elastic Moduli

## CONTENTS

## 3.1 INTRODUCTION

The next material property we will look at is the *elastic modulus*. This measures the resistance of the material to elastic—or "springy"—deformation. Low modulus materials are floppy, and stretch a lot when they are pulled (squash down a lot when pushed). High modulus materials are the opposite—they stretch very little when pulled (squash down very little when pushed). As shown in Figure 3.1, it is very easy to stretch a rubber band—it would be useless if you couldn't—but there is no way you could stretch a strip of steel of this cross-section using you bare hands. Floppy materials like rubber are ideal for things

Engineering Materials I: An Introduction to Properties, Applications, and Design, Fourth Edition

**FIGURE 3.1**
Stretching of a rubber band, showing the Poisson contraction.

like bungee-jumping ropes, but would be catastrophic for something like the deck hangers of the Sydney Harbour Bridge (Figure 3.2)—the deck would end up in the harbour! And steel—used for the Harbour Bridge precisely because it is stiff in tension and doesn't give—would kill any mountaineer who was stupid enough to use steel wire rope for climbing protection; in a fall, she/he would literally get cut in half by the shock loading.

However, before we can look at the modulus of the many and different engineering materials, we need to understand what is meant by *stress* and *strain*.

## 3.2 DEFINITION OF STRESS

Imagine a block of material to which we apply a force $F$, as in Figure 3.3(a). The force is transmitted through the block and is balanced by the equal, opposite force which the base exerts on the block (if this were not so, the block would move). We can replace the base by the equal and opposite force, $F$, which acts on all sections through the block parallel to the original surface; the whole of the block is said to be in a state of stress. The intensity of the stress, $\sigma$, is measured by the force $F$ divided by the area, $A$, of the block face, giving

$$\sigma = \frac{F}{A} \tag{3.1}$$

This particular stress is caused by a force pulling at right angles to the face; we call it the *tensile* stress.

Suppose now that the force acted not normal to the face but at an angle to it, as shown in Figure 3.3(b). We can resolve the force into two components—one,

**FIGURE 3.2**
The deck hangers of the Sydney Harbour Bridge. Note the scaffolding on the right hand side, ready for carrying the fireworks for the world famous New Year's Eve firework display, 19 December 2010.
−35 51 14.40 S 151 12 51.66E

$F_s$, normal to the face and the other, $F_s$, parallel to it. The normal component creates a tensile stress in the block. Its magnitude, as before, is $F_t/A$.

The other component, $F_s$, also loads the block, but it does so in *shear*. The shear stress, $\tau$, in the block parallel to the direction of $F_s$, is given by

$$\tau = \frac{F_s}{A} \tag{3.2}$$

The important point is that the magnitude of a stress is always equal to the magnitude of a *force* divided by the *area* of the face on which it acts.

## Ways of writing stress

Forces are measured in newtons, so stresses are measured in units of newtons per meter squared ($N\ m^{-2}$). This has also been turned into a relatively new SI unit—the pascal—written as Pa. So a stress might be written as $10\ N\ m^{-2}$ (ten newtons per square meter) or 10 Pa (ten pascals). Stresses in materials are usually sufficiently large that these basic SI units are too small, so the multiple of mega ($10^6$) usually goes in front—for example $10\ MN\ m^{-2}$

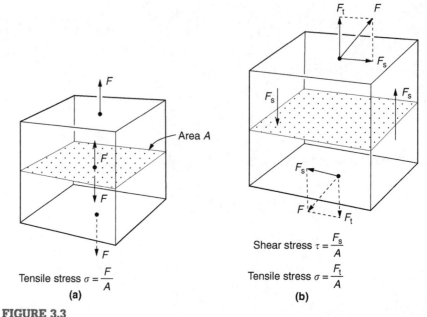

**FIGURE 3.3**

Definitions of (a) tensile stress $\sigma$ and (b) shear stress $\tau$; balancing shear required for equilibrium, as shown.

(ten mega-newtons per square meter) or 10 MPa (ten mega-pascals). Values of the elastic modulus are usually larger again, so a multiple of giga ($10^9$) is used for them—for example 200 GN m$^{-2}$ (200 giga-newtons per square meter) or 200 GPa (200 giga-pascals).

Finally, a *mega*-newton per square meter is the same as a newton per square *millimeter*—for example 10 N mm$^{-2}$ is the same as 10 MN m$^{-2}$. You can do this in your head—a square meter has an area that is larger than that of a square millimeter by $(10^3)^2 = 10^6$. **But a note of warning**—if you are using stresses in connection with calculating things such as vibration frequencies, which involve Newton's law of acceleration ($F = ma$)—you **must** use the basic SI stress unit of N m$^{-2}$ (and the basic mass/density units of kg/kg m$^{-3}$).

## Common states of stress

There are four commonly occurring states of stress, shown in Figure 3.4. The simplest is that of *simple tension* or *compression* (as in a tension member loaded by pin joints at its ends or in a pillar supporting a structure in compression). The stress is, of course, the force divided by the section area of the member or pillar. The second common state of stress is that of *biaxial tension*. If a spherical shell (e.g., a balloon) contains an internal pressure, then the skin of the shell is loaded in two directions, not one, as shown in Figure 3.4. This state

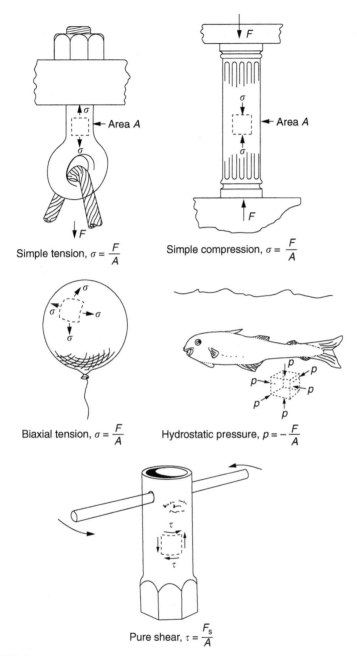

**FIGURE 3.4**
Common states of stress: tension, compression, hydrostatic pressure, and shear.

of stress is called biaxial tension (unequal biaxial tension is obviously the state in which the two tensile stresses are unequal). The third common state of stress is that of *hydrostatic pressure*. This occurs deep in the Earth's crust, or deep in the ocean, when a solid is subjected to equal compression on all sides.

There is a convention that stresses are *positive* when they *pull*, as we have drawn them in earlier figures. Pressure, however, is positive when it *pushes*, so that the magnitude of the pressure differs from the magnitude of the other stresses in its sign. Otherwise it is defined in exactly the same way as before: the force divided by the area on which it acts. The final common state of stress is that of *pure shear*. If you try to twist a tube, then elements of it are subjected to pure shear, as shown. This shear stress is simply the shearing force divided by the area of the face on which it acts.

## 3.3 DEFINITION OF STRAIN

Materials respond to stress by *straining*. Under a given stress, a stiff material (e.g., aluminum) strains only slightly; a floppy or compliant material (e.g., polyethylene) strains much more. The modulus of the material describes this property, but before we can measure it, or even define it, we must define strain properly.

The kind of stress that we called a tensile stress induces a tensile strain. If the stressed cube of side $l$, shown in Figure 3.5(a) extends by an amount $u$ parallel to the tensile stress, the *nominal tensile strain* is

$$\varepsilon_n = \frac{u}{l} \tag{3.3}$$

When it strains in this way, the cube usually gets thinner. The amount by which it shrinks inwards is described by Poisson's ratio, $\upsilon$, which is the negative of the ratio of the inward strain to the original tensile strain:

$$\upsilon = -\frac{\text{lateral strain}}{\text{tensile strain}}$$

A shear stress induces a shear strain. If a cube shears sideways by an amount $\omega$ then the *shear strain* is defined by

$$\gamma = \frac{\omega}{l} = \tan\theta \tag{3.4}$$

where $\theta$ is the angle of shear and $l$ is the edge-length of the cube (Figure 3.5(b)). Since the elastic strains are almost always very small, we may write, to a good approximation,

$$\gamma = \theta$$

Finally, hydrostatic pressure induces a volume change called *dilatation*, as shown in Figure 3.5(c). If the volume change is $\Delta V$ and the cube volume is $V$, we define the dilatation by

$$\Delta = \frac{\Delta V}{V} \tag{3.5}$$

Since strains are the ratios of two lengths or of two volumes, they are dimensionless.

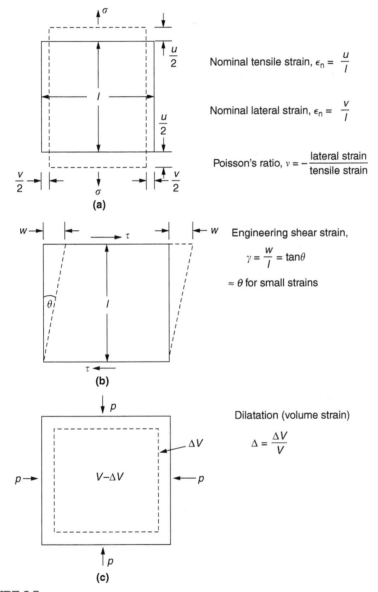

Nominal tensile strain, $\epsilon_n = \dfrac{u}{l}$

Nominal lateral strain, $\epsilon_n = \dfrac{v}{l}$

Poisson's ratio, $v = -\dfrac{\text{lateral strain}}{\text{tensile strain}}$

**(a)**

Engineering shear strain,

$$\gamma = \frac{w}{l} = \tan\theta$$

$\approx \theta$ for small strains

**(b)**

Dilatation (volume strain)

$$\Delta = \frac{\Delta V}{V}$$

**(c)**

**FIGURE 3.5**

Definitions of tensile strain, $\varepsilon_n$, shear strain, $\gamma$, and dilatation, $\Delta$.

## 3.4 HOOKE'S LAW

We can now define the elastic moduli. They are defined through Hooke's law, which is a description of the experimental observation that—when strains are small—the strain is very nearly proportional to the stress; the behavior of the solid is *linear elastic*. The nominal tensile strain, for example, is proportional to the tensile stress; for simple tension

$$\sigma = E\varepsilon_n \tag{3.6}$$

where $E$ is called *Young's modulus*. The same relationship also holds for stresses and strains in simple compression.

In the same way, the shear strain is proportional to the shear stress, with

$$\tau = G\gamma \tag{3.7}$$

where $G$ is the *shear modulus*. Finally, the negative of the dilatation is proportional to the pressure (because positive pressure causes a shrinkage of volume) so that

$$p = -K\Delta \tag{3.8}$$

where $K$ is called the *bulk modulus*. Because strain is dimensionless, the moduli have the same dimensions as those of stress.

This linear relationship between stress and strain is a very useful one when calculating the response of a solid to stress, but it must be remembered that many solids are elastic only to *very small* strains: up to about 0.002. Beyond that some break and some become plastic—and this we will discuss in later chapters. A few solids, such as rubber, are elastic up to very much larger strains of order 4 or 5, but they cease to be *linearly* elastic (that is the stress is no longer proportional to the strain) after a strain of about 0.01.

We defined Poisson's ratio as the ratio of the lateral shrinkage strain to the tensile strain. This quantity is also an elastic constant, so altogether we have four elastic constants: $E$, $G$, $K$ and $\upsilon$. In a moment when we give data for the elastic constants we list data only for $E$. But for many materials it is useful to know that

$$K \approx E, \quad G \approx \frac{3}{8}E \quad \text{and} \quad \upsilon \approx 0.33 \tag{3.9}$$

(although for some the relationship can be more complicated).

## 3.5 MEASUREMENT OF YOUNG'S MODULUS

How is Young's modulus measured? This requires both stress and strain to be measured with enough accuracy. In the case of metals, because they are stiff, either the strain needs to be measured very accurately, or there needs to be some

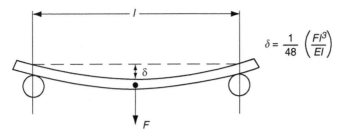

$$\delta = \frac{1}{48}\left(\frac{Fl^3}{EI}\right)$$

**FIGURE 3.6**

Three-point bend test.

way of magnifying it. So we could load a bar of material in tension, having first glued strain gauges to its surface, and use the amplified electrical signal from them to measure the strain. Or we could load a bar in bending—equations for the relationships between applied load and deflection for elastic beams having various loading geometries are given at the end of this chapter.

The three-point bend test is an especially easy geometry to adopt (Figure 3.6). The three-point bend test is also good for brittle materials (brittle metals, ceramics, polymers, composites), because if loaded in tension they may break where they are gripped by the testing machine. It is also good for natural composites, like wood or bamboo.

Floppy materials, like the lower modulus thermoplastics, rubbers and foamed polymers can be tested in compression, reading the strain directly from the movement of the testing machine. However, care needs to be taken that any deflection of the machine itself is allowed for, and also that there is no other source of non-elastic strain like creep. For such materials, the *rate* at which the specimen is strained will often have a significant effect on the modulus values calculated from the test.

Finally, we can measure the velocity of sound in the material. The velocity of longitudinal waves is given by

$$V_L = \left(\frac{E(1-v)}{\rho(1+v)(1-2v)}\right)^{1/2}$$

where $\rho$ is the material density. The velocity of shear (transverse) waves is given by

$$V_T = \left(\frac{G}{\rho}\right)^{1/2}$$

The equation

$$v = \frac{1 - 2(V_T/V_L)^2}{2 - 2(V_T/V_L)^2}$$

gives the value of Poisson's ratio. An electronic pulser-receiver is placed in contact with one end face of a short solid cylinder of the material. The times of travel of longitudinal and shear waves over the known distance are measured electronically, and used to determine $V_L$ and $V_T$.

## 3.6 DATA FOR YOUNG'S MODULUS

Now for some real numbers. Table 3.1 is a ranked list of Young's modulus of materials—we will use it later in solving problems and in selecting materials for particular applications. Diamond is at the top, with a modulus of 1000 GPa; soft rubbers and foamed polymers are at the bottom with moduli as low as 0.001 GPa.

You can, of course, make special materials with lower moduli—jelly, for instance, has a modulus of about $10^{-6}$ GPa. Practical engineering materials lie in the range $10^{-3}$ to $10^{+3}$ GPa—a range of $10^{6}$. This is the range you have to choose from when selecting a material for a given application.

A good perspective of the spread of moduli is given by the bar chart shown in Figure 3.7 (see page 41). Ceramics and metals—even the floppiest of them, like lead—lie near the top of this range. Polymers and elastomers are much more compliant, the common ones (polyethylene, PVC and polypropylene) lying several decades lower. Composites span the range between polymers and ceramics.

To understand the origin of the modulus, why it has the values it does, why polymers are much less stiff than metals, and what we can do about it, we have to examine the *structure* of materials, and *the nature of the forces* holding the atoms together. In the next two chapters we will examine these, and then return to the modulus, and to our bar chart, with new understanding.

## WORKED EXAMPLE

Siân wants to keep her large collection of history books in an alcove in her sitting room. The alcove is 1600 mm wide and you offer to install a set of pine shelves, each 1600 mm long, supported at their ends on small brackets screwed into the wall. You guess that shelves measuring 20 mm deep by 250 mm wide will work. However, professional integrity demands that you calculate the maximum stress and deflection first. For pine, $E \approx 10$ G Nm$^{-2}$ and $\sigma_y \approx 60$ M Nm$^{-2}$. A book 40 mm thick weighs about 1 kg. Equations for the elastic bending of beams are given at the end of this chapter.

## Table 3.1 Data for Young's Modulus, $E$

| Material | $E$ (GN m$^{-2}$) |
|---|---|
| Diamond | 1000 |
| Tungsten carbide, WC | 450–650 |
| Osmium | 551 |
| Cobalt/tungsten carbide cermets | 400–530 |
| Borides of Ti, Zr, Hf | 450–500 |
| Silicon carbide, SiC | 430–445 |
| Boron | 441 |
| Tungsten and alloys | 380–411 |
| Alumina, $Al_2O_3$ | 385–392 |
| Beryllia, BeO | 375–385 |
| Titanium carbide, TiC | 370–380 |
| Tantalum carbide, TaC | 360–375 |
| Molybdenum and alloys | 320–365 |
| Niobium carbide, NbC | 320–340 |
| Silicon nitride, $Si_3N_4$ | 280–310 |
| Beryllium and alloys | 290–318 |
| Chromium | 285–290 |
| Magnesia, MgO | 240–275 |
| Cobalt and alloys | 200–248 |
| Zirconia, $ZrO_2$ | 160–241 |
| Nickel | 214 |
| Nickel alloys | 130–234 |
| CFRP | 70–200 |
| Iron | 196 |
| Iron-based super-alloys | 193–214 |
| Ferritic steels, low-alloy steels | 196–207 |
| Stainless austenitic steels | 190–200 |
| Mild steel | 200 |
| Cast irons | 170–190 |
| Tantalum and alloys | 150–186 |
| Platinum | 172 |
| Uranium | 172 |
| Boron/epoxy composites | 80–160 |
| Copper | 124 |
| Copper alloys | 120–150 |
| Mullite | 145 |
| Vanadium | 130 |
| Titanium | 116 |
| Titanium alloys | 80–130 |
| Palladium | 124 |
| Brasses and bronzes | 103–124 |

*Continued*

**Table 3.1** Data for Young's Modulus, $E$—Cont'd

| Material | $E$ (GN m$^{-2}$) |
|---|---|
| Niobium and alloys | 80–110 |
| Silicon | 107 |
| Zirconium and alloys | 96 |
| Silica glass, SiO$_2$ (quartz) | 94 |
| Zinc and alloys | 43–96 |
| Gold | 82 |
| Calcite (marble, limestone) | 70–82 |
| Aluminum | 69 |
| Aluminum and alloys | 69–79 |
| Silver | 76 |
| Soda glass | 69 |
| Alkali halides (NaCl, LiF, etc.) | 15–68 |
| Granite (Westerly granite) | 62 |
| Tin and alloys | 41–53 |
| Concrete, cement | 30–50 |
| Fiberglass (glass-fiber/epoxy) | 35–45 |
| Magnesium and alloys | 41–45 |
| GFRP | 7–45 |
| Calcite (marble, limestone) | 31 |
| Graphite | 27 |
| Shale (oil shale) | 18 |
| Common woods, ‖ to grain | 9–16 |
| Lead and alloys | 16–18 |
| Alkyds | 14–17 |
| Ice, H$_2$O | 9.1 |
| Melamines | 6–7 |
| Polyimides | 3–5 |
| Polyesters | 1.8–3.5 |
| Acrylics | 1.6–3.4 |
| Nylon | 2–4 |
| PMMA | 3.4 |
| Polystyrene | 3–3.4 |
| Epoxies | 2.6–3 |
| Polycarbonate | 2.6 |
| Common woods, ⊥ to grain | 0.6–1.0 |
| Polypropylene | 0.9 |
| PVC | 0.2–0.8 |
| Polyethylene, high density | 0.7 |
| Polyethylene, low density | 0.2 |
| Rubbers | 0.01–0.1 |
| Cork | 0.01–0.03 |
| Foamed polymers | 0.001–0.01 |

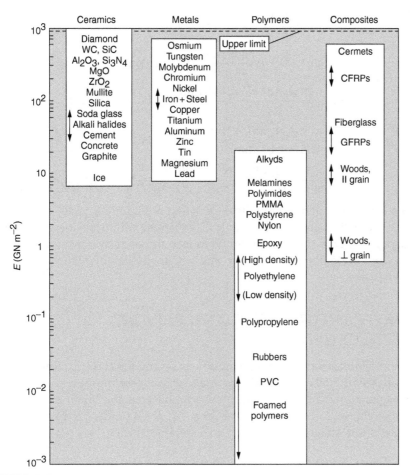

**FIGURE 3.7**
Bar chart of data for Young's modulus, $E$.

The ones to use for this problem are:

$$\delta = \frac{5}{384}\left(\frac{Fl^3}{EI}\right)$$

$$\sigma_{max} = \frac{Mc}{I}$$

$$I = \frac{bd^3}{12}$$

$$M = \frac{Fl}{8}$$

$$I = \frac{250 \times 20^3}{12} \; \text{mm}^4 = 1.67 \times 10^5 \, \text{mm}^4$$

$$F = \frac{1600 \, \text{mm}}{40 \, \text{mm}} \; \text{kgf} = 40 \, \text{kgf}$$

$$M = \frac{40 \times 9.81 \, \text{N} \times 1600 \, \text{mm}}{8} = 7.85 \times 10^4 \, \text{N mm}$$

$$\sigma_{max} = \frac{7.85 \times 10^4 \; \text{N mm} \times 10 \; \text{mm}}{1.67 \times 10^5 \; \text{mm}^4} = 4.7 \; \text{N mm}^{-2} = \underline{\underline{4.7 \; \text{MN m}^{-2}}}$$

$$\delta = \frac{5 \times 40 \times 9.81 \, \text{N} \times 1600^3 \; \text{mm}^3}{384 \times 10^4 \; \text{N mm}^{-2} \times 1.67 \times 10^5 \, \text{mm}^4} = \underline{\underline{12.5 \; \text{mm}}}$$

The shelves will not break but they will deflect noticeably. The deflection is just about acceptable aesthetically, but it may increase with time due to creep (see Chapter 20). One solution would be to turn the shelf over every six months. This will not be popular, however, and you are better advised to increase the thickness instead.

## A NOTE ON STRESSES AND STRAINS IN 3 DIMENSIONS

When we look at stresses in real components—and the elastic strains these create—it is useful to have a simple shorthand way of describing them. The following figure shows that, no matter how complicated the loadings are on the component, we can express the stresses at any given point in terms of 3 tensile stresses and 6 shear stresses acting on a small "test cube."

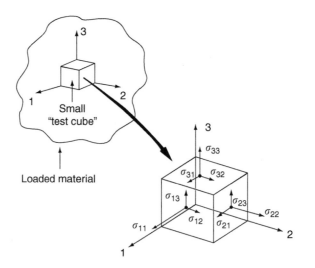

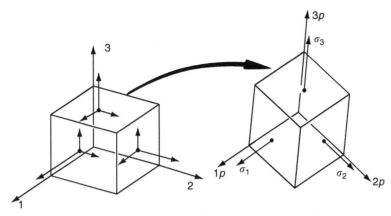

The test cube can always be rotated into one particular orientation where all the shear components vanish.

The conventions for doing this are summarized next. The shorthand form for listing them is called the *stress tensor*. You may have come across this in other courses (e.g., math or mechanics), and they have a number of important mathematical properties. But we are not using those properties here—simply using the tensor to *summarize easily* what could otherwise be cumbersome ways of writing down the stress components.

- Right set of orthogonal axes labeled 1, 2, 3; stress components $\sigma_{ij}$ where i = 1, 2, 3, and j = 1, 2, 3.
- i represents the direction of the normal to the plane on which a stress component acts.
- j respresents the direction in which the stress component acts.
- A component with i = j is a normal stress.
- A component with i $\neq$ j is a shear stress.

The stress tensor is

$$\sigma_{ij} = \begin{pmatrix} \sigma_{11} & \sigma_{12} & \sigma_{13} \\ \sigma_{21} & \sigma_{22} & \sigma_{23} \\ \sigma_{31} & \sigma_{32} & \sigma_{33} \end{pmatrix}$$

- For static equilibrium $\sigma_{21} = \sigma_{12}, \sigma_{31} = \sigma_{13}, \sigma_{32} = \sigma_{23}$. Thus tensor is symmetrical about the leading diagonal.
- Tensile components $\sigma_{11}, \sigma_{22}, \sigma_{33}$, become principal stresses $\sigma_1, \sigma_2, \sigma_3$.
- Axes 1, 2, 3 become principal directions $1_p, 2_p, 3_p$.
- Cube planes become principal planes.

The principal stress tensor is

$$\sigma_p = \begin{pmatrix} \sigma_1 & 0 & 0 \\ 0 & \sigma_2 & 0 \\ 0 & 0 & \sigma_3 \end{pmatrix}$$

Property of transformation: $\sigma_{11} + \sigma_{22} + \sigma_{33} = \sigma_1 + \sigma_2 + \sigma_3$.

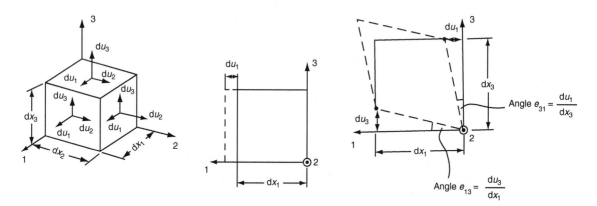

This figure shows that we can do the same for the strain components in a loaded material—the test cube has 3 axial strains and 6 shear strains, and the short-hand form for listing them is called the *strain tensor*.

- Strain components $\varepsilon_{ij}$, where i = 1, 2, 3 and j = 1, 2, 3.
- A component with i = j is an axial strain.

The three axial strains are:

$$\varepsilon_{11} = \frac{du_1}{dx_1}, \qquad \varepsilon_{22} = \frac{du_2}{dx_2}, \qquad \varepsilon_{33} = \frac{du_3}{dx_3}$$

A component with i ≠ j is a shear strain.

$$\varepsilon_{13} = \varepsilon_{31} = \frac{1}{2}(e_{13} + e_{31}) = \frac{1}{2}\left(\frac{du_3}{dx_1} + \frac{du}{dx_3}\right)$$

Thus the strain tensor is:

$$\varepsilon_{ij} = \begin{pmatrix} \varepsilon_{11} & \varepsilon_{12} & \varepsilon_{13} \\ \varepsilon_{21} & \varepsilon_{22} & \varepsilon_{23} \\ \varepsilon_{31} & \varepsilon_{32} & \varepsilon_{33} \end{pmatrix}$$

Note that the tensor is symmetrical about the leading diagonal.

The test cube can always be rotated into one particular orientation where all the shear components vanish. Then:

- Axial components $\varepsilon_{11}, \varepsilon_{22}, \varepsilon_{33}$ become principal strains $\varepsilon_1, \varepsilon_2, \varepsilon_3$.
- Axes 1, 2, 3, become principal directions $1_p, 2_p, 3_p$.
- In an isotropic material the principal strain directions are the same as the principal stress directions.

The principal strain tensor is

$$\varepsilon_p = \begin{pmatrix} \varepsilon_1 & 0 & 0 \\ 0 & \varepsilon_2 & 0 \\ 0 & 0 & \varepsilon_3 \end{pmatrix}$$

Property of transformation: $\varepsilon_{11} + \varepsilon_{22} + \varepsilon_{33} = \varepsilon_1 + \varepsilon_2 + \varepsilon_3$.

## EXAMPLES

**3.1** Uncle Albert has been given an elegant hickory (hardwood) walking stick. It is 750 mm long and 17 mm in diameter (being an engineer, you always carry a pair of vernier calipers around with you). The stick is straight, and the top end it terminates in a small ball-shaped handle. He is very pleased with his stick, but you have your doubts that it can support his full weight (90 kg on a good day). Do the calculations to see whether he will be OK. Assume that $E = 20 \text{ GN m}^{-2}$ ($2 \times 10^4 \text{ N}$ $\text{mm}^{-2}$). Select the appropriate equation from the results given at the end of this chapter for the elastic buckling of struts.

**3.2** You have a steel tuning fork, which produces a Concert A (440 Hz) when it is struck. The cantilevers of the fork are 85 mm long, and have a square cross section of side 3.86 mm. Calculate the value of Young's modulus for steel. Select the appropriate equation from the results given at the end of this chapter for the mode 1 natural vibration frequencies of beams. The density of steel is $7.85 \times 10^3 \text{ kg m}^{-3}$.

**Answer**

$204 \text{ GN m}^{-2}$

**3.3** Explain why, in the general form of the stress tensor, there are only three independent shear stresses.

**3.4** Explain what is special about principal planes, principal directions and principal stresses. What is the special characteristic of the principal stress tensor?

**3.5** Write out the principal stress tensor for the following loading conditions:
   a. A long bar subjected to a uniform tensile stress $\sigma_1$ along its length.
   b. A thin sheet subjected to unequal biaxial tension (stresses $\sigma_1$ and $\sigma_2$).
   c. A cube subjected to an external hydrostatic pressure $p$.
   d. A cube subjected to triaxial tension (tensile stresses $\sigma_1$, $\sigma_2$, and $\sigma_3$).

**3.6** A thin-walled tube is subjected to a torque (see Figure 3.4). Write out the stress tensor for the state of pure shear in the tube wall (shear stress $= \tau$). Is this stress tensor a principal stress tensor?

**3.7** The diagram on the next page shows a pressure vessel subjected to internal pressure $p$. The circumferential stress (hoop stress) is given by $\sigma = pr/t$ and the longitudinal (axial) stress is given by $\sigma = pr/2t$. Write out the principal stress tensor for this stress state. Explain why the 1, 2 and 3 axes are the principal directions.

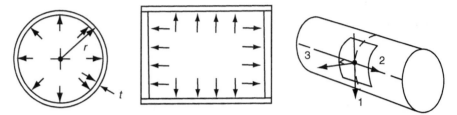

**3.8** Explain why, in the general form of the strain tensor, there are only three independent shear strains.

**3.9** Explain what is special about principal strains. What is the special characteristic of the principal strain tensor?

**3.10** Write out the strain tensor for the following straining conditions:
   **a.** A long bar subjected to a uniform tensile strain $\varepsilon_1$ along its length.
   **b.** A cube subjected to a uniform compressive strain $\varepsilon$ as the result of external hydrostatic pressure.

**3.11** Show that the engineering shear strain $\gamma$ is equal to twice the shear strain component in the strain tensor. [*Hint:* Compare Figures 3.5 and 10.3].

**3.12** Define the dilatation $\Delta$, and show that $\Delta = \varepsilon_1 + \varepsilon_2 + \varepsilon_3$.

**3.13** Show that the elastic stress–strain relations for an isotropic material are

$$\varepsilon_1 = \frac{\sigma_1}{E} - v\frac{\sigma_2}{E} - v\frac{\sigma_3}{E}$$

$$\varepsilon_2 = \frac{\sigma_2}{E} - v\frac{\sigma_1}{E} - v\frac{\sigma_3}{E}$$

$$\varepsilon_3 = \frac{\sigma_3}{E} - v\frac{\sigma_1}{E} - v\frac{\sigma_2}{E}$$

Hence derive an equation for the dilatation in terms of Poisson's ratio, Young's modulus, and the sum of the principal stresses. For what value of $v$ is the volume change zero?

**3.14** A bar of material is subjected to uniaxial tension, which produces an axial strain $\varepsilon_1$. Poisson's ratio for most metals is about 0.3. For cork, it is close to 0; for rubber it is close to 0.5. What approximate volume changes are produced in each of these materials by the strain?

**3.15** Show that Young's modulus $E$ and the bulk modulus $K$ are related by the equation

$$K = \frac{E}{3(1 - 2v)}$$

[*Hint:* Use the equation you derived for the dilatation in example 3.13, and set each of the principal stresses equal to $-p$.] Hence show that $K = E$ when $v = 1/3$.

**3.16** The sole of a shoe is to be surfaced with soft synthetic rubber having a Poisson's ratio of 0.5. The least expensive solution is to use a solid rubber slab of uniform thickness. However, a colleague suggests that the sole would give better cushioning if it were molded as shown in the diagram. Is your colleague correct? If so, why?

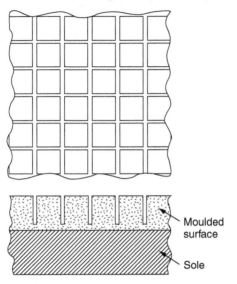

Moulded surface

Sole

**3.17** Explain why it is much easier to push a cork into a wine bottle than a rubber bung. Poisson's ratio is 0 for cork and 0.5 for rubber.

## ELASTIC BENDING OF BEAMS

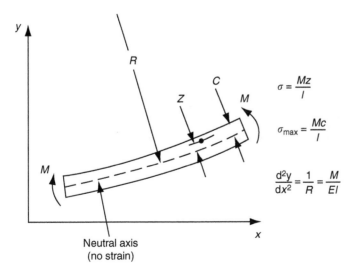

$$\sigma = \frac{Mz}{I}$$

$$\sigma_{max} = \frac{Mc}{I}$$

$$\frac{d^2y}{dx^2} = \frac{1}{R} = \frac{M}{EI}$$

Neutral axis
(no strain)

**Case 1**

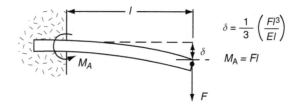

$$\delta = \frac{1}{3}\left(\frac{Fl^3}{EI}\right)$$

$$M_A = Fl$$

**Case 2**

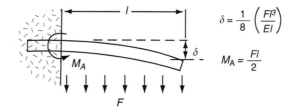

$$\delta = \frac{1}{8}\left(\frac{Fl^3}{EI}\right)$$

$$M_A = \frac{Fl}{2}$$

**Case 3**

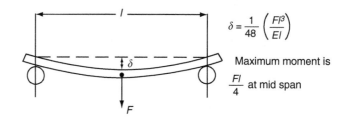

$$\delta = \frac{1}{48}\left(\frac{Fl^3}{EI}\right)$$

Maximum moment is

$\frac{Fl}{4}$ at mid span

**Case 4**

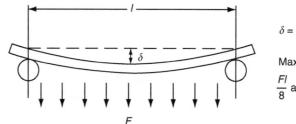

$$\delta = \frac{5}{384}\left(\frac{Fl^3}{EI}\right)$$

Maximum moment is

$\frac{Fl}{8}$ at mid span

## Case 5

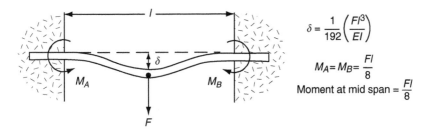

$$\delta = \frac{1}{192}\left(\frac{Fl^3}{EI}\right)$$

$$M_A = M_B = \frac{Fl}{8}$$

Moment at mid span $= \dfrac{Fl}{8}$

## Case 6

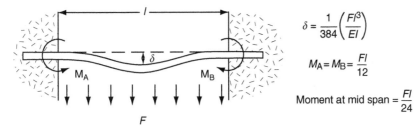

$$\delta = \frac{1}{384}\left(\frac{Fl^3}{EI}\right)$$

$$M_A = M_B = \frac{Fl}{12}$$

Moment at mid span $= \dfrac{Fl}{24}$

## Second moments of area

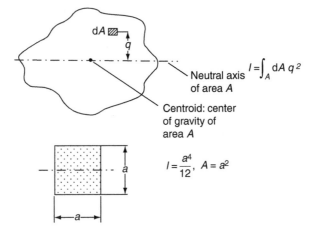

Neutral axis of area $A$

$$I = \int_A dA\, q^2$$

Centroid: center of gravity of area $A$

$$I = \frac{a^4}{12}, \quad A = a^2$$

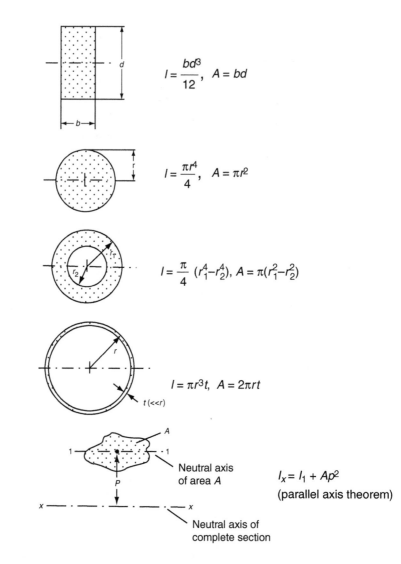

$$I = \frac{bd^3}{12}, \quad A = bd$$

$$I = \frac{\pi r^4}{4}, \quad A = \pi r^2$$

$$I = \frac{\pi}{4}(r_1^4 - r_2^4), \quad A = \pi(r_1^2 - r_2^2)$$

$$I = \pi r^3 t, \quad A = 2\pi r t$$

Neutral axis
of area $A$

$$I_x = I_1 + Ap^2$$
(parallel axis theorem)

Neutral axis of
complete section

## MODE 1 NATURAL VIBRATION FREQUENCIES

**Case 1**

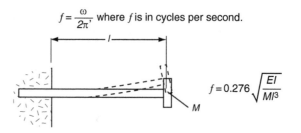

$$f = \frac{\omega}{2\pi}, \text{ where } f \text{ is in cycles per second.}$$

$$f = 0.276 \sqrt{\frac{EI}{MI^3}}$$

**Case 2**

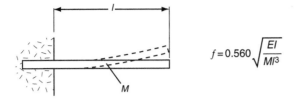

$$f = 0.560 \sqrt{\frac{EI}{Ml^3}}$$

**Case 3**

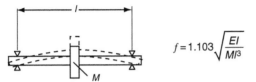

$$f = 1.103 \sqrt{\frac{EI}{Ml^3}}$$

**Case 4**

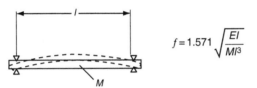

$$f = 1.571 \sqrt{\frac{EI}{Ml^3}}$$

**Case 5**

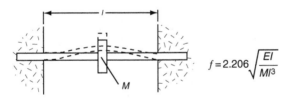

$$f = 2.206 \sqrt{\frac{EI}{Ml^3}}$$

**Case 6**

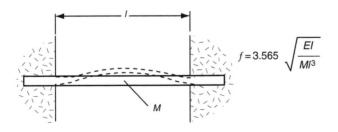

$$f = 3.565 \sqrt{\frac{EI}{Ml^3}}$$

# ELASTIC BUCKLING OF STRUTS

### Case 1

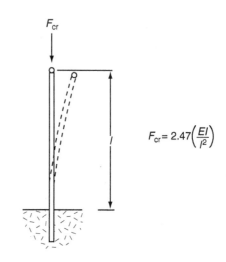

$$F_{cr} = 2.47\left(\frac{EI}{l^2}\right)$$

### Case 2

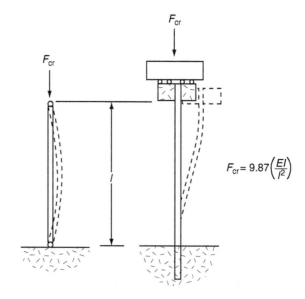

$$F_{cr} = 9.87\left(\frac{EI}{l^2}\right)$$

**Case 3**

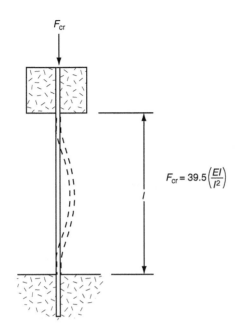

$$F_{cr} = 39.5\left(\frac{EI}{l^2}\right)$$

## CHAPTER 4

# Bonding between Atoms

## CONTENTS

## 4.1 INTRODUCTION

In order to understand the origin of material properties like *Young's modulus*, we need to focus on materials at the *atomic* level. Two things are very important in determining the modulus:

1. The forces that hold atoms together (the *interatomic bonds*) which act like little springs, linking one atom to the next in the solid state (Figure 4.1).
2. The ways in which atoms pack together (the *atom packing*), since this determines how many little springs there are per unit area, and the angle at which they are pulled (Figure 4.2).

In this chapter we look at the forces that bind atoms together—the springs. In the next chapter we examine the arrangements in which they can be packed.

**FIGURE 4.1**
The spring-like bond between two atoms.

Engineering Materials I: An Introduction to Properties, Applications, and Design, Fourth Edition
© 2012, Michael F. Ashby and David R. H. Jones. Published by Elsevier Ltd. All rights reserved.

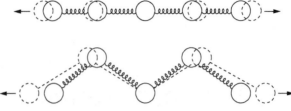

**FIGURE 4.2**
Atom packing and bond-angle.

The various ways in which atoms can be bound together involve:

1. *Primary bonds—ionic, covalent,* or *metallic* bonds, which are all relatively *strong* (they generally melt between 1000 and 4000 K)
2. *Secondary bonds—Van der Waals* and *hydrogen* bonds, which are both relatively *weak* (they melt between 100 and 500 K)

We should remember, however, when listing distinct bond types that many atoms are really bound together by bonds that are a hybrid of the simpler types (mixed bonds).

## 4.2 PRIMARY BONDS

Ceramics and metals are entirely held together by primary bonds—the ionic and covalent bond in ceramics, and the metallic and covalent bond in metals. These strong, stiff bonds give high moduli.

The *ionic bond* is the most obvious sort of electrostatic attraction between positive and negative charges. It is typified by cohesion in sodium chloride. Other alkali halides (e.g., lithium fluoride), oxides (magnesia, alumina), and components of cement (hydrated carbonates and oxides) are wholly or partly held together by ionic bonds.

Let us start with the *sodium atom.* It has a nucleus of 11 *protons*, each with a $+$ charge (and 12 neutrons with no charge at all) surrounded by 11 *electrons* each carrying a minus ($-$) charge (Figure 4.3).

The electrons are attracted to the nucleus by electrostatic forces and therefore have negative energies. But the energies of the electrons are not all the same. Those furthest from the nucleus naturally have the highest (least negative) energy. The electron that we can most easily remove from the sodium atom

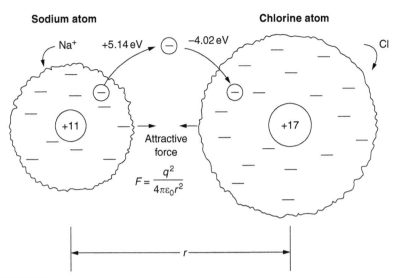

**Sodium atom**   **Chlorine atom**

**FIGURE 4.3**

The formation of an ionic bond—in this case, between a sodium atom and a chlorine atom, making sodium chloride.

is therefore the outermost one: we can remove it by expending 5.14 eV*of work. This electron can most profitably be transferred to a vacant position on a distant chlorine atom, giving us back 4.02 eV of energy. We can make isolated $Na^+$ and $Cl^-$ by doing 5.14 eV – 4.02 eV = 1.12 eV of work, $U_i$.

So far, we have had to *do* work to create the ions that will go to make up the ionic bond: not a very good start. However, the + and – charges attract each other and if we now bring them together, the force of attraction does work. The force is simply that between two opposite point charges

$$F = q^2/4\pi\varepsilon_0 r^2 \tag{4.1}$$

where $q$ is the charge on each ion, $\varepsilon_0$ is the permittivity of vacuum, and $r$ is the separation of the ions. The work done as the ions are brought to a separation $r$ (from infinity) is:

$$U = \int_r^\infty F dr = q^2/4\pi\varepsilon_0 r \tag{4.2}$$

Figure 4.4 shows how the energy of the pair of ions falls as $r$ decreases, until, at $r \approx 1$ nm for a typical ionic bond, we have paid off the 1.12 eV of work borrowed to form $Na^+$ and $Cl^-$ in the first place. For $r < 1$ nm (1 nm = $10^{-9}$ m), it is all gain, and the ionic bond now becomes more and more stable.

---

* The eV is a convenient unit for energy when dealing with atoms because the values generally lie in the range 1–10. 1 eV is equal to $1.6 \times 10^{-19}$ J.

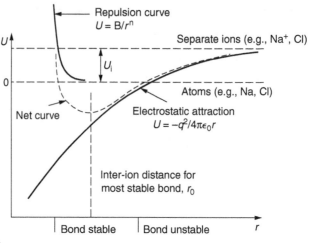

**FIGURE 4.4**
The formation of an ionic bond, viewed in terms of energy.

Why does $r$ not decrease indefinitely, releasing more and more energy, and ending in the *fusion* of the two ions? When the ions get close enough together, the electronic charge distributions start to overlap one another, and this causes a very large repulsion. Figure 4.4 shows the potential energy increase that this causes. Clearly, the ionic bond is most stable at the minimum in the $U(r)$ curve, which is well approximated by

$$U(r) = U_i - \underbrace{\frac{q^2}{4\pi\varepsilon_0 r}}_{\text{attractive part}} + \underbrace{\frac{B}{r^n}}_{\text{repulsive part}} \qquad (4.3)$$

$n$ is a large power—typically about 12.

How much can we bend this bond? The electrons of each ion occupy complicated three-dimensional regions (or "orbitals") around the nuclei. But at an approximate level we can assume the ions to be spherical, and there is then considerable freedom in the way we pack them round each other. The ionic bond therefore *lacks directionality*, although in packing ions of opposite signs, it is obviously necessary to make sure that the total charge (+ and −) adds up to zero, and that positive ions (which repel each other) are always separated by negative ions.

*Covalent bonding* appears in its pure form in diamond, silicon, and germanium—all materials with large moduli (that of diamond is the highest known). It is the dominant bond type in silicate ceramics and glasses (stone, pottery, brick, all common glasses, components of cement) and contributes to the bonding of the high-melting-point metals (tungsten, molybdenum, tantalum, etc.). It appears, too, in polymers, linking carbon atoms to each other along the

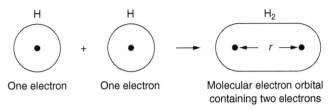

One electron          One electron          Molecular electron orbital
                                            containing two electrons

**FIGURE 4.5**

The formation of a covalent bond—in this case, between two hydrogen atoms, making a hydrogen molecule.

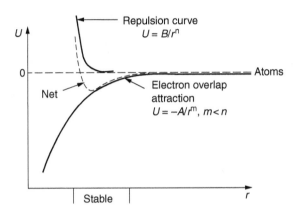

**FIGURE 4.6**

The formation of a covalent bond, viewed in terms of energy.

polymer chain; but because polymers also contain bonds of other, much weaker, types (see what follows) their moduli are usually small.

The simplest example of covalent bonding is the hydrogen molecule. The proximity of the two nuclei creates a new electron orbital, shared by the two atoms, into which the two electrons go (Figure 4.5). This sharing of electrons leads to a reduction in energy, and a stable bond, as Figure 4.6 shows. The energy of a covalent bond is well described by the empirical equation

$$U = -\frac{A}{r^m} + \frac{B}{r^n} \quad (m < n) \tag{4.4}$$

$$\underbrace{\phantom{-\frac{A}{r^m}}}_{\text{attractive part}} \quad \underbrace{\phantom{\frac{B}{r^n}}}_{\text{repulsive part}}$$

Hydrogen is hardly an engineering material. A more relevant example of the covalent bond is that of diamond, one of several solid forms of carbon. It is more of an engineering material than you might think, finding wide application for rock-drilling bits, cutting tools, polishing wheels and precision bearings. The shared electrons occupy regions that point to the corners of a tetrahedron, as shown in Figure 4.7(a). The unsymmetrical shape of these

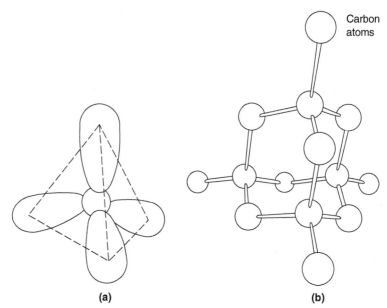

**FIGURE 4.7**
Directional covalent bonding in diamond.

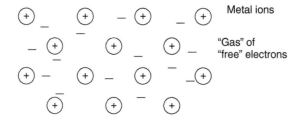

**FIGURE 4.8**
Bonding in a metal—metallic bonding.

orbitals leads to a very *directional* form of bonding in diamond, as Figure 4.7(b) shows. All covalent bonds have directionality which, in turn, influences the ways in which atoms pack together to form crystals.

The *metallic bond*, as the name says, is the dominant (though not the only) bond in metals and their alloys. In a solid (or, for that matter, a liquid) metal, the highest energy electrons tend to leave the parent atoms (which become ions) and combine to form a "sea" of freely wandering electrons, not attached to any ion in particular (Figure 4.8). This gives an energy curve that is very similar to that for covalent bonding; it is well described by Equation (4.4) and has a shape like that of Figure 4.6.

The easy movement of the electrons gives the high electrical conductivity of metals. The metallic bond has no directionality, so that metal ions tend to pack to give simple, high-density structures, like ball bearings shaken down in a box.

## 4.3 SECONDARY BONDS

Although much weaker than primary bonds, secondary bonds are still very important. They provide the links between polymer molecules in polyethylene (and other polymers) which make them solids. Without them, water would boil at $-80°C$, and life as we know it on earth would not exist.

*Van der Waals bonding* describes a dipolar attraction between *uncharged* atoms. The electronic charge on an atom is in motion; one can think of the electrons as little charged blobs whizzing round the nucleus like the moon around the earth. Averaged over time, the electron charge has spherical symmetry, but at any given instant it is unsymmetric relative to the nucleus. The instantaneous distribution has a dipole moment; this moment induces a like moment on a nearby atom and the two dipoles attract (Figure 4.9).

Dipoles attract such that their energy varies as $1/r^6$. Thus the energy of the Van der Waals bond has the form

$$U = - \underbrace{\frac{A}{r^6}}_{\text{attractive part}} + \underbrace{\frac{B}{r^n}}_{\text{repulsive part}} \quad (n \approx 12) \tag{4.5}$$

A good example is liquid nitrogen, which liquifies at $-198°C$. It is glued together by Van der Waals forces between the covalently bonded $N_2$ molecules. The thermal agitation produced when liquid nitrogen is poured on the floor at room temperature is more than enough to break the Van der Waals bonds, showing how weak they are. But without these bonds, most gases would not

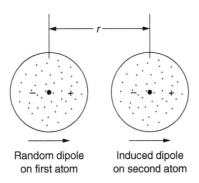

Random dipole          Induced dipole
on first atom          on second atom

**FIGURE 4.9**
Van der Waals bonding; the atoms are held together by the dipole charge distribution.

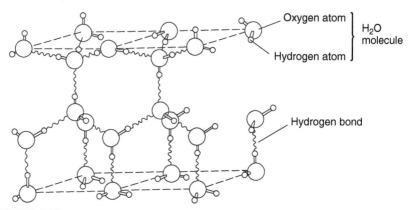

**FIGURE 4.10**

The arrangement of $H_2O$ molecules in the common form of ice, showing the hydrogen bonds. The hydrogen bonds keep the molecules well apart, which is why ice has a lower density than water.

liquefy at attainable temperatures, and we should not be able to separate industrial gases from the atmosphere.

*Hydrogen bonds* keep water liquid at room temperature, and bind polymer chains together to give solid polymers. Ice (Figure 4.10) is hydrogen bonded. Each hydrogen atom shares its charge with the nearest oxygen atom. The hydrogen, having lost part of its share, acquires a +ve charge; the oxygen, having a share in more electrons than it should, is −ve. The positively charged H atom acts as a bridging bond between neighboring oxygen ions—the charge redistribution gives each $H_2O$ molecule a dipole moment, which attracts other $H_2O$ dipoles.

## 4.4 THE CONDENSED STATES OF MATTER

It is because these primary and secondary bonds form that matter condenses from the gaseous state to give liquids and solids. Five distinct *condensed states of matter*, differing in their structure and the state of their bonding, can be identified (Table 4.1). The bonds in ordinary liquids have melted, and for this reason the liquid resists compression, but not shear; the bulk modulus, $K$, is large compared to the gas because the atoms are in contact, so to speak; but the shear modulus, $G$, is zero because they can slide past each other.

The other states of matter, listed in Table 4.1, are distinguished by the state of their bonding (molten versus solid) and by their structure (crystalline versus noncrystalline). These differences are reflected in the relative magnitudes of their bulk modulus and of their shear modulus—the more liquid the material becomes, the smaller is its ratio of $G/K$.

**Table 4.1** Condensed States of Matter

| State | Bonds | | Moduli | |
| --- | --- | --- | --- | --- |
| | Molten | Solid | K | G and E |
| 1. Liquids | * | | Large | Zero |
| 2. Liquid crystals | * | | Large | Some nonzero but very small |
| 3. Rubbers | * (secondary) | * (primary) | Large | Small (E ≈ K) |
| 4. Glasses | | * | Large | Large (E ≈ K) |
| 5. Crystals | | * | Large | Large (E ≈ K) |

## 4.5 INTERATOMIC FORCES

Having established the various types of bonds that can form between atoms, and the shapes of their potential energy curves, we are now in a position to explore the *forces* between atoms. Starting with the $U(r)$ curve, we can find this force $F$ for any separation of the atoms, $r$, from the relationship

$$F = \frac{dU}{dr}$$

(4.6)

Figure 4.11 shows the shape of the *force*/distance curve that we get from a typical *energy*/distance curve in this way. Points to note are:

1. $F$ is zero at the equilibrium point $r = r_0$; however, if the atoms are pulled apart by distance $(r - r_0)$ a resisting force appears. For *small* $(r - r_0)$ the resisting force is proportional to $(r - r_0)$ for all materials, in both tension and compression.
2. The *stiffness*, $S$, of the bond is given by

$$S = \frac{dF}{dr} = \frac{d^2U}{dr^2}$$

(4.7)

When the stretching is small, $S$ is constant and equal to

$$S_0 = \left(\frac{d^2U}{dr^2}\right)_{r=r_0}$$

(4.8)

The bond behaves in linear elastic manner—this is the physical origin of Hooke's law.

To conclude, the concept of bond stiffness, based on the energy–distance curves for the various bond types, goes a long way towards explaining the origin of the elastic modulus. But we need to find out how individual atom bonds build up to form whole pieces of material before we can fully explain experimental data for the modulus. The nature of the bonds we have mentioned influences the *packing* of atoms in engineering materials. This is the subject of the next chapter.

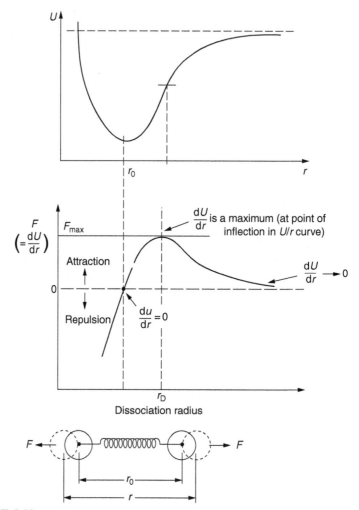

**FIGURE 4.11**
The energy–distance curve (*top*), when differentiated (Equation (4.6)) gives the force–distance curve (*center*).

## EXAMPLES

**4.1** The potential energy $U$ of two atoms, a distance $r$ apart, is

$$U = -\frac{A}{r^m} + \frac{B}{r^n}, \quad m = 2, \quad n = 10$$

Given that the atoms form a stable molecule at a separation of 0.3 nm with an energy of $-4$ eV, calculate $A$ and $B$. Also find the force required to break the molecule, and the critical separation at which the molecule breaks. You should sketch an energy/distance curve for the atom, and sketch beneath this curve the appropriate *force*/distance curve.

**Answers**
A: $7.2 \times 10^{-20}$ J nm$^2$; B: $9.4 \times 10^{-25}$ J nm$^{10}$; Force: $2.39 \times 10^{-9}$ N at 0.352 nm

**4.2** The potential energy $U$ of a pair of atoms in a solid can be written as

$$U = -\frac{A}{r^m} + \frac{B}{r^n}$$

where $r$ is the separation of the atoms, and $A$, $B$, $m$, and $n$ are positive constants. Indicate the physical significance of the two terms in this equation.

**4.3** The table below gives the Young's modulus, $E$, the atomic volume, $\Omega$, and the melting temperature, $T_M$, for a number of metals. If

$$E \approx \frac{\tilde{A}kT_M}{\Omega}$$

(where $k$ is Boltzmann's constant and $\tilde{A}$ is a constant), calculate and tabulate the value of the constant $\tilde{A}$ for each metal. Hence find an arithmetic mean of $\tilde{A}$ for these metals.

Use the equation, with the average $\tilde{A}$, to calculate the approximate Young's modulus of (a) diamond and (b) ice. Compare these with the experimental values of $1.0 \times 10^{12}$ N m$^{-2}$ and $7.7 \times 10^9$ N m$^{-2}$, respectively. Watch the units!

| Material | $\Omega \times 10^{29}$ (m$^3$) | $T_M$ (K) | $E$ (GN m$^{-2}$) |
|---|---|---|---|
| Nickel | 1.09 | 1726 | 214 |
| Copper | 1.18 | 1356 | 124 |
| Silver | 1.71 | 1234 | 76 |
| Aluminum | 1.66 | 933 | 69 |
| Lead | 3.03 | 600 | 14 |
| Iron | 1.18 | 1753 | 196 |
| Vanadium | 1.40 | 2173 | 130 |
| Chromium | 1.20 | 2163 | 289 |
| Niobium | 1.80 | 2741 | 100 |
| Molybdenum | 1.53 | 2883 | 360 |
| Tantalum | 1.80 | 3271 | 180 |
| Tungsten | 1.59 | 3683 | 406 |

Data for ice and for diamond.

| Ice | Diamond |
|---|---|
| $\Omega = 3.27 \times 10^{-29}$ m$^3$ | $\Omega = 5.68 \times 10^{-30}$ m$^3$ |
| $T_M = 273$ K | $T_M = 4200$ K |
| $E = 7.7 \times 10^9$ N m$^{-2}$ | $E = 1.0 \times 10^{12}$ N m$^{-2}$ |

**Answers**

Mean $\tilde{A} = 88$. Calculated moduli: diamond, $9.0 \times 10^{11}$ N m$^{-2}$; ice, $1.0 \times 10^{10}$ N m$^{-2}$

**4.4** Starting with Equation (4.3), show that the value of the constant $B$ for an ionic bond is given by

$$B = \frac{q^2 r_0^{n-1}}{4\pi n\varepsilon_0}$$

Hence show that, for an ionic bond, the stiffness $S_0$ for small displacements from the equilibrium separation is given by

$$S_0 = \frac{\alpha q^2}{4\pi\varepsilon_0 r_0^3}$$

where $\alpha = (n - 1)$.

**4.5** How does the type of bonding in metals explain the following?

    **a.** Atoms in solid metals tend to pack into simple high-density structures.

    **b.** Most metals have a comparatively high electrical conductivity.

**4.6** How does the nature of the bonding in ice explain the fact that ice is less dense than water?

**4.7** How does the shape of the energy-distance curve between atoms explain the linear elastic behavior of many materials?

## CHAPTER 5

# Packing of Atoms in Solids

## CONTENTS

## 5.1 INTRODUCTION

In the previous chapter we examined the stiffnesses of the bonds holding atoms together. But bond stiffness alone does not explain the stiffness of solids; the way in which the atoms are packed together is equally important. In this chapter we examine how atoms are arranged in the main engineering solids.

Engineering Materials I: An Introduction to Properties, Applications, and Design, Fourth Edition

## 5.2 ATOM PACKING IN CRYSTALS

Many engineering materials (almost all metals and ceramics, for instance) are made up entirely of small crystals or *grains* in which atoms are packed in regular, repeating, three-dimensional patterns; the grains are stuck together, meeting at *grain boundaries*, which we will describe later. We focus now on the individual crystals, which can best be understood by thinking of the atoms as *hard spheres* (although, from what we said in the previous chapter, it is clear that this is a big simplification). To make things even simpler, for the moment consider a material that is *pure*—with only one size of hard sphere—which has *nondirectional bonding*, so we can arrange the spheres subject only to geometrical constraints. Pure copper is a good example of a material satisfying these conditions.

To build up a three-dimensional packing pattern, it is easier conceptually to begin by

(i) packing atoms two-dimensionally in *atomic planes*,
(ii) stacking these planes on top of one another to give *crystals*.

## 5.3 CLOSE-PACKED STRUCTURES AND CRYSTAL ENERGIES

An example of how we might pack atoms in a *plane* is shown in Figure 5.1; it is the arrangement in which the reds are set up on a billiard table before starting a game of snooker. The balls are packed in a triangular fashion so as to take up the least possible space on the table. This type of plane is thus called a *close-packed plane*, and contains three *close-packed directions*; they are the directions along which the balls touch. The figure shows only a small region of close-packed plane—if we had more reds we could extend the plane sideways and could, if we wished, fill the whole billiard table. The important thing to notice is the way in which the balls are packed in a *regularly repeating two-dimensional pattern*.

How could we add a second layer of atoms to our close-packed plane? As Figure 5.1 shows, the depressions where the atoms meet are ideal "seats" for the next layer of atoms. By dropping atoms into alternate seats, we can generate a second close-packed plane lying on top of the original one and having an identical packing pattern. Then a third layer can be added, and a fourth, and so on until we have made a sizeable piece of crystal—with, this time, a *regularly repeating pattern of atoms in three dimensions*. The particular structure we have produced is one in which the atoms take up the *least volume* and is therefore called a *close-packed structure*. The atoms in many solid metals are packed in this way.

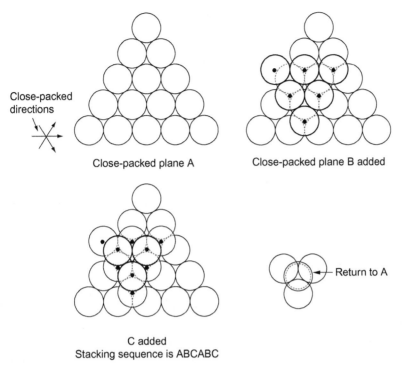

Close-packed directions

Close-packed plane A

Close-packed plane B added

Return to A

C added
Stacking sequence is ABCABC

**FIGURE 5.1**
The close packing of hard-sphere atoms. The ABC stacking gives the "face-centered cubic" (f.c.c.) structure.

There is a complication to this apparently simple story. There are two alternative and different *sequences* in which we can stack the close-packed planes on top of one another. If we follow the stacking sequence in Figure 5.1 rather more closely, we see that, by the time we have reached the *fourth* atomic plane, we are placing the atoms directly above the original atoms (although, naturally, separated from them by the two interleaving planes of atoms). We then carry on adding atoms as before, generating an ABCABC... sequence.

In Figure 5.2 we show the alternative way of stacking, in which the atoms in the *third* plane are now directly above those in the first layer. This gives an ABAB... sequence. These two different stacking sequences give two different three-dimensional packing structures—*face-centered cubic (f.c.c.)* and *close–packed hexagonal (c.p.h.)* respectively. Many common metals (e.g., Al, Cu and Ni) have the f.c.c. structure and many others (e.g., Mg, Zn, and Ti) have the c.p.h. structure.

Why should Al choose to be f.c.c. while Mg chooses to be c.p.h.? The answer to this is that materials choose the crystal structure that gives minimum *energy*. This structure may not necessarily be close packed or, indeed, very simple

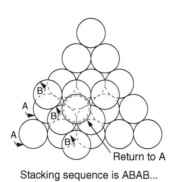

Return to A

Stacking sequence is ABAB...

**FIGURE 5.2**

Close packing of hard-sphere atoms—an alternative arrangement, giving the "close-packed hexagonal" structure.

geometrically—although to be a crystal it must still have some sort of three–dimensional repeating pattern.

The difference in energy between alternative structures is often slight. Because of this, the crystal structure that gives the minimum energy at one temperature may not do so at another. Thus tin changes its crystal structure if it is cooled enough; and, incidentally, becomes much more brittle in the process (causing the tin-alloy coat buttons of Napoleon's army to fall apart during the harsh Russian winter; and the soldered cans of paraffin on Scott's South Pole expedition to leak, with tragic consequences). Cobalt changes its structure at 450°C, transforming from a c.p.h. structure at lower temperatures to an f.c.c. structure at higher temperatures. More important, pure iron transforms from a b.c.c. structure (defined in the following) to one that is f.c.c. at 911°C, a process which is important in the heat treatment of steels.

## 5.4 CRYSTALLOGRAPHY

We have not yet explained why an ABCABC sequence is called "f.c.c." or why an ABAB sequence is referred to as "c.p.h." And we have not even begun to describe the features of the more complicated crystal structures like those of ceramics such as alumina. To explain things such as the geometric differences between f.c.c. and c.p.h. or to ease the conceptual labor of constructing complicated crystal structures, we need an appropriate descriptive language. The methods of *crystallography* provide this language, and give us an essential shorthand way of describing crystal structures.

Let us illustrate the crystallographic approach in the case of f.c.c. Figure 5.3 shows that the *atom centers* in f.c.c. can be placed at the corners of a cube and in the centers of the cube faces. The cube, of course, has no physical significance but is merely a constructional device. It is called a *unit cell*. If we look

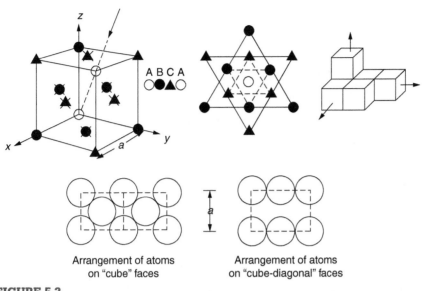

ABCA

Arrangement of atoms
on "cube" faces

Arrangement of atoms
on "cube-diagonal" faces

**FIGURE 5.3**
The face-centered cubic structure.

along the cube diagonal, we see the view shown in Figure 5.3 (*top center*): a tri-angular pattern which, with a little effort, can be seen to be bits of close-packed planes stacked in an ABCABC sequence.

This unit–cell visualization of the atomic positions is thus exactly equivalent to our earlier approach based on stacking of close-packed planes, but is much more powerful as a descriptive aid. For example, we can see how our complete f.c.c. crystal is built up by attaching further unit cells to the first one (like assembling a set of children's building cubes) so as to fill space without leaving awkward gaps—something you cannot so easily do with 5-sided shapes (in a plane) or 7-sided shapes (in three dimensions).

Beyond this, inspection of the unit cell reveals planes in which the atoms are packed other than in a close-packed way. On the "cube" faces the atoms are packed in a square array, and on the cube-diagonal planes in separated rows, as shown in Figure 5.3. Obviously, properties like the shear modulus might well be different for close-packed planes and cube planes, because the number of bonds attaching them per unit area is different. This is one of the reasons that it is important to have a method of describing various planar packing arrangements.

Let us now look at the c.p.h. unit cell as shown in Figure 5.4. A view looking down the vertical axis reveals the ABA stacking of close-packed planes. We build up our c.p.h. crystal by adding hexagonal building blocks to one another: hexagonal blocks also stack so that they fill space. Here, again, we can use the unit cell concept to "open up" views of the various types of planes.

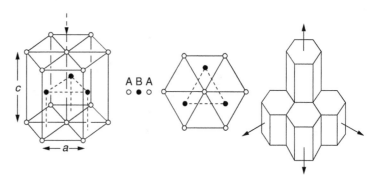

**FIGURE 5.4**
The close-packed hexagonal structure.

## 5.5 PLANE INDICES

We could make scale drawings of the many types of planes that we see in unit cells; but the concept of a unit cell also allows us to describe any plane by a set of numbers called *Miller indices*. The two examples given in Figure 5.5 should enable you to find the Miller index of any plane in a cubic unit cell, although they take a little getting used to. The indices (for a plane) are the *reciprocals* of the intercepts the plane makes with the three axes, reduced to the smallest integers (reciprocals are used simply to avoid infinities when planes are parallel to axes).

As an example, the six individual *"cube" planes* are called (100), (010), (001). Collectively this type of plane is called {100}, with curly brackets. Similarly the six cube *diagonal* planes are (110), (1$\bar{1}$0), (101), ($\bar{1}$01), (011), and (0$\bar{1}$1), or, collectively, {110}. (Here the sign $\bar{1}$ means an intercept of –1.) As a final example, our original close-packed planes—the ones of the ABC stacking—are of {111} type. Obviously the unique structural description of "{111} f.c.c." is a good deal more succinct than a scale drawing of close-packed billiard balls.

Different indices are used in hexagonal cells (we build a c.p.h. crystal up by adding bricks in four directions, not three as in cubic). We do not need them here—the crystallography books listed under "References" at the end of the book describe them fully.

## 5.6 DIRECTION INDICES

Properties like Young's modulus may well vary with *direction* in the unit cell; for this (and other) reasons we need a succinct description of crystal directions. Figure 5.6 shows the method and illustrates some typical directions. The indices

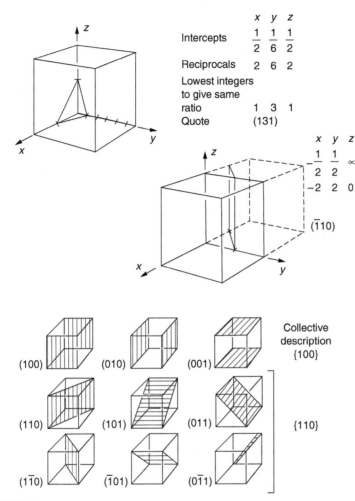

| | x | y | z |
|---|---|---|---|
| Intercepts | $\frac{1}{2}$ | $\frac{1}{6}$ | $\frac{1}{2}$ |
| Reciprocals | 2 | 6 | 2 |
| Lowest integers to give same ratio | 1 | 3 | 1 |
| Quote | (131) | | |

| | x | y | z |
|---|---|---|---|
| | $\frac{1}{2}$ | $\frac{1}{2}$ | $\infty$ |
| | $-2$ | 2 | 0 |
| | ($\bar{1}$10) | | |

(100)  (010)  (001)    Collective description {100}

(110)  (101)  (011)    {110}

($1\bar{1}0$)  ($\bar{1}01$)  ($0\bar{1}1$)

**FIGURE 5.5**
Miller indices for identifying crystal planes, showing how the (1 3 1) plane and the ($\bar{1}$ 1 0) planes are defined. The lower part of the figure shows the family of {1 0 0} and of {1 1 0} planes.

of direction are the components of a vector (*not* reciprocals, because infinities do not crop up here), starting from the origin, along the desired direction, again reduced to the smallest integer set. A single direction (like the "111" direction which links the origin to the opposite corner of the cube is given square brackets (i.e., [111]), to distinguish it from the Miller index of a plane. The family of directions of this type (illustrated in Figure 5.6) is identified by angled brackets: ⟨111⟩.

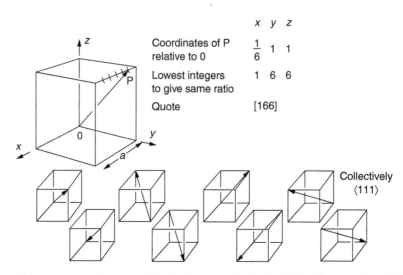

Note: In cubic systems only [111] is the normal to (111); [100] is the normal to (100), and so on.

**FIGURE 5.6**
Direction indices for identifying crystal directions, showing how the [1 6 6] direction is defined. The lower part of the figure shows the family of ⟨111⟩ directions.

## 5.7 OTHER SIMPLE IMPORTANT CRYSTAL STRUCTURES

Figure 5.7 shows a new crystal structure, and an important one: it is the body-centered cubic (b.c.c.) structure of tungsten, chromium, iron, and many steels. The ⟨111⟩ directions are close packed (that is to say: the atoms touch along this direction) but there are no close-packed planes. The result is that b.c.c. packing is less dense than either f.c.c. or c.p.h. It is found in materials that have *directional bonding*: the directionality distorts the structure, preventing the atoms from dropping into one of the two close-packed structures we have just described. There are other structures involving only one sort of atom which are not close packed, for the same reason, but we don't need them here.

In compound materials—in the ceramic sodium chloride, for instance—there are two (sometimes more) species of atoms, packed together. The crystal structures of such compounds can still be simple. Figure 5.8(a) shows that the ceramics NaCl, KCl, and MgO, for example, also form a cubic structure. When two species of atoms are not in the ratio 1 : 1, as in compounds like the nuclear fuel $UO_2$ (a ceramic too) the structure is more complicated, as shown in Figure 5.8(b), although this too has a cubic unit cell.

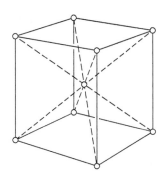

**FIGURE 5.7**
The body-centered cubic structure.

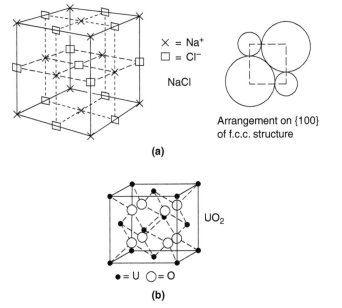

X = Na⁺
□ = Cl⁻

NaCl

Arrangement on {100}
of f.c.c. structure

**(a)**

UO₂

● = U  ○ = O

**(b)**

**FIGURE 5.8**
(a) Packing of unequally sized ions of sodium chloride to give the f.c.c. structure; KCl and MgO pack in the same way. (b) Packing of ions in uranium dioxide; this is more complicated than in NaCl because the U and O ions are not in a 1:1 ratio.

## 5.8 ATOM PACKING IN POLYMERS

Polymers are much more complex structurally than metals, and because of this they have special mechanical properties. The extreme elasticity of a rubber band is one; the formability of polyethylene is another.

Polymers are huge chain-like molecules (huge, that is, by the standards of an atom) in which the atoms forming the backbone of the chain are linked by *covalent* bonds. The chain backbone is usually made from carbon atoms (although a limited range of silicon–based polymers can be synthesized—they are called "silicones"). A typical high polymer ("high" means "of large molecular weight") is polyethylene. It is made by the catalytic polymerization of ethylene, shown on the left, to give a chain of ethylenes, minus the double bond:

$$
\begin{array}{cc}
\text{H} \ \ \text{H} \\
| \ \ \ | \\
\text{C} = \text{C} \\
| \ \ \ | \\
\text{H} \ \ \text{H}
\end{array}
\rightarrow
\begin{array}{cccccc}
\text{H} & \text{H} & \text{H} & \text{H} & \text{H} & \text{H} \\
| & | & | & | & | & | \\
-\text{C} & -\text{C} & -\text{C} & -\text{C} & -\text{C} & -\text{C} - \text{etc.} \\
| & | & | & | & | & | \\
\text{H} & \text{H} & \text{H} & \text{H} & \text{H} & \text{H}
\end{array}
$$

Polystyrene, similarly, is made by the polymerization of styrene (*left*), again by sacrificing the double bond to provide the bonds that produce the chain:

$$
\begin{array}{cc}
\text{H} \ \ \text{C}_6\text{H}_5 \\
| \ \ \ \ \ | \\
\text{C} = \text{C} \\
| \ \ \ \ \ | \\
\text{H} \ \ \text{H}
\end{array}
\rightarrow
\begin{array}{cccccc}
\text{H} & \text{C}_6\text{H}_5 & \text{H} & \text{H} & \text{H} & \text{C}_6\text{H}_5 \\
| & | & | & | & | & | \\
-\text{C} & -\text{C} & -\text{C} & -\text{C} & -\text{C} & -\text{C} - \text{etc.} \\
| & | & | & | & | & | \\
\text{H} & \text{H} & \text{H} & \text{C}_6\text{H}_5 & \text{H} & \text{H}
\end{array}
$$

A copolymer is made by polymerization of two monomers, adding them randomly (a random copolymer) or in an ordered way (a block copolymer). An example is styrene–butadiene rubber, SBR. Styrene, extreme left, loses its double bond; butadiene, richer in double bonds to start with, keeps one.

$$
\begin{array}{cc}
\text{H} \ \ \text{C}_6\text{H}_5 \ \ \text{H} \\
| \ \ \ \ \ | \ \ \ \ \ | \\
\text{C} = \text{C} + \text{C} = \text{C} - \text{C} = \text{C} \\
| \ \ \ \ \ | \ \ \ \ \ | \\
\text{H} \ \ \text{H} \ \ \text{H}
\end{array}
\rightarrow
\begin{array}{cccccc}
\text{H} & \text{C}_6\text{H}_5 & \text{H} & \text{H} & \text{H} & \text{H} \\
| & | & | & | & | & | \\
-\text{C} & -\text{C} & -\text{C} & -\text{C} & = \text{C} & -\text{C} - \text{etc.} \\
| & | & | & | & | & | \\
\text{H} & \text{H} & \text{H} & & \text{H} & \text{H}
\end{array}
$$

Molecules such as these form long, flexible, spaghetti-like chains (Figure 5.9). Figure 5.10 shows how they pack to form bulk material. In some polymers the chains can be folded carefully backwards and forwards over one another so as to look like the firework called the "jumping jack". The regularly repeating symmetry of this chain folding leads to crystallinity, so polymers can be crystalline. More usually the chains are arranged *randomly* and *not* in regularly repeating three-dimensional patterns. These polymers are thus *noncrystalline*, or *amorphous*. Many contain both amorphous and crystalline regions, as shown in Figure 5.10—they are *partly crystalline*.

There is a whole science called *molecular architecture* devoted to making all sorts of chains and trying to arrange them in all sorts of ways to make the final material. There are currently thousands of different polymeric materials, all having different properties. This sounds complicated, but we need only a few: six basic polymers account for almost 95% of all current production.

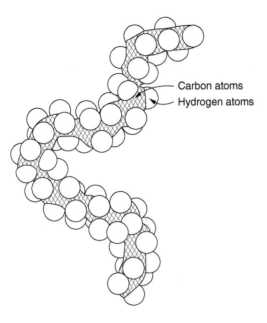

Carbon atoms
Hydrogen atoms

**FIGURE 5.9**
The three–dimensional appearance of a short bit of a polyethylene molecule.

## 5.9 ATOM PACKING IN INORGANIC GLASSES

Inorganic glasses are mixtures of oxides, almost always with silica, $SiO_2$, as the major ingredient. The atoms in glasses are packed in a noncrystalline (or amorphous) way. Figure 5.11(a) shows the structure of silica glass, which is solid to well over 1000°C because of the strong covalent bonds linking the Si to the O atoms. Adding soda ($Na_2O$) breaks up the structure and lowers the *softening temperature* (at which the glass can be worked) to about 600°C. This soda glass (Figure 5.11(b)) is the material from which milk bottles and window panes are made. Adding boron oxide ($B_2O_3$) instead gives boro–silicate glasses (pyrex is one) which withstand thermal shocks better than ordinary window glass.

## 5.10 THE DENSITY OF SOLIDS

The densities of common engineering materials are listed in Table 5.1 and shown in Figure 5.12 (see page 80). These reflect the mass and diameter of the atoms that make them up and the efficiency with which they are packed to fill space. Metals, most of them, have high densities because the atoms are heavy and closely packed. Polymers are much less dense because the atoms of which they are made (C, H, O) are light, and the structures are not close packed.

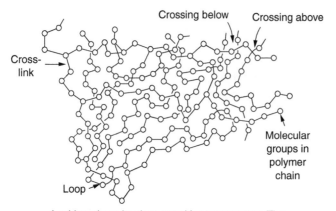

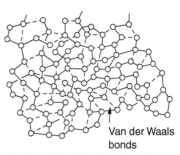

A rubber above its glass-transition temperature. The structure is entirely amorphous. The chains are held together only by occasional *covalent* cross-linking.

**(a)**

A rubber below its glass-transition temperature. In addition to occasional covalent cross-linking the molecular groups in the polymer chains attract by Van der Waals bonding, tying the chains closely to one another.

**(b)**

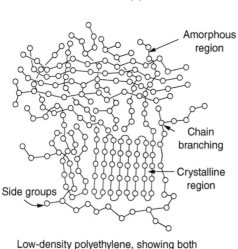

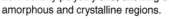

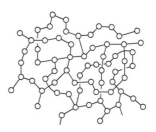

Low-density polyethylene, showing both amorphous and crystalline regions.

**(c)**

A polymer (e.g., epoxy resin) where the chains are tied tightly together by frequent *covalent* cross-links.

**(d)**

**FIGURE 5.10**
How the molecules are packed together in polymers.

Ceramics—even the ones in which atoms are packed closely—are, on average, a little less dense than metals because most of them contain light atoms such as O, N, and C. Composites have densities that are an average of the materials from which they are made.

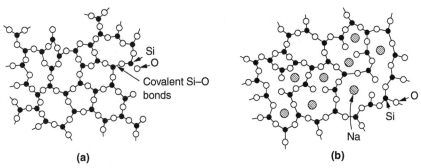

**FIGURE 5.11**
(a) Atom packing in amorphous (glassy) silica. (b) How the addition of soda breaks up the bonding in amorphous silica, giving soda glass.

## EXAMPLES

**5.1 a.** Calculate the density of an f.c.c. packing of spheres of unit density.

  **b.** If these same spheres are packed to form a *glassy* structure, the arrangement is called "dense random packing" (d.r.p.) and has a density of 0.636. If crystalline f.c.c. nickel has a density of 8.90 Mg m$^{-3}$, calculate the density of glassy nickel.

**Answers**
(a) 0.740; (b) 7.65 Mg m$^{-3}$

**5.2 a.** Sketch three-dimensional views of the unit cell of a b.c.c. crystal, showing a (100), a (110), a (111) and a (210) plane.

  **b.** The slip planes of b.c.c. iron are the {110} planes: sketch the atom arrangement in these planes, and mark the ⟨111⟩ slip directions.

  **c.** Sketch three-dimensional views of the unit cell of an f.c.c. crystal, showing a [100], a [110], a [111] and a [211] direction.

  **d.** The slip planes of f.c.c. copper are the {111} planes: sketch the atom arrangement in these planes and mark the ⟨110⟩ slip directions.

**5.3 a.** The atomic diameter of an atom of nickel is 0.2492 nm. Calculate the lattice constant *a* of f.c.c. nickel.

  **b.** The atomic weight of nickel is 58.71 kg kmol$^{-1}$. Calculate the density of nickel. (Calculate first the mass per atom, and the number of atoms in a unit cell.)

  **c.** The atomic diameter of an atom of iron is 0.2482 nm. Calculate the lattice constant *a* of b.c.c. iron.

  **d.** The atomic weight of iron is 55.85 kg kmol$^{-1}$. Calculate the density of iron.

**Answers**
(a) 0.352 nm; (b) 8.91 Mg m$^{-3}$; (c) 0.287 nm; (d) 7.88 Mg m$^{-3}$

**5.4** Crystalline copper and magnesium have f.c.c and c.p.h structures respectively.

  **a.** Assuming that the atoms can be represented as hard spheres, calculate the percentage of the volume occupied by atoms in each material.

## Table 5.1 Data for Density, $\rho$

| Material | $\rho$ (Mg m$^{-3}$) |
| --- | --- |
| Osmium | 22.7 |
| Platinum | 21.4 |
| Tungsten and alloys | 13.4–19.6 |
| Gold | 19.3 |
| Uranium | 18.9 |
| Tungsten carbide, WC | 14.0–17.0 |
| Tantalum and alloys | 16.6–16.9 |
| Molybdenum and alloys | 10.0–13.7 |
| Cobalt/tungsten–carbide cermets | 11.0–12.5 |
| Lead and alloys | 10.7–11.3 |
| Silver | 10.5 |
| Niobium and alloys | 7.9–10.5 |
| Nickel | 8.9 |
| Nickel alloys | 7.8–9.2 |
| Cobalt and alloys | 8.1–9.1 |
| Copper | 8.9 |
| Copper alloys | 7.5–9.0 |
| Brasses and bronzes | 7.2–8.9 |
| Iron | 7.9 |
| Iron-based super-alloys | 7.9–8.3 |
| Stainless steels, austenitic | 7.5–8.1 |
| Tin and alloys | 7.3–8.0 |
| Low-alloy steels | 7.8–7.85 |
| Mild steel | 7.8–7.85 |
| Stainless steel, ferritic | 7.5–7.7 |
| Cast iron | 6.9–7.8 |
| Titanium carbide, TiC | 7.2 |
| Zinc and alloys | 5.2–7.2 |
| Chromium | 7.2 |
| Zirconium carbide, ZrC | 6.6 |
| Zirconium and alloys | 6.6 |
| Titanium | 4.5 |
| Titanium alloys | 4.3–5.1 |
| Alumina, $Al_2O_3$ | 3.9 |
| Alkali halides | 3.1–3.6 |
| Magnesia, MgO | 3.5 |
| Silicon carbide, SiC | 2.5–3.2 |
| Silicon nitride, $Si_3N_4$ | 3.2 |
| Mullite | 3.2 |
| Beryllia, BeO | 3.0 |
| Common rocks | 2.2–3.0 |

**Table 5.1** Cont'd

| Material | $\rho$ (Mg m$^{-3}$) |
|---|---|
| Calcite (marble, limestone) | 2.7 |
| Aluminum | 2.7 |
| Aluminum alloys | 2.6–2.9 |
| Silica glass, SiO$_2$ (quartz) | 2.6 |
| Soda glass | 2.5 |
| Concrete/cement | 2.4–2.5 |
| GFRPs | 1.4–2.2 |
| Carbon fibers | 2.2 |
| PTFE | 2.3 |
| Boron fiber/epoxy | 2.0 |
| Beryllium and alloys | 1.85–1.9 |
| Magnesium and alloys | 1.74–1.88 |
| Fiberglass (GFRP/polyester) | 1.55–1.95 |
| Graphite, high strength | 1.8 |
| PVC | 1.3–1.6 |
| CFRPs | 1.5–1.6 |
| Polyesters | 1.1–1.5 |
| Polyimides | 1.4 |
| Epoxies | 1.1–1.4 |
| Polyurethane | 1.1–1.3 |
| Polycarbonate | 1.2–1.3 |
| PMMA | 1.2 |
| Nylon | 1.1–1.2 |
| Polystyrene | 1.0–1.1 |
| Polyethylene, high-density | 0.94–0.97 |
| Ice, H$_2$O | 0.92 |
| Natural rubber | 0.83–0.91 |
| Polyethylene, low-density | 0.91 |
| Polypropylene | 0.88–0.91 |
| Common woods | 0.4–0.8 |
| Cork | 0.1–0.2 |
| Foamed plastics | 0.01–0.6 |

b. Calculate, from first principles, the dimensions of the unit cell in copper and in magnesium. (The densities of copper and magnesium are 8.96 Mg m$^{-3}$ and 1.74 Mg m$^{-3}$.)

**Answers**

(a) 74% for both; (b) copper: $a = 0.361$ nm, magnesium: $a = 0.320$ nm, $c = 0.523$ nm

5.5 What are the factors that determine the densities of solids? Why are metals more dense than polymers?

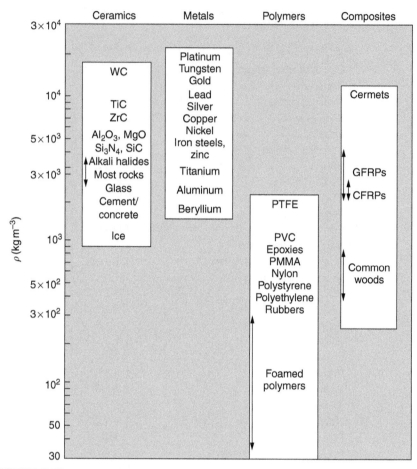

**FIGURE 5.12**
Bar chart of data for density, $\rho$.

**5.6** The density of a crystal of polyethylene (PE) is 1.014 (all densities are in Mg m$^{-3}$ at 20°C). The density of amorphous PE is 0.84. Estimate the percentage crystallinity in:

**a.** a low-density PE with a density of 0.92,

**b.** a high-density PE with a density of 0.97.

**Answers**

(a) 46%; (b) 75%

## CHAPTER 6

# The Physical Basis of Young's Modulus

## CONTENTS

## 6.1 INTRODUCTION

We can now bring together the factors underlying the moduli of materials. First, look back to Figure 3.7, the bar chart showing the moduli of materials. Recall that most ceramics and metals have moduli in a relatively narrow range: 30–300 GN m$^{-2}$. Cement and concrete (45 GN m$^{-2}$) are near the bottom of that range. Aluminum (69 GN m$^{-2}$) is higher up; and steels (200 GN m$^{-2}$) are near the top. Special materials lie outside it—diamond and tungsten lie above; ice and lead lie a little below—but most crystalline materials lie in that fairly narrow range. Polymers are quite different: all of them have moduli that are smaller, some by several orders of magnitude. Why is this? What determines the general level of the moduli of solids? And is there the possibility of producing stiff polymers?

## 6.2 MODULI OF CRYSTALS

We showed in Chapter 4 that atoms in crystals are held together by bonds that behave like little springs. We defined the stiffness of these bonds as

Engineering Materials I: An Introduction to Properties, Applications, and Design, Fourth Edition

$$S_0 = \left(\frac{d^2 U}{dr^2}\right)_{r=r0} \tag{6.1}$$

For *small* strains, $S_0$ stays constant (it is the *spring constant* of the bond). This means that the force between a pair of atoms, stretched apart to a distance $r$ ($r \approx r_0$), is

$$F = S_0(r - r_0) \tag{6.2}$$

Imagine a solid held together by these little springs, linking the atoms between two planes within the material, as shown in Figure 6.1. For simplicity we put atoms at the corners of cubes of side $r_0$.

The total force exerted across *unit area*, if the two planes are pulled apart a distance $(r - r_0)$ is the stress $\sigma$ with

$$\sigma = NS_0(r - r_0) \tag{6.3}$$

$N$ is the number of bonds/unit area, equal to $1/r_0^2$ (since $r_0^2$ is the average area per atom). We convert displacement $(r - r_0)$ into strain $\varepsilon_n$ by dividing by the *initial* spacing, $r_0$, giving

$$\sigma = \left(\frac{S_0}{r_0}\right)\varepsilon_n \tag{6.4}$$

Young's modulus is

$$E = \frac{\sigma}{\varepsilon_n} = \left(\frac{S_0}{r_0}\right) \tag{6.5}$$

$S_0$ can be calculated from the theoretically derived $U(r)$ curves described in Chapter 4. This is the realm of the solid-state physicist and quantum chemist,

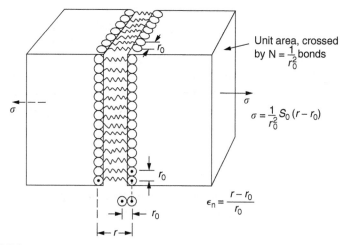

FIGURE 6.1

The method of calculating Young's modulus from the stiffnesses of individual bonds.

but we shall consider one example: the ionic bond, for which $U(r)$ is given in Equation (4.3). We showed in Example 4.4 that

$$S_0 = \frac{\alpha q^2}{4\pi\varepsilon_0 r_0^3} \qquad (6.6)$$

for the ionic bond. The coulombic attraction is a *long-range* interaction (it varies as $1/r$; an example of a short-range interaction is one that varies as $1/r^{10}$). Because of this, a given $Na^+$ ion not only interacts (attractively) with its shell of six neighboring $Cl^-$ ions, it also interacts (repulsively) with the 12 slightly more distant $Na^+$ ions, with the eight $Cl^-$ ions beyond that, and with the six $Na^+$ ions that form the shell beyond *that*. To calculate $S_0$ properly, we must sum over all these bonds, taking attractions and repulsions properly into account. The result is identical with Equation (6.7), but with $\alpha = 0.58$.

The Table of Physical Constants on the inside front cover gives values for $q$ and $\varepsilon_0$. $r_0$, the atom spacing, is close to $2.5 \times 10^{-10}$ m. Inserting these values gives:

$$S_0 = \frac{0.58(1.6 \times 10^{-19})^2}{4\pi \times 8.85 \times 10^{-12}(2.5 \times 10^{-10})^3} = 8.54 \; N\,m^{-1}$$

The stiffnesses of other bond types are calculated in a similar way (in general, the cumbersome sum just described is not needed because the interactions are of *short range*). The resulting hierarchy of bond stiffnesses is as shown in Table 6.1.

A comparison of these predicted values of $E$ with the measured values plotted in the bar chart of Figure 3.7 shows that, for metals and ceramics, the values of $E$ we calculate are about right: the bond-stretching idea explains the stiffness of these solids. We can be happy that we can explain the moduli of these classes of solid. But a paradox remains: *there exists a whole range of polymers and rubbers that have moduli that are lower—by up to a factor of 100—than the lowest we have calculated.* Why is this? What determines the moduli of these floppy polymers if it is not the springs between the atoms?

**Table 6.1 Hierarchy of Bond Stiffnesses**

| Bond Type | $S_0$ (N m$^{-1}$) | E(GPa); from Equation (6.5) (with $r_0 = 2.5 \times 10^{-10}$ m) |
|---|---|---|
| Covalent, e.g., C–C | 50–180 | 200–1000 |
| Metallic, e.g., Cu–Cu | 15–75 | 60–300 |
| Ionic, e.g., Na–Cl | 8–24 | 32–96 |
| H-bond, e.g., $H_2O$–$H_2O$ | 2–3 | 8–12 |
| Van der Waals, e.g., polymers | 0.5–1 | 2–4 |

## 6.3 RUBBERS AND THE GLASS TRANSITION TEMPERATURE

All polymers, if really solid, should have moduli above the lowest level we have calculated—about 2 GN m$^{-2}$—since they are held together partly by Van der Waals and partly by covalent bonds. If you take ordinary rubber tubing (a polymer) and cool it down in liquid nitrogen, it becomes stiff—its modulus rises rather suddenly from around $10^{-2}$ GN m$^{-2}$ to a "proper" value of 2 GN m$^{-2}$. But if you warm it up again, its modulus drops back to $10^{-2}$ GN m$^{-2}$.

This is because rubber, like many polymers, is composed of long spaghetti-like chains of carbon atoms, all tangled together, as we showed in Chapter 5. In the case of rubber, the chains are also lightly cross-linked, as shown in Figure 5.10. There are covalent bonds along the carbon chain, and where there are occasional cross-links. These are very stiff, but they contribute very little to the overall modulus because when you load the structure it is the flabby Van der Waals bonds *between* the chains that stretch, and it is these that determine the modulus.

Well, that is the case at the low temperature, when the rubber has a "proper" modulus of a few GPa. As the rubber warms up to room temperature, the Van der Waals bonds *melt*. (In fact, the stiffness of the bond is proportional to its melting point: that is why diamond, which has the highest melting point of any material, also has the highest modulus.) The rubber remains solid because of the cross-links, which form a sort of skeleton: but when you load it, the chains now slide over each other in places where there are no cross-linking bonds. This, of course, gives extra strain, and the modulus goes down (remember, $E = \sigma/\varepsilon_n$).

Many of the most floppy polymers have half-melted in this way at room temperature. The temperature at which this happens is called the *glass temperature*, $T_G$. Some polymers, which have no cross-links, melt completely at temperatures above $T_G$, becoming viscous liquids. Others, containing cross-links, become *leathery* (e.g., PVC) or rubbery (as polystyrene butadiene does). Some typical values for $T_G$ are: polymethylmethacrylate (PMMA, or perspex), 100°C; polystyrene (PS), 90°C; polyethylene (low-density form), –20°C; *natural* rubber, –40°C.

To summarize, above $T_G$, the polymer is leathery, rubbery or molten; below, it is a true solid with a modulus of at least 2 GN m$^{-2}$. This behavior is shown in Figure 6.2 which also shows how the stiffness of polymers increases as the covalent cross-link density increases, towards the value for diamond (which is simply a polymer with 100 percent of its bonds cross-linked, Figure 4.7). Stiff polymers, then, *are* possible; the stiffest now available have moduli comparable with that of aluminum.

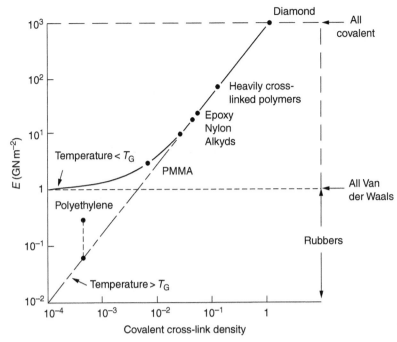

**FIGURE 6.2**

How Young's modulus increases with increasing density of covalent cross-links in polymers, including rubbers above the glass temperature. Below $T_G$, the modulus of rubbers increases markedly because the Van der Waals bonds kick in. Above $T_G$ they melt, and the modulus drops.

## 6.4 COMPOSITES

Is it possible to make polymers stiffer than the Van der Waals bonds that usually hold them together? The answer is yes—if we mix into the polymer a second, stiffer, material. Good examples of materials stiffened in this way are:

- GFRP—glass-fiber reinforced polymers, where the polymer is stiffened or reinforced by long fibers of soda glass;
- CFRP—carbon-fiber reinforced polymers, where the reinforcement is achieved with fibers of graphite;
- KFRP—Kevlar-fiber reinforced polymers, using Kevlar fibers (a unique polymer with a high density of covalent bonds oriented along the fiber axis);
- Filled polymers—polymers into which glass powder or silica flour has been mixed;
- Wood—a natural composite of lignin (an amorphous polymer) stiffened with fibers of cellulose.

The bar chart of moduli (Figure 3.7) shows that composites can have moduli much higher than those of their matrices. And it also shows that they can be *very* anisotropic, meaning that the modulus is higher in some directions than others. Wood is an example: its modulus, measured parallel to the fibers, is about 10 GN m$^{-2}$; at right angles to this, it is less than 1 GN m$^{-2}$.

There is a simple way to estimate the modulus of a fiber-reinforced composite. Suppose we stress a composite, containing a volume fraction $V_f$ of fibers, parallel to the fibers (see Figure 6.3(a)). Loaded in this direction, the strain, $\varepsilon_n$, in the fibers and the matrix is the same. The stress carried by the composite is

$$\sigma = V_f \sigma_f + (1 - V_f)\sigma_m$$

where the subscripts f and m refer to the fiber and matrix, respectively.

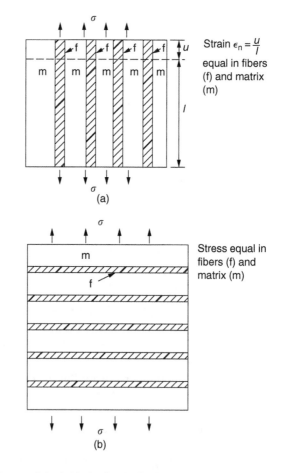

**FIGURE 6.3**

A fiber-reinforced composite loaded in the direction in which the modulus is (a) a maximum, (b) a minimum.

Since $\sigma = E\varepsilon_n$, we can rewrite this as:

$$\sigma = E_f V_f \varepsilon_n + E_m (1 - V_f)\varepsilon_n$$

But

$$E_{composite} = \sigma / \varepsilon_n$$

so

$$E_{composite} = V_f E_f + (1 - V_f)E_m \tag{6.7}$$

This gives us an upper estimate for the modulus of our fiber-reinforced compos-ite. The modulus cannot be greater than this, since the strain in the stiff fibers can never be greater than that in the matrix.

How is it that the modulus can be less? Suppose we had loaded the composite in the opposite way, at right angles to the fibers (as in Figure 6.3(b)). It is now reasonable to assume that the *stresses*, not the strains, in the two components are equal. If this is so, then the total nominal strain $\varepsilon_n$ is the weighted sum of the individual strains:

$$\varepsilon_n = V_f \varepsilon_{n,f} + (1 - V_f)\varepsilon_{n,m}$$

Using $\varepsilon_n = \sigma / E$ gives

$$\varepsilon_n = \frac{V_f \sigma}{E_f} + \left(\frac{1 - V_f}{E_m}\right)\sigma$$

The modulus is still $\sigma / \varepsilon_n$, so

$$E_{composite} = 1 / \left\{ \frac{V_f}{E_f} + \frac{(1 - V_f)}{E_m} \right\} \tag{6.8}$$

This is a lower limit for the modulus—it cannot be less than this.

The two estimates, if plotted, look as shown in Figure 6.4. This explains why fiber-reinforced composites like wood and GFRP are so stiff along the rein-forced direction (the upper line of the figure) and yet so floppy at right angles to the direction of reinforcement (the lower line), that is, it explains their *anisotropy*. Anisotropy is sometimes what you want—as in the shaft of a squash racquet or a vaulting pole. Sometimes it is not, and then the layers of fibers can be *laminated* in a criss-cross way, as they are in the skins of aircraft flaps.

Not all composites contain fibers. Materials can also be stiffened by small *particles*. The theory is, as one might imagine, more difficult than for fiber-reinforced composites. But it is useful to know that the moduli of these so-called *particulate* composites lie between the upper and lower limits of Equations (6.7) and (6.8), nearer the lower one than the upper one, as shown in Figure 6.4. It is much less expensive to mix sand into a polymer than to

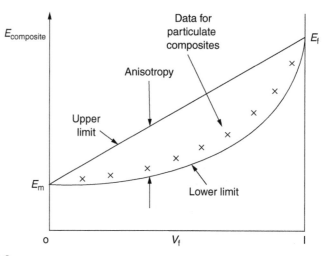

**FIGURE 6.4**

Composite modulus for various volume fractions of stiffener, showing the upper and lower limits of Equations (6.7) and (6.8).

carefully align specially produced glass fibers in the same polymer. Thus, the modest increase in stiffness given by particles is economically worthwhile.

The resulting particulate composite is *isotropic*, rather than *anisotropic* (as would be the case for the fiber-reinforced composites) and this, too, can be an advantage. These *filled polymers* can be formed and molded by normal methods (most fiber composites cannot) and so are cheap to fabricate. Many of the polymers of everyday life—bits of cars and bikes, household appliances and so on—are filled in this way.

## WORKED EXAMPLE

The photograph on the next page shows an oboe—one of the woodwind instruments of the classical symphony orchestra; see *http://en.wikipedia.org/wiki/Oboe*.

The close-up shows the most critical component of the instrument—the *reed* (see page 90). The player places the reed in his/her mouth, and blows through it in order to drive the natural frequencies of air inside the main tube of the instrument. The "reed" is actually made from a species of *cane—Arundo Donax*. This grows in marshy coastal areas in the south of France, as well as other places (California and Argentina, for example) where there is the right soil and climate—essential to growing the best cane.

A piece of the tubular cane is first split along its length, and the inside is "gouged" to produce the right thickness. The gouged piece is then cut to the correct length and width, folded in half, and tied on to a small brass tube

(the "staple"). After tying-on, the folded end is cut off, and the reed is "scraped" to the correct thickness profile using a reed-scraping knife. The final scraping is critical to the correct functioning of the reed. In fact, most professional oboeists spend much more time making and adjusting reeds than they ever do playing the instrument!

The two separated halves of the reed behave like cantilever springs, and when the oboeist blows through the reed, the cantilevers beat rapidly against one another. This is what powers the vibrations inside the instrument. But this is just the start—with as little stress as possible, the player must be able to play loudly and softly, start notes quietly over the whole range of the instrument, play staccato and legato, make a nice sound, and so on. A well-made and adjusted cane reed does all these things well, because it has just the right combination of properties—stiffness, density, damping, water uptake (conditioning). But a

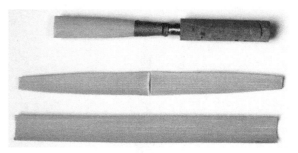

critical factor is that Young's modulus for cane is *anisotropic*—Young's modulus along the tube axis is approximately 10 times that transverse to the axis. The performance of any synthetic substitutes will be wholly inadequate unless they can match the anisotropy of natural cane.

## EXAMPLES

**6.1**    The table lists Young's modulus, $E_{composite}$, for a glass-filled epoxy composite. The material consists of a volume fraction $V_f$ of glass particles (Young's modulus, $E_f$, 80 GN m$^{-2}$) dispersed in a matrix of epoxy (Young's modulus, $E_m$, 5 GN m$^{-2}$). Calculate the upper and lower values for the modulus of the composite material, and plot them, together with the data, as a function of $V_f$. Which set of values most nearly describes the results? Why? How does the modulus of a random chopped-fiber composite differ from those of an aligned continuous-fiber composite?

| Volume Fraction of Glass, $V_f$ | $E_{composite}$ (GN m$^{-2}$) |
|---|---|
| 0 | 5.0 |
| 0.05 | 5.5 |
| 0.10 | 6.4 |
| 0.15 | 7.8 |
| 0.20 | 9.5 |
| 0.25 | 11.5 |
| 0.30 | 14.0 |

**6.2**    A composite material consists of parallel fibers of Young's modulus $E_f$ in a matrix of Young's modulus $E_m$. The volume fraction of fibers is $V_f$. Derive an expression for $E_c$, Young's modulus of the composite along the direction of the fibers, in terms of $E_f$, $E_m$ and $V_f$. Obtain an analogous expression for the density of the composite, $\rho_c$. Using material parameters given in the next table, find $\rho_c$ and $E_c$ for the following composites: (a) carbon fiber–epoxy resin ($V_f = 0.5$), (b) glass fiber–polyester resin ($V_f = 0.5$), (c) steel–concrete ($V_f = 0.02$).

| Material | Density (Mg m$^{-3}$) | Young's modulus (GN m$^{-2}$) |
|---|---|---|
| Carbon fiber | 1.90 | 390 |
| Glass fiber | 2.55 | 72 |
| Epoxy resin ⎫ Polyester resin ⎬ | 1.15 | 3 |
| Steel | 7.90 | 200 |
| Concrete | 2.40 | 45 |

**Answers**

p0270 $E_c = E_f V_f + (1 - V_f)E_m$; $\rho_c = \rho_f V_f + (1 - V_f)\rho_m$

    a. $\rho_c = 1.53 \text{ Mg m}^{-3}$, $E_c = 197 \text{ GN m}^{-2}$

    b. $\rho_c = 1.85 \text{ Mg m}^{-3}$, $E_c = 37.5 \text{ GN m}^{-2}$

    c. $\rho_c = 2.51 \text{ Mg m}^{-3}$, $E_c = 48.1 \text{ GN m}^{-2}$

**6.3** A composite material consists of flat, thin metal plates of uniform thickness glued one to another with a thin, epoxy-resin layer (also of uniform thickness) to form a "multi-decker-sandwich" structure. Young's modulus of the metal is $E_1$, that of the epoxy resin is $E_2$ (where $E_2 < E_1$) and the volume fraction of metal is $V_1$. Find the ratio of the maximum composite modulus to the minimum composite modulus in terms of $E_1$, $E_2$ and $V_1$. Which value of $V_1$ gives the largest ratio?

**Answer**

Largest ratio when $V_1 = 0.5$

**6.4** a. Define a high polymer.

    b. What is the order of magnitude of the number of carbon atoms in a single molecule of a high polymer?

    c. List three engineering polymers.

**6.5** a. List the monomers of polyethylene (PE), polyvinyl chloride (PVC) and polystyrene (PS).

    b. What is a copolymer?

**6.6** a. Distinguish between a glassy polymer, a crystalline polymer and a rubber.

    b. Distinguish between a cross-linked and a non-cross-linked polymer.

    c. Define a thermoplastic and a thermoset.

**6.7** a. What is the glass transition temperature, $T_G$?

    b. What is the temperature range in which $T_G$ lies for most polymers?

    c. Explain the change of moduli of polymers at $T_G$.

**6.8** How would you increase the modulus of a polymer?

## CHAPTER 7

# Case Studies in Modulus-Limited Design

## CONTENTS

## 7.1 CASE STUDY 1: SELECTING MATERIALS FOR RACING YACHT MASTS

Figure 7.1 shows a typical mast from a modern racing yacht. It is built in the form of a thin-walled tube, and is fabricated from carbon fiber reinforced plastic—CFRP (which is why it appears black). To win long-haul ocean races, these yachts need to have every advantage that high-performance materials can give, in terms of maximum strength and stiffness, plus minimum weight and aerodynamic drag. The clip shows conditions on a typical ocean race—the Sydney to Hobart. The yacht is travelling at over 10 knots, and gear is being worked hard to move along as fast as possible. When storms are encountered (as they were shortly after this clip was taken), gear such as masts and rigging is exposed to very high loadings indeed, and masts can even be lost overboard; see *http:// www.youtube.com/watch?v=MzePFwYk88c&feature=related*

A good way to compare the performance of different materials in this type of application is to look at the mechanics of a cantilever beam in bending (see Figure 7.2). Equations for the elastic bending of beams are given at the end

Engineering Materials I: An Introduction to Properties, Applications, and Design, Fourth Edition
© 2012, Michael F. Ashby and David R. H. Jones. Published by Elsevier Ltd. All rights reserved.

**FIGURE 7.1**
Carbon fibre mast on an ocean racing yacht. *(Cruising Yacht Club of Australia, Rushcutters Bay, Sydney.)*
−33 52 25.34 S 151 13 55.08 E

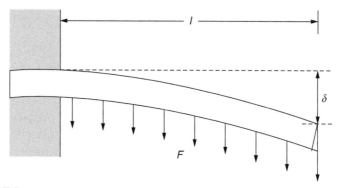

**FIGURE 7.2**
The elastic deflection δ of a cantilever beam of length *l* under a uniformly distributed force *F*.

of Chapter 3 (pp. 47–50)—the closest result is for a cantilever subjected to a uniformly distributed loading (see also Case Study 2). This is because the boom and the mainsail impose a lateral loading on the mast that is roughly uniform along the length of the mast. The following are the equations to use:

$$\delta = \frac{1}{8}\left(\frac{Fl^3}{EI}\right) \tag{7.1}$$

$$I = \pi r^3 t \tag{7.2}$$

Combining these two equations gives

$$\frac{F}{\delta} = \frac{8\pi E r^3 t}{l^3} \tag{7.3}$$

for the *bending stiffness* of the tube. The mass of the beam is given by

$$M = 2\pi r t l \rho \tag{7.4}$$

The length and radius of the tube are fixed by the design, leaving the wall thickness $t$ as the only dimensional variable. Substituting for $t$ from Equation (7.4) into Equation (7.3) gives

$$M = \frac{1}{4} \left( \frac{l^4}{r^2} \right) \left( \frac{F}{\delta} \right) \left( \frac{\rho}{E} \right) \tag{7.5}$$

The mass of tube for a specified bending stiffness is therefore minimized by selecting a material with the minimum value of the *material index*

$$M_1 = \left( \frac{\rho}{E} \right) \tag{7.6}$$

Table 7.1 gives data for some candidate materials for this application.

With the exception of CFRP, there is clearly not much to choose between any of these materials. Historically, masts for sailing ships were made from wood; that was the only suitable material that was available, and was very good at doing the job. With the advent of larger ships and steel at the end of the nineteenth century, masts were increasingly made from steel—equally good at the job, but available in much larger sizes to order, and with much better consistency of properties.

From the 1960s, masts—especially for yachts and sailing dinghies—have almost always been made from aluminum alloy, mainly because of the corrosion problems with steel in thin sections. But CFRP gives a dramatic improvement in performance—a weight saving of a factor of 5 or more over the alternatives. It is little wonder that it is universally used in large ocean racers. But why is it not used more extensively? Price is the major factor here—CFRP costs 30 times

**Table 7.1** Data for Tube of Given Stiffness

| Material | $\rho$ (Mg m$^{-3}$) | $E$ (GN m$^{-2}$) | $M_1$ |
|---|---|---|---|
| Steel | 7.8 | 200 | 0.039 |
| Aluminum alloy | 2.7 | 69 | 0.039 |
| Titanium alloy | 4.5 | 120 | 0.038 |
| GFRP | 2.0 | 40 | 0.05 |
| Wood | 0.6 | 12 | 0.05 |
| CFRP | 1.5 | 200 | 0.0075 |

**FIGURE 7.3**

Bamboo scaffolding. Nilambagh Palace, Bhavnagar, Gujurat, India. − 21 45 57.07 N 72 07 50.07 E

more than aluminum alloys (see Table 2.1), and is only justified when there is plenty of money to pay for it. Finally, it is not surprising in view of this analysis that tubular members made from *natural* composite materials have been used for millenia—bamboo for buildings, cane for furniture, reed for roofing, and so on. Figure 7.3 shows an example of such a material still in use today.

## 7.2 CASE STUDY 2: DESIGNING A MIRROR FOR A LARGE REFLECTING TELESCOPE

Figure 7.4 shows a typical large reflecting telescope—the U.K.'s infrared telescope (UKIRT), situated on the summit plateau of Mauna Kea, Hawaii—13,796 ft above sea level. A list of the world's largest optical telescopes can be found at *http://astro.nineplanets.org/bigeyes.html*

The classic big reflector is the Hale 200" (5.1 m), located on the Palomar Mountain, in California. Completed in 1948, until 1993 it was the world's largest correctly operating optical telescope. The following links give details of the construction, in particular the mirror and its supporting mechanism—we will return to these critical design solutions at the end of the case study. See also 33 21 22.54 N 116 51 53.61 W in Google Earth (*http://www.astro.caltech.edu/palomar/*; *http://articles.adsabs.harvard.edu//full/1950PASP...62...91B/0000091.000.html*).

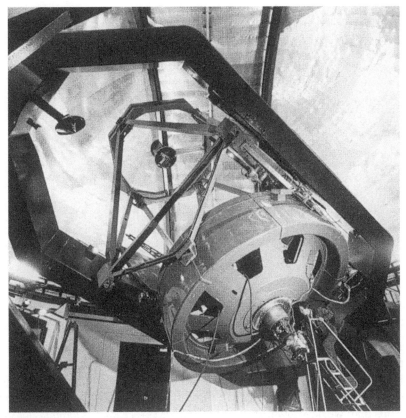

**FIGURE 7.4**
The British infrared telescope at Mauna Kea, Hawaii. The picture shows the housing for the 3.8-m diameter mirror, the supporting frame, and the interior of the aluminum dome with its sliding "window." *(Courtesy of Photo Labs, Royal Observatory, Edinburgh.)* – 19 49 32.2 N 155 28 23.6 W

In simple terms, the mirror is an inert backing (of low-expansion glass, weighing 14.5 tons) for a very thin layer of vapour-deposited aluminum, 100 nm thick. Before coating, it is ground and polished to the shape of a parabola of revolution. The final dimensions must be incredibly accurate, otherwise the optical image seen through the telescope will not be sharp. In service, the mirror must hold an accuracy of 2 millionths of an inch (0.05 μm)—a huge technical challenge.

Not surprisingly, the cost of such a telescope is enormous. However, much of the cost is associated with the steel framework that supports the mirror and the other optical components, and makes them move to follow the stars. The scaling laws of these structural elements is such that the cost of the telescope increases at a much faster rate than just the mirror diameter. So there is a big incentive to select materials for mirrors that reduce the mirror mass as much as possible.

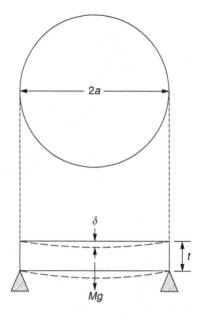

**FIGURE 7.5**
The elastic deflection of a telescope mirror under its own weight.

At its simplest, a telescope mirror is a circular disc of diameter $2a$ and thickness $t$ simply supported at its periphery (Figure 7.5). When horizontal, it will deflect under its own weight $M$, but when vertical it will not deflect significantly. The equation for the elastic deflection needs to be found from a much more extensive source of results than we have given at the end Chapter 3—see *Roark's formulas for stress and strain* (Young and Budynas—in the References). It is

$$\delta = \frac{0.67}{\pi}\left(\frac{Mga^2}{Et^3}\right) \tag{7.7}$$

for a material with Poisson's ratio fairly close to 0.33. The mass of the mirror is given by

$$M = \pi a^2 t \rho \tag{7.8}$$

Substituting for $t$ from equation (7.8) into equation (7.7) gives

$$M = \left(\frac{0.67g}{\delta}\right)^{1/2} \pi a^4 \left(\frac{\rho^3}{E}\right)^{1/2} \tag{7.9}$$

$\delta$ and $a$ are fixed by the design, so the only variable remaining on the right-hand side of Equation (7.9) is the materials index

$$M_2 = \left(\frac{\rho^3}{E}\right)^{1/2} \tag{7.10}$$

Table 7.2 gives data for some candidate materials for this application.

**Table 7.2** Data for Reflecting Telescope Mirror

| Material | $\rho$ (Mg m$^{-3}$) | E (GN m$^{-2}$) | $M_2$ |
|---|---|---|---|
| Steel | 7.8 | 200 | 1.54 |
| Concrete | 2.5 | 47 | 0.58 |
| Aluminum alloy | 2.7 | 69 | 0.53 |
| Glass | 2.5 | 69 | 0.48 |
| Titanium alloy | 4.5 | 120 | 0.87 |
| GFRP—quasi isotropic laminate | 2.0 | 15 | 0.73 |
| Wood—quasi isotropic laminate | 0.6 | 4 | 0.23 |
| CFRP—unidirectional | 1.5 | 200 | 0.13 |
| CFRP—quasi isotropic laminate | 1.5 | 70 | 0.22 |

CFRP is no longer clearly the lightest material—we cannot exploit its high unidirectional modulus because we must have the same modulus in all directions in the plane of the mirror, and this means we have to use a *quasi isotropic laminate*, with a much lower modulus. In fact, wood is now equal to CFRP—but we cannot use it because it is far too unstable dimensionally. Next up is glass, which also has the advantages that it is very stable, and can be ground and polished to a mirror finish—so it is not surprising that nearly all telescope mirrors continue to use it, even the Hubble space telescope! *http://en.wikipedia.org/wiki/Hubble_Space_Telescope*

Materials selection based on the materials index is a useful first step towards identifying potential new ways of doing things—CFRP yacht masts are a good example of the huge weight savings possible by using a new material. But in some cases, the traditional material is the best. And the materials index is sometimes too simplistic to incorporate other design-based solutions to the problem.

Two clever engineering solutions are used in the Hale telescope to minimize the deflection of the mirror under gravitational loading. First, the mirror is not a plain disc—its underside is deeply ribbed in order to reduce weight while retaining a stiff structural shape (the weight saving is 42%). Second, mirror sag is offset by applying mechanical forces to the underside of the mirror at 36 separate locations (Figure 7.6).

These forces are provided by a system of jacks which are actuated by weighted levers—each jack is unique to each of the 36 locations. As the mirror is tilted (and the sag decreases), the jacks back off automatically; due to clever mechanical design, the gravitational pull on the weighted levers is made to back off at the same rate as the pull on the mirror. This is an entirely passive control mechanism, and needs no external actuators or control systems.

Modern reflectors, on the other hand, use highly sophisticated computer controlled actuators to do this job—the signals to the actuators are even driven in real time by aberrations in the optical images, so the telescope is continually

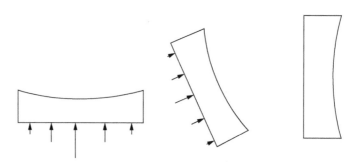

**FIGURE 7.6**

The distortion of the mirror under its own weight can be corrected by applying forces (shows as arrows) to the back surface.

correcting itself. This has permitted optical telescopes to be built with mirrors as large as 10 m—the Keck telescopes on Mauna Kea each comprise 36 separate glass mirror segments, all of which are kept precisely aligned by this type of control system. Even larger reflectors are planned using this system. The message here is clear—in many applications, control engineering solutions are now outclassing solutions based on materials selection. And here is a solution to the mirror problem that does not even require us to use a solid material! (*http:// en.wikipedia.org/wiki/Liquid_mirror*)

## 7.3 CASE STUDY 3: THE *CHALLENGER* SPACE SHUTTLE DISASTER

On 28 January 1986, the U.S. space shuttle *Challenger* caught fire and broke up 73 seconds after lift-off. Together with the break-up of the space shuttle Columbia during re-entry on 1 February 2003, these are the worst accidents in the history of space travel. The circumstances of the disaster are well known, and are extensively documented in the report of the Presidential Commission (the Rogers report) and many other sources:

- *http://history.nasa.gov/rogersrep/genindex.htm*
- *http://en.wikipedia.org/wiki/Space_Shuttle_Challenger_disaster*
- *http://en.wikipedia.org/wiki/Space_Shuttle_Columbia_disaster*
- *http://www.cbsnews.com/htdocs/space_place/framesource_challenger.html*
- *http://www.msnbc.msn.com/id/11031097/ns/technology_and_science-space/*
- *http://www.onlineethics.org/cms/7123.aspx*
- *http://challenger-o-ring.com/*
- *http://en.wikipedia.org/wiki/Space_Shuttle_Solid_Rocket_Booster*
- *http://en.wikipedia.org/wiki/Solid-fuel_rocket*

The disaster started with a blow-by of hot combustion gases through a joint in the steel casing of one of the two solid rocket boosters, or SRBs. You can see this blow-by immediately after lift-off on the YouTube clip—at 4 minutes 36 seconds. (*http://www.youtube.com/watch?v=zk_wi4QD5WE*)

There is a puff of black smoke from the lowest (aft end) joint of the right-hand SRB—in fact, there was a short sequence of repeated smoke puffs from T + 0.678 s to T + 2.733 s. The blow-by then stopped, but started up again after 59 s, resulting in burn-through of the steel casing at this location, breaching of the huge external propellant tank, fracture of the lower connection between the SRB and the propellant tank and rapid disintegration of the whole assembly.

Each SRB was made from seven lengths of large diameter steel tube, 12 ft (144" or 3658 mm) outside diameter, and 0.479" (12.17 mm) wall thickness. The overall length of the SRB was 149 ft (45.5 m). The six upper (forward) lengths of tube were joined end-to-end in pairs at the factory in Promontory, Utah, making four separate sections to be transported by rail to the launch site at the Kennedy Space Center (28 36 31.04 N 80 36 15.36 W). These four sections were then assembled at the launch site, which involved making three "field" joints. It was the lowest (aftermost) of these field joints which failed during launch.

Figure 7.7 shows a simplified cross section through this joint. The end of one tube (the "tang") plugs into a groove (the "clevis") in the end of the next tube. The joint is made pressure-tight by two synthetic rubber "O" rings, which sit in circumferential grooves machined in the clevis. After assembly, the O rings are compressed slightly ("nipped") because the as-manufactured diameter of the O ring (0.280") is slightly greater than the gap between the tang and the bottom of the O-ring groove. The joint is tested for pressure tightness after assembly by pressurising the cavity between the two O rings with air introduced through the test port. If the air pressure in the pressurising system holds the initial 50 psi for long enough, then the joint is deemed to be correctly assembled. However—something we will return to in a moment—the pressure test will push the O rings in opposite directions: the upper O ring will be pushed upwards until it sits against the top face of the groove, and the opposite will happen for the lower O ring.

After ignition, the pressure of the combustion gases in the SRB climbs to its steady value of 912 psi (6.303 MN m$^{-2}$) within 0.600 s. This is supposed to force the heat-resistant putty along its channel, compress the air in the space above the O ring, and push the O ring down so it forms a seal against the tang and the lower face of the groove.

In practice, things did not work out this way with real joints. Figure 7.7 shows that the wall thickness of the SRB is much greater at the joint than in the main run of the tube—1.591" as compared to 0.479", a factor of 3.32 times thicker. Under the internal pressure of 912 psi, the main run of the tube expands elastically. The worked example at the end of this chapter shows how to calculate this expansion. At 912 psi, it is 0.283" (7.18 mm) on tube *radius*. The joint also

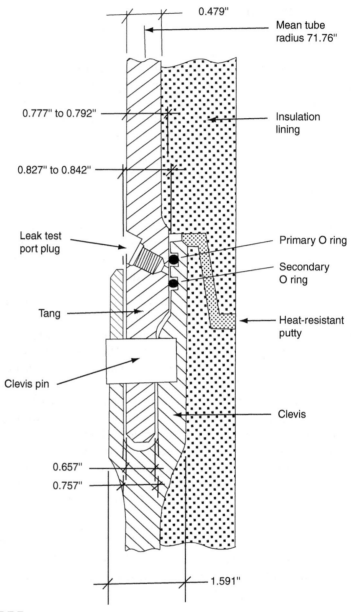

**FIGURE 7.7**
Joint in solid rocket booster casing.

expands elastically, but because it is much thicker, it expands a lot less—by 0.183" (4.64 mm) radially. This leads to what is termed "joint rotation"— shown schematically in Figure 7.8. You can model this effect very easily with a party balloon and a rubber band, as shown in Figure 7.9.

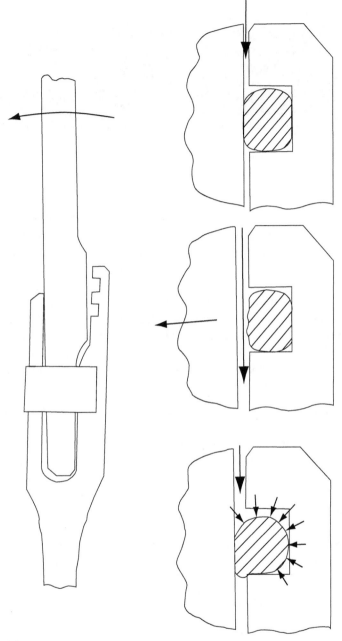

**FIGURE 7.8**
Joint rotation, blow-by of hot gases, and reseating of O ring.

**FIGURE 7.9**
The party balloon expands less where it is stiffened by the rubber band.

What joint rotation does is to increase the gap between the tang and the clevis. Tests showed that the gap would open by an average value of 0.029" over the timescale of 0.600 s to reach the steady operating pressure. During this time, the O ring lost its initial compression, and lost its seal. As we explained in Chapter 6, when rubbers are warmed up above the glass temperature $T_G$ the Van der Waals bonds melt, and this allows the polymer chains to slide over one another when the rubber is strained. But this chain sliding takes time, and the rate at which it takes place is very dependent on temperature—after 0.6 s, the O ring will recover 60% of its compression at 24°C, 40% of its compression at 10°C, but only 20% of its compression at 0°C. This was elegantly demonstrated to the Commission by the Nobel prize-winning physicist Richard Feynman: *http://www.youtube.com/watch?v=8qAi_9quzUY*

If the launch is carried out in warm weather, the chances are that the O ring will rapidly reseat itself under the action of the pressurized putty. However, at lower temperatures reseating takes much longer, and during this time, combustion gases can make their way down to the O ring—and even leak past it—before the O ring has had a chance to reseat. The hot gases burn back the surface of the O ring, so there is a race on for the O ring to re-seat before it gets eroded too much. The *Challenger* launch took place in cold weather—the joint temperature was estimated to have been around 0°C—and the O ring could not re-seat before it burnt out. At this stage, the second O ring should have come into operation as a back-up, but this also would not reseat in time and burnt out, leaving a clear leak path through the joint.

Engineers had been aware for some time of this flaw in the joint design. Because the SRBs were designed to be recoverable, post-mission inspections of joints had been made, and the tell-tale signs of blow-by had been found in a number of SRBs. Not surprisingly, these were more common at the lower launch temperatures. But the launch on 28 January 1986 was at the lowest temperature ever, by some margin. Engineers at the manufacturers were strongly opposed to the launch because of this, but were overruled—on the basis that they did not have enough information on the behavior of the O rings at low temperatures to prove that the launch was dangerous!

## Postscript

A very basic knowledge of elastic behavior has shown how the apparently simple use of an O ring seal can go badly wrong—because of the textbook elastic behavior of a pressurized tube of varying wall thickness, and the textbook time-dependent behavior of rubbers. To end, here are a few quotes from the Rogers report, which make compelling reading for any engineer involved in managing risk.

> [Mr. Ray], when the joint was first designed, the analysis produced by Thiokol says the joint would close, the extrusion gap would actually close. We had quite a debate about that until we did a test on the first couple of segments that we received from the manufacturer, which in fact showed that the joint did open. At that time, we really nailed it down. We got some very accurate numbers on joint rotation, and we know for a fact that during these tests the joint rotated. The primary O-ring was extruded up into the joint. The secondary O ring did in fact detach from the seat.
>
> One of the most astute summaries of the cause came from a lawyer, not an engineer. David Acheson, one of members of the Commission, said: "A lot of material we have received, one reads that the designers, presumably both the corporate designers and the NASA supervisors, believe the joint was designed to compress and seal in the gas tight under combustion pressure. And it turns out very quickly in the joint history that it did the opposite. It opened up. I just don't understand why the program then decided to go into a lot of little fixes to see if you could compensate for the fundamental error."
>
> The Space Shuttle's Solid Rocket Booster problem began with the faulty design of its joint and increased as both NASA and contractor management first failed to recognize it as a problem, then failed to fix it and finally treated it as an acceptable flight risk. Morton Thiokol, Inc., the contractor, did not accept the implication of tests early in the program that the design had a serious and unanticipated flaw. NASA did not accept the judgment of its engineers that the design was unacceptable, and as the joint problems grew in number and severity NASA minimized them in management briefings and reports. Thiokol's stated position was that "the condition is not desirable but is acceptable." Neither Thiokol nor NASA expected the rubber O-rings sealing the joints to be touched by hot gases of motor ignition, much less to be partially burned. However, as tests and then flights confirmed damage to the sealing rings, the reaction by both NASA and Thiokol was to increase the amount of damage

considered "acceptable." At no time did management either recommend a redesign of the joint or call for the Shuttle's grounding until the problem was solved.

NASA and Thiokol accepted escalating risk apparently because they "got away with it last time." As Commissioner Feynman observed, "the decision making was: a kind of Russian roulette. [The Shuttle] flies [with O-ring erosion] and nothing happens. Then it is suggested, therefore, that the risk is no longer so high for the next flights. We can lower our standards a little bit because we got away with it last time.... You got away with it but it shouldn't be done over and over again like that."

## WORKED EXAMPLE

In Chapter 3, we looked at two things that are useful for analysing the strains that the internal pressure generates in the solid rocket booster casing. Looking again at Example 3.7, we can see that the applied stress state in the wall of the casing consists of a hoop stress given by $\sigma_1 = pr/t$, an axial stress given by $\sigma_2 = pr/2t$, and a free surface stress $\sigma_3 = 0$. Because the stress state is not uniaxial, we need to use the equations for stresses and strains in 3 dimensions to work out the hoop strain in the SRB casing. These were given in Example 3.13:

$$\varepsilon_1 = \frac{\sigma_1}{E} - v\frac{\sigma_2}{E} - v\frac{\sigma_3}{E}$$

$$\varepsilon_2 = \frac{\sigma_2}{E} - v\frac{\sigma_1}{E} - v\frac{\sigma_3}{E}$$

$$\varepsilon_3 = \frac{\sigma_3}{E} - v\frac{\sigma_1}{E} - v\frac{\sigma_2}{E}$$

The equation that is relevant to the hoop strain is obviously

$$\varepsilon_1 = \frac{\sigma_1}{E} - v\frac{\sigma_2}{E} - v\frac{\sigma_3}{E}$$

Poisson's ratio is 1/3 for steel, so the hoop strain becomes

$$\varepsilon_1 = \frac{\sigma_1}{E} - \frac{1}{3}\frac{\sigma_2}{E} = \frac{\sigma_1}{E} - \frac{1}{3}\frac{\sigma_1}{2E} = \frac{\sigma_1}{E}\left(1 - \frac{1}{6}\right) = \frac{5}{6}\frac{\sigma_1}{E}$$

The casing has an outer diameter of 12 ft (144"), and a wall thickness of 0.479". The inner diameter of the tube is 144"—2 × 0.479" = 143.04". The mean diameter of the tube is 143.52", so the mean radius of the tube is half this = 71.76". NASA conducted a pressure test on an SRB casing—this used water as the pressurising medium, so a fracture of the casing would not cause an explosion (it would if air had been used!). The test pressure used was 1004 psi (6.939 MN m$^{-2}$). At this pressure, the hoop stress is given by

$$\sigma_1 = \frac{6.939 \times 71.76}{0.479} = 1040 \, \text{M Nm}^{-2}$$

Young's modulus for the steel is $200.4 \, \text{GN m}^{-2}$, so the hoop strain is

$$\varepsilon_1 = \frac{5}{6} \frac{\sigma_1}{E} = \frac{5}{6} \times \frac{1040 \, \text{M Nm}^{-2}}{200.4 \, \text{G Nm}^{-2}} = 0.00433$$

This produces a radial expansion of the casing of $0.00433 \times 71.76" = 0.311"$.

NASA measured a radial expansion during the test which is exactly the same—0.311". Not bad for a simple analysis!

## EXAMPLES

**7.1** Refer to the equations for the elastic bending of beams at the end of Chapter 3 (pp. 47–50). For each of the cases listed,

$$\delta = C\left(\frac{Fl^3}{EI}\right)$$

Write down the value of $C$ for each case.

**7.2** Obtain a materials index for the elastic bending of beams for the following cross sections. Both the length and stiffness of the beam are fixed by the design.
   **a.** Square
   **b.** Rectangle (breadth $b$ not a variable)
   **c.** Solid circle
   **d.** Thin-walled tube (radius $r$ not a variable)

**Answers**
(a) $(\rho^2/E)^{1/2}$; (b) $(\rho^3/E)^{1/3}$; (c) $(\rho^2/E)^{1/2}$; (d) $(\rho/E)$

**7.3** Refer to the equations for the elastic buckling of struts at the end of Chapter 3. For each of the cases listed,

$$F_{cr} = C\left(\frac{EI}{l^2}\right)$$

Write down the value of $C$ for each case.

**7.4** Obtain a materials index for the elastic buckling of struts for the following cross sections. Both the length and critical buckling load of the strut are fixed by the design.
   **a.** Square
   **b.** Rectangle (breadth $b$ not a variable)
   **c.** Solid circle
   **d.** Thin-walled tube (radius $r$ not a variable)

**Answers**

(a) $(\rho^2/E)^{1/2}$; (b) $(\rho^3/E)^{1/3}$; (c) $(\rho^2/E)^{1/2}$; (d) $(\rho/E)$

7.5    You have been asked to prepare an outline design for the pressure hull of a deep-sea submersible vehicle capable of descending to the bottom of the Mariana Trench in the Pacific Ocean—the deepest water on the planet (*Challenger* Deep, 11 20 57 N 142 13 00 E, 11,000 m depth). The external pressure at this depth is approximately 100 MN m$^{-2}$, and the design pressure is to be taken as 200 MN m$^{-2}$. The pressure hull is to have the form of a thin-walled sphere with a specified radius $r$ of 1 m and a uniform thickness $t$. The sphere can fail by external-pressure buckling at a pressure $p_b$ given by

$$P_b = 0.3\, E\left(\frac{t}{r}\right)^2$$

where $E$ is Young's modulus.

The basic design requirement is that the pressure hull shall have the minimum possible mass compatible with surviving the design pressure. By eliminating $t$ from the equations, show that the minimum mass of the hull is given by the expression

$$M_b = 22.9 r^3 p_b^{0.5}\left(\frac{\rho^2}{E}\right)^{1/2}$$

Hence obtain a materials index to meet the design requirement for the failure mechanism. [You may assume that the surface area of the sphere is $4\pi r^2$.]

**Answer**

$(\rho^2/E)^{1/2}$

7.6    The diagram that follows shows a prop shaft from a truck. This is supported at either end by a universal joint, which allows the ends of the shaft to flex relative to the axes of the driving and driven shafts. As a consequence of this pin-jointed end support condition, when the rotational speed of the shaft increases, there is an increasing tendency for the shaft to bow out sideways under centrifugal force. Eventually, a critical speed is reached at which the bowing takes off, and the shaft goes out of control. This phenomenon is called *whirling*. The critical speed $f_{cr}$ (in revolutions per second) for whirling is (by mathematical coincidence) the same as the mode 1 vibration frequency (in cycles per second, or hertz)—this is given for various support conditions and mass distributions in the equations for mode 1 natural vibration frequencies at the end of Chapter 3.

Assuming that the prop shaft is a thin-walled tube of radius $r$ and thickness $t$, derive an equation for the mass of the shaft showing how it depends on $t$, $I$, $f_{cr}$, $\rho$, and $E$. If $I$, $f_{cr}$, $\rho$, and $E$ are fixed by design and material choice, comment on the dependence of the mass on $t$. In practice, what additional requirement for a prop shaft will restrict the minimum mass attainable?

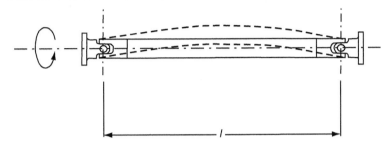

# Yield Strength, Tensile Strength, and Ductility

## CHAPTER 8

# Yield Strength, Tensile Strength, and Ductility

## CONTENTS

## 8.1 INTRODUCTION

All solids have an *elastic limit* beyond which something happens. A totally brittle solid will fracture suddenly (e.g., glass). Most engineering materials do something different; they deform *plastically* or change their shapes in a *permanent* way. It is important to know when, and how, they do this—so we can design structures to withstand normal service loads without any permanent deformation, and rolling mills, sheet presses and forging machinery that will be strong enough to deform the materials we wish to form.

To study this, we pull carefully prepared samples in a tensile testing machine, or compress them in a compression machine, and record the *stress* required to produce a given *strain*.

Engineering Materials I: An Introduction to Properties, Applications, and Design, Fourth Edition
© 2012, Michael F. Ashby and David R. H. Jones. Published by Elsevier Ltd. All rights reserved.

## 8.2 LINEAR AND NONLINEAR ELASTICITY

Figure 8.1 shows the *stress–strain* curve of a material exhibiting *perfectly linear elastic* behavior. This is the behavior that is characterized by Hooke's law (Chapter 3). All solids are linear elastic at small strains—by which we usually mean less than 0.001, or 0.1%. The slope of the stress–strain line, which is the same in compression as in tension, is Young's modulus, $E$. The area (shaded) is the elastic energy stored, per unit volume: we can get it all back if we unload the solid, which behaves like a spring.

Figure 8.2 shows a *nonlinear* elastic solid. *Rubbers* have a stress–strain curve like this, extending to very large strains (of order 5 or even 8). The material is still

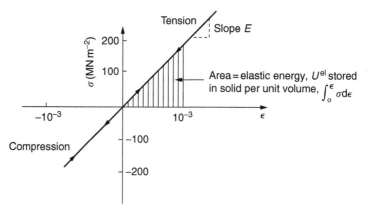

**FIGURE 8.1**

Stress–strain behavior for a *linear elastic solid.* The axes are calibrated for a material such as steel.

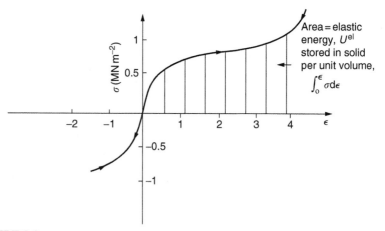

**FIGURE 8.2**

Stress–strain behavior for a *nonlinear elastic solid.* The axes are calibrated for a material such as rubber.

elastic: if unloaded much of the energy stored during loading is recovered—that is why catapults can be as lethal as they are.

## 8.3 LOAD–EXTENSION CURVES FOR NONELASTIC (PLASTIC) BEHAVIOR

Rubbers are exceptional in behaving elastically to high strains; as we said, *almost all materials, when strained by more than about* 0.001 (0.1%), *do something irreversible*: and most engineering materials deform *plastically* to change their shape *permanently*. If we load a piece of ductile metal (e.g., copper), in tension, we get the following relationship between the load and the extension (Figure 8.3). This can be demonstrated nicely by pulling a piece of plasticine modelling clay (a ductile non-metallic material). Initially, the plasticine deforms elastically, but at a small strain begins to deform plastically, so if the load is removed, the piece of plasticine is permanently longer than it was at the beginning of the test: it has undergone *plastic* deformation (Figure 8.4).

If you continue to pull, it continues to get longer, at the same time getting thinner because in plastic deformation *volume is conserved* (matter is just flowing from place to place). Eventually, the plasticine becomes unstable and begins to *neck* at the maximum load point in the force-extension curve (see Figure 8.3). Necking is an *instability* which we shall look at in more detail in Chapter 11. The neck grows rapidly, and the load that the specimen can carry through the neck decreases until breakage takes place. The two pieces produced *after* breakage have a total length that is slightly *less* than the length *just* before breakage by the amount of the *elastic* extension produced by the terminal load.

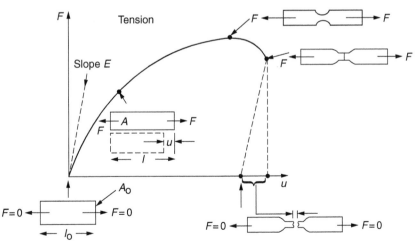

**FIGURE 8.3**
Load-extension curve for a bar of ductile metal (e.g., annealed copper) pulled in tension.

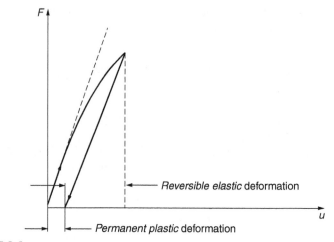

**FIGURE 8.4**
Permanent plastic deformation after a sample has yielded and been unloaded.

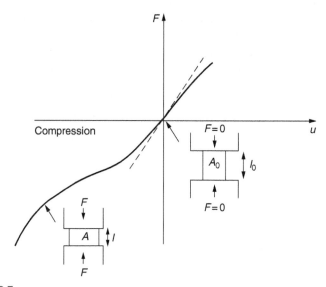

**FIGURE 8.5**
Squashing of the specimen increases the load needed to keep it flowing.

If we load a material in *compression,* the force-displacement curve is simply the reverse of that for tension *at small strains,* but it becomes different at larger strains. As the specimen squashes down, becoming shorter and fatter to conserve volume, the load needed to keep it flowing rises (Figure 8.5). No instability such as necking appears, and the specimen can be squashed almost indefinitely, only being limited by severe cracking in the specimen or plastic flow of the compression plates.

Why this great difference in behavior? After all, we are dealing with the same material in either case.

## 8.4 TRUE STRESS–STRAIN CURVES FOR PLASTIC FLOW

The apparent difference between the curves for tension and compression is due solely to the geometry of testing. If, instead of plotting *load*, we plot *load divided by the actual area of the specimen, A, at any particular elongation or compression*, the two curves become much more like one another. In other words, we simply plot *true stress* (see Chapter 3) as our vertical coordinate (Figure 8.6). This method of plotting allows for the *thinning* of the material when pulled in tension, or the *fattening* of the material when compressed.

But the two curves still do not exactly match, as Figure 8.6 shows. The reason is that a displacement of (for example) $u = l_0/2$ in tension and in compression gives different *strains;* it represents a drawing out of the tensile specimen from $l_0$ to $1.5l_0$, but a squashing down of the compressive specimen from $l_0$ to $0.5l_0$. The material of the compressive specimen has thus undergone *much* more plastic deformation than the material in the tensile specimen, and can hardly be expected to be in the same state, or to show the same resistance to plastic deformation. The two conditions can be compared properly by taking small *strain increments*

$$\delta\varepsilon = \frac{\delta u}{l} = \frac{\delta l}{l} \tag{8.1}$$

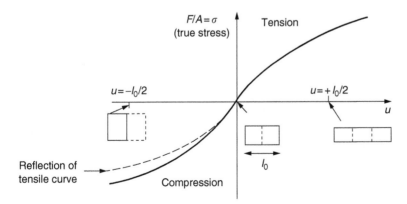

**FIGURE 8.6**

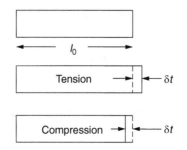

**FIGURE 8.7**

about which the state of the material is the same for either tension or compression (Figure 8.7).

This is the same as saying that a decrease in length from 100 mm ($l_0$) to 99 mm ($l$), or an increase in length from 100 mm ($l_0$) to 101 mm ($l$) both represent a 1% change in the state of the material. Actually, they do not give *exactly* the same state in both cases, but they *do* in the limit

$$d\varepsilon = \frac{dl}{l} \tag{8.2}$$

Then, if the stresses in compression and tension are plotted against

$$\varepsilon = \int_{l_0}^{l} \frac{dl}{l} = \ln\left(\frac{l}{l_0}\right) \tag{8.3}$$

the two curves *exactly* mirror one another (Figure 8.8). The quantity $\varepsilon$ is called the *true* strain (to be contrasted with the *nominal* strain $u/l_0$ defined in Chapter 3) and the matching curves are *true stress/true strain* ($\sigma/\varepsilon$) curves.

Now, a final catch. From our original load-*extension* or load-*compression* curves we can easily calculate $\varepsilon$, simply by knowing $l_0$ and taking natural logs. But how do we calculate $\sigma$? Because volume is conserved during plastic deformation we can write, at any strain,

$$A_0 l_0 = A l$$

provided the extent of plastic deformation is much greater than the extent of elastic deformation (volume is only conserved during *elastic* deformation if Poisson's ratio $v = 0.5$; and it is near 0.33 for most materials). Thus

$$A = \frac{A_0 l_0}{l} \tag{8.4}$$

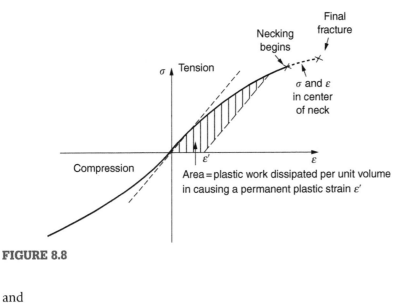

FIGURE 8.8

and

$$\sigma = \frac{F}{A} = \frac{Fl}{A_0 l_0} \tag{8.5}$$

all of which we know or can measure easily.

## 8.5 PLASTIC WORK

When metals are rolled or forged, or drawn to wire, or when polymers are injection molded or pressed or drawn, energy is absorbed. The work done on a material to change its shape permanently is called the *plastic work*; its value, per unit volume, is the area of the crosshatched region shown in Figure 8.8; it may easily be found (if the stress–strain curve is known) for any amount of permanent plastic deformation, $\varepsilon'$. Plastic work is important in metal and polymer forming operations because it determines the forces that the rolls, or press or molding machine must exert on the material.

## 8.6 TENSILE TESTING

The plastic behavior of a material is usually measured by conducting a tensile test. Tensile testing equipment is standard in engineering laboratories. Such equipment produces a load/displacement ($F/u$) curve for the material, which is then converted to a nominal stress/nominal strain, or $\sigma_n/\varepsilon_n$, curve (see Figure 8.9), where

$$\sigma_n = \frac{F}{A_0} \tag{8.6}$$

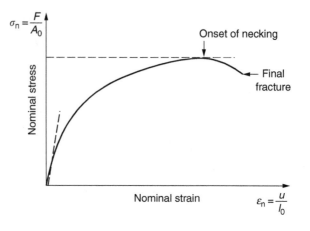

**FIGURE 8.9**

and

$$\varepsilon_n = \frac{u}{l_0} \qquad (8.7)$$

Naturally, because $A_0$ and $l_0$ are constant, the *shape* of the $\sigma_n/\varepsilon_n$ curve is identical to that of the load-extension curve. But the $\sigma_n/\varepsilon_n$ plotting method allows us to compare data for specimens having different (though now standardized) $A_0$ and $l_0$, and thus to examine the properties of *material*, unaffected by specimen size. The advantage of keeping the stress in *nominal* units and not converting to *true* stress (as shown before) is that the onset of necking can clearly be seen on the $\sigma_n/\varepsilon_n$ curve.

Now, we define the quantities usually listed as the results of a *tensile test*. The easiest way to do this is to show them on the $\sigma_n/\varepsilon_n$ curve itself (Figure 8.10). They are:

- $\sigma_y$ = *yield strength* ($F/A_0$ at onset of plastic flow)
- $\sigma_{0.1\%}$ = *0.1% proof stress* ($F/A_0$ at a permanent strain of 0.1%) (0.2% proof stress is often quoted instead; proof stress is useful for characterizing yield of a material that yields gradually, and does not show a distinct yield point)
- $\sigma_{TS}$ = *tensile strength* ($F/A_0$ at onset of necking)
- $\varepsilon_f$ = (plastic) *strain after fracture*, or tensile ductility; the broken pieces are put together and measured, and $\varepsilon_f$ calculated from $(l - l_0)/l_0$, where $l$ is the length of the assembled pieces

## 8.7 DATA

Data for *yield strength, tensile strength,* and *tensile ductility* are given in Table 8.1 and shown on the bar chart (Figure 8.11). Like moduli, they span a range of

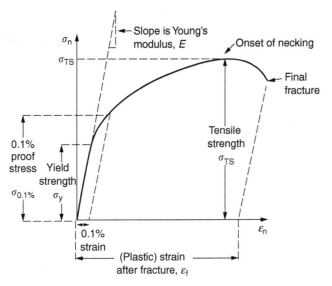

FIGURE 8.10

about $10^6$: from about 0.1 MN m$^{-2}$ (for polystyrene foams) to nearly $10^5$ MN m$^{-2}$ (for diamond).

Most ceramics have enormous yield stresses. In a tensile test, at room temperature, ceramics almost all fracture long before they yield: this is because their fracture toughness, which we will discuss later, is very low. Because of this, you cannot measure the yield strength of a ceramic by using a tensile test. Instead, you have to use a test that somehow suppresses fracture: a compression test, for instance. The best and easiest is the hardness test: the data shown here are obtained from hardness tests, which we shall discuss in a moment.

Pure metals are very soft indeed, and have a high ductility. This is what, for centuries, has made them so attractive at first for jewellery and weapons, and then for other implements and structures: they can be *worked* to the shape that you want them in; furthermore, their ability to work-harden means that, after you have finished, the metal is much stronger than when you started. By alloying, the strength of metals can be further increased, though—in yield strength—the strongest metals still fall short of most ceramics.

Polymers, in general, have lower yield strengths than metals. The very strongest barely reach the strength of aluminum alloys. They can be strengthened, however, by making composites out of them: GFRP has a strength only slightly inferior to aluminum, and CFRP is substantially stronger.

**Table 8.1** Yield Strength, $\sigma_y$, Tensile Strength, $\sigma_{TS}$, and Tensile Ductility, $\varepsilon_f$

| Material | $\sigma_y$ (MN m$^{-2}$) | $\sigma_{TS}$ (MN m$^{-2}$) | $\varepsilon_f$ |
|---|---|---|---|
| Diamond | 50,000 | – | 0 |
| Boron carbide, B$_4$C | 14,000 | (330) | 0 |
| Silicon carbide, SiC | 10,000 | (200–800) | 0 |
| Silicon nitride, Si$_3$N$_4$ | 9600 | (200–900) | 0 |
| Silica glass, SiO$_2$ | 7200 | (110) | 0 |
| Tungsten carbide, WC | 5600–8000 | (80–710) | 0 |
| Niobium carbide, NbC | 6000 | – | 0 |
| Alumina, Al$_2$O$_3$ | 5000 | (250–550) | 0 |
| Beryllia, BeO | 4000 | (130–280) | 0 |
| Syalon (Si-Al-O-N ceramic) | 4000 | (945) | 0 |
| Mullite | 4000 | (128–140) | 0 |
| Titanium carbide, TiC | 4000 | – | 0 |
| Zirconium carbide, ZrC | 4000 | – | 0 |
| Tantalum carbide, TaC | 4000 | – | 0 |
| Zirconia, ZrO$_2$ | 4000 | (100–700) | 0 |
| Soda glass (standard) | 3600 | (50–70) | 0 |
| Magnesia, MgO | 3000 | (100) | 0 |
| Cobalt and alloys | 180–2000 | 500–2500 | 0.01–6 |
| Low-alloy steels (water-quenched and tempered) | 500–1900 | 680–2400 | 0.02–0.3 |
| Pressure-vessel steels | 1500–1900 | 1500–2000 | 0.3–0.6 |
| Stainless steels, austenitic | 286–500 | 760–1280 | 0.45–0.65 |
| Boron/epoxy composites | – | 725–1730 | – |
| Nickel alloys | 200–1600 | 400–2000 | 0.01–0.6 |
| Nickel | 70 | 400 | 0.65 |
| Tungsten | 1000 | 1510 | 0.01–0.6 |
| Molybdenum and alloys | 560–1450 | 665–1650 | 0.01–0.36 |
| Titanium and alloys | 180–1320 | 300–1400 | 0.06–0.3 |
| Carbon steels (water-quenched and tempered) | 260–1300 | 500–1880 | 0.2–0.3 |
| Tantalum and alloys | 330–1090 | 400–1100 | 0.01–0.4 |
| Cast irons | 220–1030 | 400–1200 | 0.–0.18 |
| Copper alloys | 60–960 | 250–1000 | 0.01–0.55 |
| Copper | 60 | 400 | 0.55 |
| Cobalt/tungsten carbide cermets | 400–900 | 900 | 0.02 |
| CFRPs | – | 670–640 | – |
| Brasses and bronzes | 70–640 | 230–890 | 0.01–0.7 |
| Aluminum alloys | 100–627 | 300–700 | 0.05–0.3 |
| Aluminum | 40 | 200 | 0.5 |
| Stainless steels, ferritic | 240–400 | 500–800 | 0.15–0.25 |
| Zinc alloys | 160–421 | 200–500 | 0.1–1.0 |
| Concrete, steel reinforced | – | 50–200 | 0.02 |
| Alkali halides | 200–350 | – | 0 |

**Table 8.1  Cont'd**

| Material | $\sigma_y$ (MN m$^{-2}$) | $\sigma_{TS}$ (MN m$^{-2}$) | $\varepsilon_f$ |
|---|---|---|---|
| Zirconium and alloys | 100–365 | 240–440 | 0.24–0.37 |
| Mild steel | 220 | 430 | 0.18–0.25 |
| Iron | 50 | 200 | 0.3 |
| Magnesium alloys | 80–300 | 125–380 | 0.06–0.20 |
| GFRPs | – | 100–300 | – |
| Beryllium and alloys | 34–276 | 380–620 | 0.02–0.10 |
| Gold | 40 | 220 | 0.5 |
| PMMA | 60–110 | 110 | 0.03–0.05 |
| Epoxies | 30–100 | 30–120 | – |
| Polyimides | 52–90 | – | – |
| Nylons | 49–87 | 100 | – |
| Ice | 85 | (6) | 0 |
| Pure ductile metals | 20–80 | 200–400 | 0.5–1.5 |
| Polystyrene | 34–70 | 40–70 | – |
| Silver | 55 | 300 | 0.6 |
| ABS/polycarbonate | 55 | 60 | – |
| Common woods (∥ to grain) | – | 35–55 | – |
| Lead and alloys | 11–55 | 14–70 | 0.2–0.8 |
| Acrylic/PVC | 45–48 | – | – |
| Tin and alloys | 7–45 | 14–60 | 0.3–0.7 |
| Polypropylene | 19–36 | 33–36 | – |
| Polyurethane | 26–31 | 58 | – |
| Polyethylene, high density | 20–30 | 37 | – |
| Concrete, non-reinforced | 20–30 | (1–5) | 0 |
| Natural rubber | – | 30 | 8.0 |
| Polyethylene, low density | 6–20 | 20 | – |
| Common woods (⊥ to grain) | – | 4–10 | – |
| Ultrapure f.c.c. metals | 1–10 | 200–400 | 1.–2 |
| Foamed polymers, rigid | 0.2–10 | 0.2–10 | 0.1–1 |
| Polyurethane foam | 1 | 1 | 0.1–1 |

Note: *Bracketed $\sigma_{TS}$ data for brittle materials refer to the modulus of rupture $\sigma_r$ (see Chapter 15).*

## 8.8 A NOTE ON THE HARDNESS TEST

We said in Section 8.7 that the hardness test is used for estimating the yield strengths of hard brittle materials. It is also widely used as a simple non-destructive test on material or finished components to check whether they meet the specified mechanical properties. Figure 8.12 shows how one type of test works (the diamond pyramid test). A very hard (diamond) pyramid "indenter" (with a standardized angle of 136° between opposite faces) is pressed into the

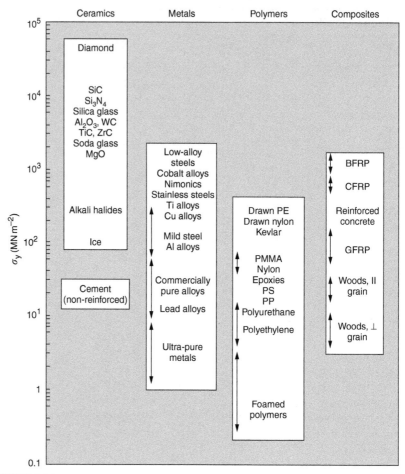

**FIGURE 8.11**

Bar chart of data for yield strength, $\sigma_y$.

surface of the material with a force $F$. Material flows away from underneath the indenter, so when the indenter is removed again, a permanent "indent" is left. The diagonals of the indent are measured with a microscope, and the total surface (or contact) area $A$ of the indent is found from look-up tables. The hardness $H$ is then given by $H = F/A$. Obviously, the softer the material, the larger the indents.

The yield strength can be estimated from the relation (derived in Chapter 11)

$$H = 3\sigma_y \tag{8.8}$$

When flowing away from the indenter, the material experiences an average nominal strain of 0.08. This means that Equation (8.8) is only accurate for

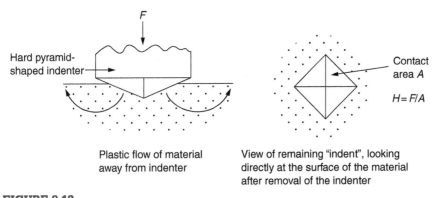

**FIGURE 8.12**
The hardness test.

materials which do not work harden. For materials that do, the hardness test gives the yield strength of material that has been work hardened by 8%. Good correlations can also be obtained between hardness and *tensile strength*. For example,

$$\sigma_{TS} = 0.33H \qquad (8.9)$$

holds well (to $\pm 10\%$) for all types of steel.

The *indentation hardness* is defined slightly differently—still as $H = F/A$, but $A$ is now the area of the indent measured at the surface of the material (which is slightly less than that measured on the faces of the indentation).

## EXAMPLES

8.1  Nine strips of pure, fully annealed copper were deformed plastically by being passed between a pair of rotating rollers so that the strips were made thinner and longer. The increases in length produced were 1, 10, 20, 30, 40, 50, 60, 70, and 100%, respectively. The diamond–pyramid hardness of each piece was measured after rolling. The results are given in the following table:

| Nominal strain | 0.01 | 0.1 | 0.2 | 0.3 | 0.4 | 0.5 | 0.6 | 0.7 | 1.0 |
|---|---|---|---|---|---|---|---|---|---|
| Hardness (MN m$^{-2}$) | 423 | 606 | 756 | 870 | 957 | 1029 | 1080 | 1116 | 1170 |

Assuming that a diamond–pyramid hardness test creates a further nominal strain, on average, of 0.08, and that the hardness value is 3.0 times the true stress, construct the curve of *nominal* stress against nominal strain.
[*Hint*: Add 0.08 to each value of nominal strain in the table.]

**8.2**  Using the plot of nominal stress against nominal strain from Example 8.1, find:
  a.  The tensile strength of copper
  b.  The strain at which tensile failure commences
  c.  The percentage reduction in cross-sectional area at this strain
  d.  The work required to initiate tensile failure in a cubic metre of annealed copper

**Answers**

(a) 217 MN m$^{-2}$; (b) 0.6 approximately; (c) 38%, (d) 109 MJ.

**8.3**  Why can copper survive a much higher extension during rolling than during a tensile test?

**8.4**  The following data were obtained in a tensile test on a specimen with 50 mm gauge length and a cross-sectional area of 160 mm$^2$.

| Extension (mm) | 0.050 | 0.100 | 0.150 | 0.200 | 0.250 | 0.300 | 1.25 | 2.50 | 3.75 | 5.00 | 6.25 | 7.50 |
|---|---|---|---|---|---|---|---|---|---|---|---|---|
| Load (kN) | 12 | 25 | 32 | 36 | 40 | 42 | 63 | 80 | 93 | 100 | 101 | 90 |

The total elongation of the specimen just before final fracture was 16%, and the reduction in area at the fracture was 64%.

Find the maximum allowable working stress if this is to equal

  a.  0.25 × tensile strength
  b.  0.6 × 0.1% proof stress

**Answers**

(a) 160 MN m$^{-2}$; (b) 131 MN m$^{-2}$

**8.5**  One type of hardness test involves pressing a hard sphere (radius $r$) into the test material under a fixed load $F$, and measuring the *depth*, $h$, to which the sphere sinks into the material, plastically deforming it. Derive an expression for the indentation hardness, $H$, of the material in terms of $h$, $F$, and $r$. Assume $h \ll r$.

**Answer**

$$H = \frac{F}{2\pi r h}$$

**8.6** The following diagram shows the force-extension characteristics of a bungee rope. The length of the rope under zero load is 15 m. One end of the rope is attached to a

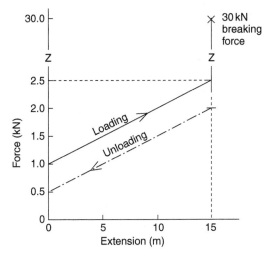

bridge deck, and the other end is attached to a person standing on the deck. The person then jumps off the deck, and descends vertically until arrested by the rope.

**a.** Find the maximum mass of the person for a successful arrest, $m_{max}$.

**b.** The diagram shows that the unloading line is 0.5 kN below the loading line. Comment on the practical significance of this to bungee jumping.

**Answer**

(a) 89.2 kg

# REVISION OF TERMS AND USEFUL RELATIONS

$\sigma_n$, nominal stress $\sigma_n = F/A_0$

$\sigma$, true stress $\sigma = F/A$

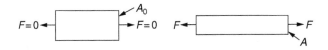

$\varepsilon_n$, nominal strain

$$\varepsilon_n = \frac{u}{l_0}, \text{ or } \frac{l - l_0}{l_0} \text{ or } \frac{l}{l_0} - 1$$

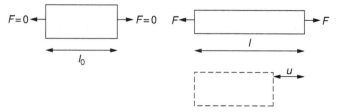

### Relations between $\sigma_n$, $\sigma$, and $\varepsilon_n$

Assuming constant volume (valid if $v = 0.5$ or, if not, plastic deformation $\gg$ elastic deformation):

$$A_0 l_0 = Al; \quad A_0 = \frac{Al}{l_0} = A(1 + \varepsilon_n)$$

Thus

$$\sigma = \frac{F}{A} = \frac{F}{A_0}(1 + \varepsilon_n) = \sigma_n(1 + \varepsilon_n)$$

$\varepsilon$, true strain and the relation between $\varepsilon$ and $\varepsilon_n$

$$\varepsilon = \int_{l_0}^{l} \frac{dl}{l} = \ln\left(\frac{l}{l_0}\right)$$

Thus

$$\varepsilon = \ln(1 + \varepsilon_n)$$

### Small strain condition

For small $\varepsilon_n$

$$\varepsilon \approx \varepsilon_n, \text{ from } \varepsilon = \ln(1 + \varepsilon_n)$$
$$\sigma \approx \sigma_n, \text{ from } \sigma = \sigma_n(1 + \varepsilon_n)$$

Thus, when dealing with most *elastic* strains (but not in rubbers), it is immaterial whether $\varepsilon$ or $\varepsilon_n$, or $\sigma$ or $\sigma_n$, are chosen.

### Energy

The energy expended in deforming a material *per unit volume* is given by the area under the stress–strain curve.

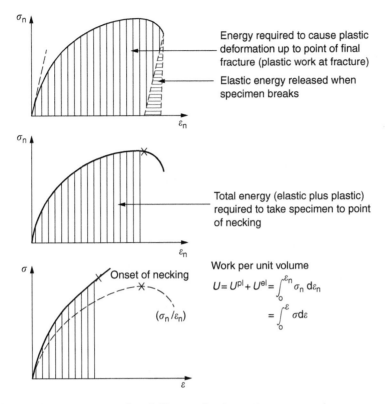

Energy required to cause plastic deformation up to point of final fracture (plastic work at fracture)

Elastic energy released when specimen breaks

Total energy (elastic plus plastic) required to take specimen to point of necking

Onset of necking

$(\sigma_n / \varepsilon_n)$

Work per unit volume

$$U = U^{pl} + U^{el} = \int_0^{\varepsilon_n} \sigma_n \, d\varepsilon_n$$

$$= \int_0^{\varepsilon} \sigma d\varepsilon$$

For *linear elastic strains*, and *only* linear elastic strains

$$\frac{\sigma_n}{\varepsilon_n} = E, \text{ and } U^{el} = \int \sigma_n d\varepsilon_n = \int \sigma_n \frac{d\sigma_n}{E} = \left\{ \frac{\sigma_n^2}{2E} \right\}$$

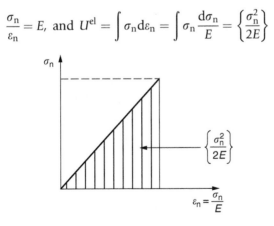

$$\left\{ \frac{\sigma_n^2}{2E} \right\}$$

$$\varepsilon_n = \frac{\sigma_n}{E}$$

### Elastic limit

In a tensile test, as the load increases, the specimen at first is strained *elastically*, that is reversibly. Above a limiting stress—the elastic limit—some of the strain is permanent; this is *plastic* deformation.

> **Yielding** The change from elastic to measurable plastic deformation.
>
> **Yield strength** The nominal stress at yielding. In many materials this is difficult to spot on the stress–strain curve and in such cases it is better to use a proof stress.
>
> **Proof stress** The stress that produces a permanent strain equal to a specified percentage of the specimen length. A common proof stress is one that corresponds to 0.2% permanent strain.

### Strain hardening (work-hardening)

The increase in stress needed to produce further strain in the plastic region. Each strain increment strengthens or hardens the material so that a larger stress is needed for further strain.

### $\sigma_{TS}$, tensile strength (ultimate tensile strength, or UTS)

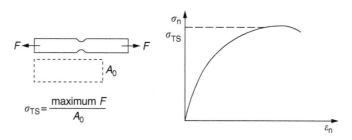

$$\sigma_{TS} = \frac{\text{maximum } F}{A_0}$$

### $\varepsilon_f$, strain after fracture, or tensile ductility

The permanent extension in length (measured by fitting the broken pieces together) expressed as a percentage of the original gauge length.

$$\left\{ \frac{l_{break} - l_0}{l_0} \right\} \times 100$$

### Reduction in area at break

The maximum decrease in cross-sectional area at the fracture expressed as a percentage of the original cross-sectional area.

Strain after fracture and percentage reduction in area are used a measures of ductility, that is the ability of a material to undergo large plastic strain under stress before it fractures.

$$\left\{ \frac{A_0 - A_{break}}{A_0} \right\} \times 100$$

## CHAPTER 9

# Dislocations and Yielding in Crystals

## CONTENTS

## 9.1 INTRODUCTION

In the previous chapter we looked at data for the yield strengths of materials. But what would we expect? From our understanding of the structure of solids and the stiffness of the bonds between the atoms, can we estimate what the yield strength should be? A simple calculation (given in the next section) over-estimates it grossly. This is because real crystals contain defects—*dislocations*—which move easily. When they move, the crystal deforms; the stress needed to move them is the yield strength. Dislocations are the *carriers* of deformation, much as electrons are the carriers of charge.

## 9.2 THE STRENGTH OF A PERFECT CRYSTAL

As we showed in Equation 6.1 in Chapter 6, the slope of the interatomic force–distance curve at the equilibrium separation is proportional to Young's modulus $E$. Interatomic forces typically drop off to negligible values at a distance of separation of atom centers of $2r_0$. The maximum in the force–distance curve is typically reached at $1.25r_0$ separation, and if the stress applied to the material

Engineering Materials I: An Introduction to Properties, Applications, and Design, Fourth Edition
© 2012, Michael F. Ashby and David R. H. Jones. Published by Elsevier Ltd. All rights reserved.

exceeds this maximum force per *bond*, fracture is bound to occur. We call the stress at which this bond rupture takes place $\tilde{\sigma}$, the *ideal strength*—a material cannot be stronger than this. From Figure 9.1

$$y\sigma = E\varepsilon$$

$$2\tilde{\sigma} \approx E \frac{0.25 r_0}{r_0} \approx \frac{E}{4}$$

$$\tilde{\sigma} \approx \frac{E}{8}$$

(9.1)

More refined estimates of $\tilde{\sigma}$ are possible, using real interatomic potentials—they give about $E/15$ instead of $E/8$.

Let us now see whether materials really show this strength. The bar chart (Figure 9.2) shows values of $\sigma_y/E$ for materials. The heavy broken line at the top is drawn at the level $\sigma/E = 1/15$. Glasses, and some ceramics, lie close to this line—they show their ideal strength, and we could not expect them to be stronger than this. Most polymers, too, lie near the line—although they have low yield strengths, these are low because the *moduli* are low.

Metals, on the other hand, have yield strengths far below the levels predicted by our calculation—as much as a factor of $10^5$ less. Even many ceramics yield at stresses that are a factor of 10 below their ideal strength. Why is this?

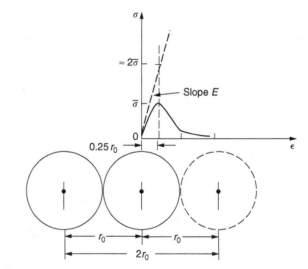

**FIGURE 9.1**
The ideal strength, $\tilde{\sigma}$.

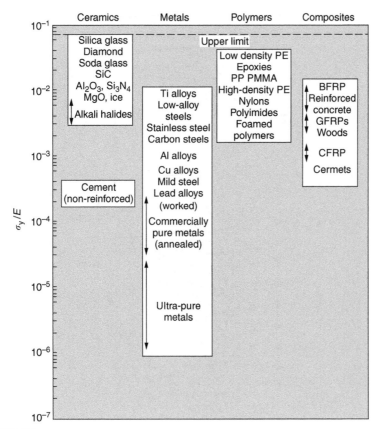

**FIGURE 9.2**

Data for normalized yield strength, $\sigma_y/E$.

## 9.3 DISLOCATIONS IN CRYSTALS

In Chapter 5 we said that many important engineering materials (e.g., metals) were normally made up of crystals, and explained that a perfect crystal was an assembly of *atoms packed together in a regularly repeating pattern.*

But crystals are not perfect—they have *defects* in them. Just as the strength of a chain is determined by the strength of the weakest link, so the strength of a crystal—and thus of the material—is usually limited by the defects that are present in it. The *dislocation* is a particular type of defect that has the effect of allowing materials to deform plastically (yield) at stress levels much less than $\tilde{\sigma}$.

Figure 9.3(a) shows an *edge dislocation* from a continuum viewpoint (ignoring the atoms). Such a dislocation is made in a block of material by cutting the block up to the line marked $\perp - \perp$, displacing the material below the cut relative to that above by a distance $b$ (the distance between adjacent planes of atoms),

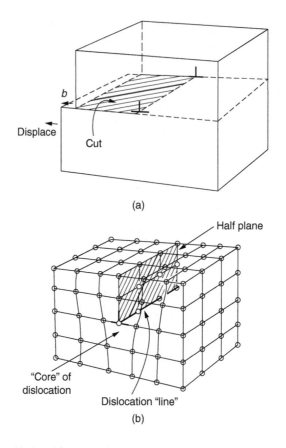

**FIGURE 9.3**

An edge dislocation, (a) viewed from a continuum standpoint (ignoring the atoms) and (b) showing the positions of the atoms near the dislocation. Only the atom centers are shown.

and finally gluing the cut surfaces back together. The result, on an atomic scale, is shown in Figure 9.3(b); the material in the middle of the block now contains a *half plane* of atoms, with its lower edge lying along the line ⊥ – ⊥ (the *dislocation line*). This defect is called an edge dislocation because it is formed by the edge of the half plane of atoms, and it is written for short using the symbol ⊥.

Dislocation motion produces plastic strain. Figure 9.4 shows how the atoms rearrange as the dislocation moves through the crystal, and how when a dislocation moves through a crystal, the lower part is displaced under the upper by distance $b$ (called the Burgers vector). The same process is drawn without the atoms, and using the symbol ⊥ for the position of the dislocation line, in Figure 9.5. The way in which this dislocation works can be likened to the way in which a large carpet can be moved across a floor simply by moving rucks through the carpet—a much easier process than pulling the whole carpet across at one go (Figure 9.6)

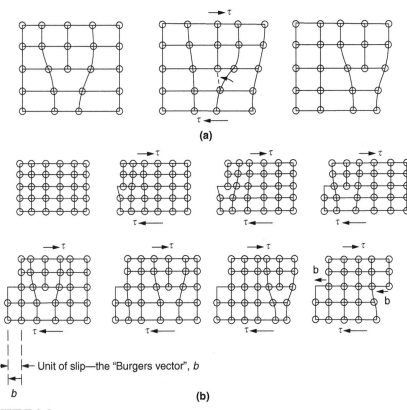

**FIGURE 9.4**

How an edge dislocation moves through a crystal. (a) How the atomic bonds at the center of the dislocation break and reform to allow the dislocation to move. (b) A complete sequence for the introduction of a dislocation into a crystal from the left side, its migration through the crystal, and its expulsion on the right side; this process causes the lower half of the crystal to slip by a distance $b$ under the upper half.

Referring back to Figure 9.3 we could, after making the cut, have displaced the lower part of the crystal under the upper part in a direction *parallel* to the bottom of the cut, instead of normal to it. Figure 9.7 shows the result—it, too, is a dislocation, called a *screw dislocation* (because it converts the planes of atoms into a helical surface, or screw). Like an edge dislocation, it produces plastic strain when it moves (Figures 9.8 and 9.9). Its geometry is a little more complicated but its properties are otherwise just like those of the edge. Any dislocation, in a real crystal, is either a screw or an edge—or can be thought of as little steps of each. Dislocations can be seen in the electron microscope. Figure 9.10 shows an example (see page 141).

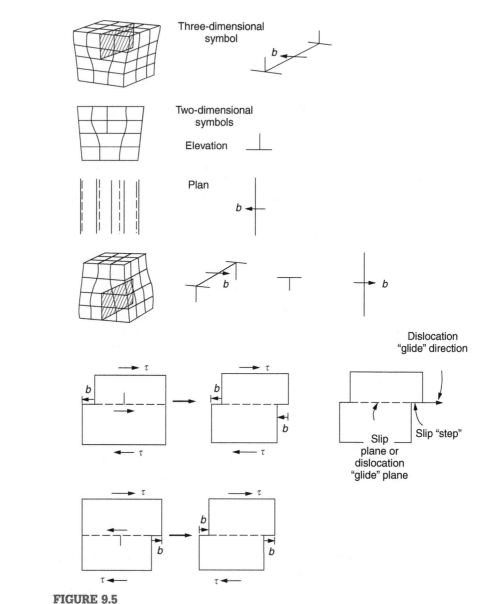

**FIGURE 9.5**
Edge dislocation notation.

## 9.4 THE FORCE ACTING ON A DISLOCATION

A shear stress ($\tau$) exerts a force on a dislocation, pushing it through the crystal. For yielding to take place, this force must be great enough to overcome the *resistance* to the motion of the dislocation. This resistance is due to intrinsic

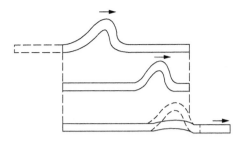

**FIGURE 9.6**
The "carpet-ruck" analogy of an edge dislocation.

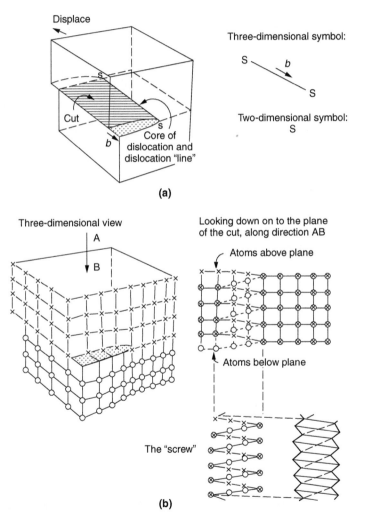

**(a)**

**(b)**

**FIGURE 9.7**
A screw dislocation, (a) viewed from a continuum standpoint and (b) showing the atom positions.

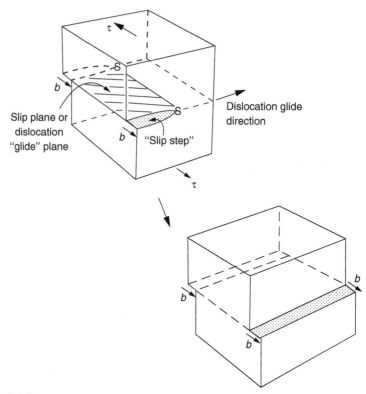

**FIGURE 9.8**
Screw-dislocation notation.

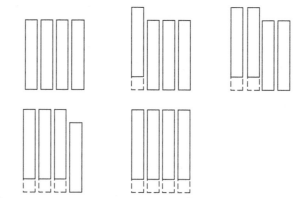

**FIGURE 9.9**
The "planking" analogy for the screw dislocation. Imagine four planks resting side by side on a factory floor. It is much easier to slide them across the floor one at a time than all at the same time.

**FIGURE 9.10**
An electron microscope picture of dislocation lines in stainless steel. The picture was taken by firing electrons through a very thin slice of steel about 100 nm thick. The dislocation lines here are only about 1000 atom diameters long because they have been "chopped off" where they meet the top and bottom surfaces of the thin slice. But a sugar-cube-sized piece of any engineering alloy contains about $10^5$ km of dislocation line. *(Courtesy of Dr Peter Southwick.)*

friction opposing dislocation motion, plus contributions from alloying or work-hardening; these are discussed in detail in the next chapter. Here we show that the magnitude of the force is $\tau b$ per unit length of dislocation.

We prove this by a virtual work calculation. We equate (a) the work done by the applied stress when the dislocation moves through the crystal, and (b) the work done against the resistance opposing its motion (Figure 9.11). The upper part is displaced relative to the lower by the distance $b$, so the applied stress does work $(\tau l_1 l_2) \times b$. In moving through the crystal, the dislocation travels a distance $l_2$, doing work against a resistance $f$ *per unit length*—$fl_1 \times l_2$. Equating the two gives

$$\tau b = f \qquad (9.2)$$

This result holds for any dislocation—edge, screw or a mixture of both.

## 9.5 OTHER PROPERTIES OF DISLOCATIONS

The two remaining properties of dislocations that are important in understanding the plastic deformation of materials are the following:

- Dislocations always glide on crystallographic planes, as we might imagine from our earlier drawings of edge dislocation motion. In f.c.c. crystals, for example, the dislocations glide on {111} planes, and therefore plastic shearing takes place on {111} in f.c.c. crystals.
- The atoms around a dislocation are displaced from their proper positions and thus have a higher energy. In order to keep the total energy as low as possible, the dislocation tries to be as short as possible—it behaves as if it had a *line tension, T,* like a rubber band (see Figure 9, 12). The line tension is energy per unit length (just as surface tension is energy per unit area), and is given by

$$T \approx \frac{Gb^2}{2} \tag{9.3}$$

In absolute terms, $T$ is small (we should need $\approx 10^8$ dislocations to hold an apple up) but it is large in relation to the size of a dislocation. In addition, it has an important bearing on the way in which obstacles obstruct the motion of dislocations.

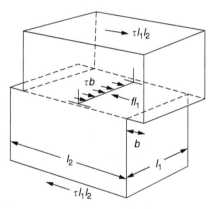

**FIGURE 9.11**
The force acting on a dislocation.

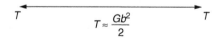

**FIGURE 9.12**
The line tension in a dislocation (*G* is the shear modulus).

# EXAMPLES

**9.1** Explain what is meant by the *ideal strength* of a material.

**9.2** The energy per unit length of a dislocation is $\frac{1}{2}Gb^2$. Dislocations dilate a close-packed crystal structure very slightly, because at their cores the atoms are not close packed: the dilatation is $\frac{1}{4}b^2$ per unit length of dislocation. It is found that the density of a bar of copper changes from 8.9323 Mg m$^{-3}$ to 8.9321 Mg m$^{-3}$ when it is very heavily deformed. Calculate:

  **a.** the dislocation density introduced into the copper by the deformation.

  **b.** the energy associated with that density.

Compare these results with the latent heat of melting of copper (1833 MJ m$^{-3}$) ($b$ for copper is 0.256 nm; $G$ is $\frac{3}{8}E$).

**Answers**

(a) $1.4 \times 10^{15}$ m$^{-2}$; (b) 2.1 MJ m$^{-3}$

**9.3** Explain briefly what is meant by a *dislocation*. Show with diagrams how the motion of (a) an edge dislocation and (b) a screw dislocation can lead to the plastic deformation of a crystal under an applied shear stress.

**9.4** How can the presence of dislocations in crystalline metals explain the observation that their strengths can be as much as $10^5$ times less than their ideal strengths?

**9.5** Explain why a dislocation has a line tension.

**9.6** Prove the relationship $f = \tau b$ for the force acting on a dislocation, and explain its relevance to the yielding of a crystal.

**9.7** Why do dislocations always glide on crystallographic planes? What planes do they glide on in f.c.c. crystals?

## CHAPTER 10

# Strengthening Methods and Plasticity of Polycrystals

## CONTENTS

## 10.1 INTRODUCTION

We showed in the last chapter that.

a. Crystals contain dislocations.
b. A shear stress $\tau$, applied to the slip plane of a dislocation, exerts a force $\tau b$ per unit length of the dislocation trying to push it forward.
c. When dislocations move, the crystal deforms plastically (yields).

In this chapter, we examine ways of increasing the resistance to motion of a dislocation—which determines the *dislocation yield strength* of a single isolated crystal of a metal or ceramic. Bulk engineering materials, however, are aggregates of many crystals, or *grains*. To understand the plasticity of such an

Engineering Materials I: An Introduction to Properties, Applications, and Design, Fourth Edition

aggregate, we also have to examine how the individual crystals interact with each other. This lets us calculate the *polycrystal yield strength*—the quantity that enters engineering design.

## 10.2 STRENGTHENING MECHANISMS

A crystal yields when the force $\tau b$ (per unit length) exceeds $f$, the *resistance* (force per unit length) opposing the motion of a dislocation. This defines the dislocation yield strength

$$\tau_y = \frac{f}{b} \tag{10.1}$$

Most crystals have a certain *intrinsic* strength, caused by the bonds between the atoms which have to be broken and reformed as the dislocation moves. Covalent bonding, particularly, gives a very large *intrinsic lattice resistance*, $f_i$ per unit length of dislocation. It is this that causes the enormous strength and hardness of diamond, and the carbides, oxides, nitrides, and silicates that are used for abrasives and cutting tools. But pure metals are very soft: they have a very low lattice resistance. Then it is useful to increase $f$ by *solid solution strengthening*, by *precipitate* or *dispersion* strengthening, or by *work-hardening*, or by any combination of the three. Remember, however, that there is an upper limit to the yield strength: it can never exceed the ideal strength. In practice, only a few materials have strengths that even approach it.

## 10.3 SOLID SOLUTION HARDENING

A good way of hardening a metal is simply to make it impure. Impurities go into solution in a solid metal just as sugar dissolves in coffee. A good example is the addition of zinc (Zn) to copper (Cu) to make the *alloy* called brass. The zinc atoms replace copper atoms to form a *random substitutional solid solution*. At room temperature Cu will dissolve up to 30% Zn in this way. The Zn atoms are bigger than the Cu atoms, and, in squeezing into the Cu structure, generate stresses.

These stresses "roughen" the slip plane, making it harder for dislocations to move; they increase the resistance $f$, and thereby increase the dislocation yield strength, $\tau_y$ (Equation (10.1)). If the contribution to $f$ given by the solid solution is $f_{ss}$ then $\tau_y$ is increased by $f_{ss}/b$. In a solid solution of concentration $C$, the spacing of dissolved atoms on the slip plane varies as $C^{-1/2}$; and the smaller the spacing, the "rougher" is the slip plane. As a result, $\tau_y$ increases parabolically (i.e., as $C^{1/2}$) with solute concentration (Figure 10.1). Single-phase brass, bronze, and stainless steels, as well as many other metallic alloys, derive their strength in this way.

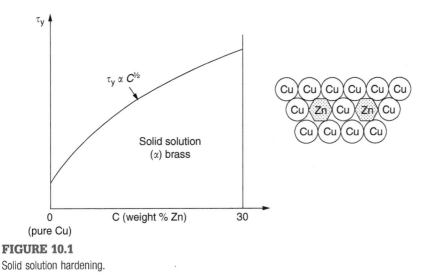

**FIGURE 10.1**
Solid solution hardening.

## 10.4 PRECIPITATE AND DISPERSION STRENGTHENING

If an impurity is dissolved in a metal or ceramic at a high temperature, and the alloy is cooled to room temperature, the impurity may *precipitate* as small particles, much as sugar will crystallize from a saturated solution when it is cooled. An alloy of Al containing 4% Cu ("Duralumin") treated in this way gives small, closely spaced precipitates of the hard compound $CuAl_2$. Most steels are strengthened by precipitates of carbides.

Small particles can be introduced into metals or ceramics in other ways. The most obvious is to mix a dispersoid (such as an oxide) into a powdered metal (aluminum and lead are both treated in this way), and then compact and sinter the mixed powders.

Either approach distributes small, hard particles in the path of a moving dislocation. Figure 10.2 shows how they obstruct its motion. The stress $\tau$ has to push the dislocation between the obstacles. It is like blowing up a balloon in a bird cage: a very large pressure is needed to bulge the balloon between the bars, though once a large enough bulge is formed, it can easily expand further. The *critical configuration* is the semicircular one (Figure 10.2(c)): here the force $\tau b L$ on one segment is just balanced by the force $2T$ due to the line tension, acting on either side of the bulge. The dislocation escapes (and yielding occurs) when

$$\tau_y = \frac{2T}{bL} \approx \frac{Gb}{L} \qquad (10.2)$$

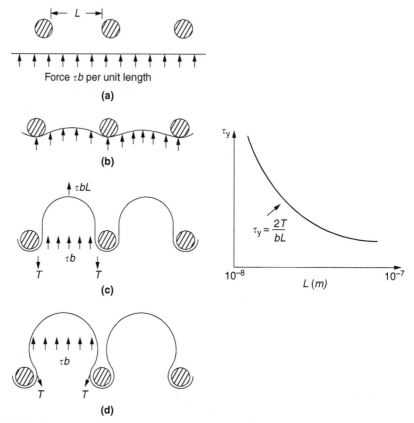

**FIGURE 10.2**
How dispersed precipitates help prevent the movement of dislocations and help prevent plastic flow of materials: (a) approach situation, (b) subcritical situation, (c) critial situation, and (d) escape situation.

The obstacles thus exert a resistance of $f_0 = 2T/L$. Obviously, the greatest hardening is produced by *strong, closely spaced* precipitates or dispersions (Figure 10.2).

## 10.5 WORK-HARDENING

When crystals yield, dislocations move through them. Most crystals have several slip planes: the f.c.c. structure, which slips on {111} planes, has four, for example. Dislocations on these intersecting planes interact, and obstruct each other, and accumulate in the material.

The result is *work-hardening*: the steeply rising stress–strain curve after yield, shown in Chapter 8. All metals work-harden. It can be a nuisance: if you want to roll thin sheet, work-hardening quickly raises the yield strength so much that you have to stop and *anneal* the metal (heat it up to remove the accumulated dislocations)

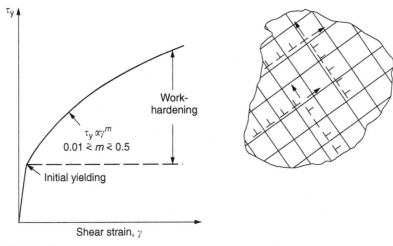

**FIGURE 10.3**
Collision of dislocations leads to work-hardening.

before you can go on. But it is also useful—it is a potent strengthening method, which can be added to the other methods to produce strong materials.

The analysis of work-hardening is difficult. Its contribution $f_{wh}$ to the total dislocation resistance $f$ is considerable and increases with strain (Figure 10.3).

## 10.6 THE DISLOCATION YIELD STRENGTH

It is adequate to assume that the strengthening methods contribute in an additive way to the strength. Then

$$\tau_y = \frac{f_i}{b} + \frac{f_{ss}}{b} + \frac{f_o}{b} + \frac{f_{wh}}{b} \tag{10.3}$$

Strong materials either have a high intrinsic strength, $f_i$ (e.g., diamond), or they rely on the superposition of solid solution strengthening $f_{ss}$, obstacles $f_o$, and work-hardening $f_{wh}$ (e.g., high-tensile steels). But before we can use this information, one problem remains: we have calculated the yield strength of an *isolated crystal* in *shear*. We want the yield strength of a *polycrystalline aggregate* in *tension*.

## 10.7 YIELD IN POLYCRYSTALS

The crystals, or *grains*, in a polycrystal fit together exactly but their crystal orientations differ (Figure 10.4). Where they meet, at *grain boundaries*, the crystal structure is disturbed, but the atomic bonds across the boundary are numerous and strong enough that the boundaries do not usually weaken the material.

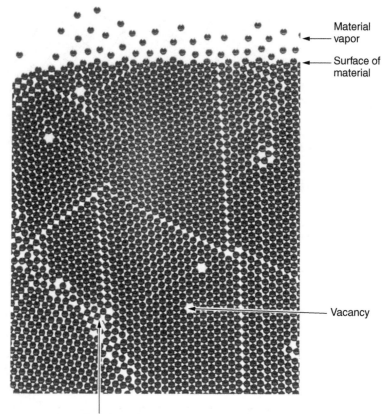

Material
vapor

Surface of
material

Vacancy

Grain boundary

**FIGURE 10.4**
Ball bearings can be used to simulate how atoms are packed together in solids. The photograph shows a
ball-bearing model set up to show what the *grain boundaries* look like in a polycrystalline material.
The model also shows up another type of defect—the vacancy—which is caused by a missing atom.

Let us now look at what happens when a polycrystalline component begins to
yield (Figure 10.5). Slip begins in grains where there are slip planes as nearly
parallel to $\tau$ as possible, for example, grain (1). Slip later spreads to grains like
(2) which are not so favorably oriented, and lastly to the worst oriented grains
like (3). Yielding does not take place all at once, therefore, and there is no sharp
polycrystalline yield point on the stress–strain curve.

Further, gross (total) yielding does not occur at the dislocation yield strength $\tau_y$,
because not all the grains are oriented favorably for yielding. The gross yield
strength is higher, by a factor called the Taylor factor, which is calculated (with
difficulty) by averaging the stress over all possible slip planes—it is close to 1.5.

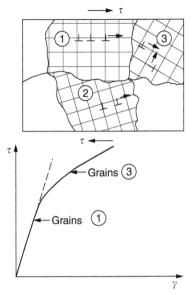

**FIGURE 10.5**
The progressive nature of yielding in a polycrystalline material.

But we want the tensile yield strength, $\sigma_y$. A tensile stress $\sigma$ creates a shear stress in the material that has a maximum value of $\tau = \sigma/2$. (We show this in Chapter 11 where we resolve the tensile stress onto planes within the material.) To calculate $\sigma_y$ from $\tau_y$, we combine the Taylor factor with the resolution factor to give

$$\sigma_y = 3\tau_y \tag{10.4}$$

$\sigma_y$ is the quantity we want: the yield strength of bulk, polycrystalline solids. It is larger than the dislocation yield strength $\tau_y$ (by the factor 3) but is proportional to it. So all the statements we have made about increasing $\tau_y$ apply unchanged to $\sigma_y$.

## Grain-boundary strengthening (Hall–Petch effect)
The presence of grain boundaries in a polycrystalline material has an additional consequence—they contribute to the yield strength because grain boundaries act as obstacles to dislocation movement. Because of this, dislocations pile up against grain boundaries, as shown in Figure 10.6.

The number of dislocations that can form in a pile-up is given by

$$n = \frac{\alpha \tau d}{Gb}$$

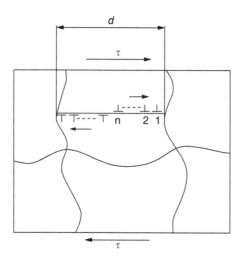

FIGURE 10.6
How dislocations pile up against grain boundaries.

where $\alpha$ is a constant and $d$ is the grain size. If there were only one dislocation piled-up against the grain boundary, the force on it would be $f = \tau b$, as shown in Equation (9.2). However, in a pile-up of $n$ dislocations, each dislocation exerts a force on the one in front, so the force on the leading (number 1) dislocation is $n\tau b$. If this force exceeds a critical value, $F_{GB}$, then the leading dislocation will move through the grain boundary, and yield will occur. The yield condition is given by

$$F_{GB} = n\tau b = \left(\frac{\alpha\tau d}{Gb}\right)\tau b = \frac{\alpha\tau^2 d}{G}$$

The contribution to the dislocation yield strength $\tau_y$ is therefore

$$\tau_y = \left(\frac{F_{GB}G}{\alpha}\right)^{1/2} d^{-1/2} = \beta d^{-1/2} \tag{10.5}$$

where $\beta$ is a constant. The interesting consequence of this result is that fine-grained materials (large $d^{-1/2}$) have a higher yield strength than coarse-grained materials (small $d^{-1/2}$). However, the value of $\beta$ is specific to the material, and is less for some alloys than others.

## 10.8 FINAL REMARKS

A whole science of alloy design for high strength has grown up in which alloys are blended and heat-treated to achieve maximum $\tau_y$. Important components that are strengthened in this way range from lathe tools ("high-speed" steels) to turbine blades ("Nimonic" alloys based on nickel). We shall have more to say about strong solids when we come to look at how materials are *selected* for a particular application.

## EXAMPLES

**10.1** Show how dislocations can account for the following observations:

    **a.** Cold working makes aluminum harder.

    **b.** An alloy of 20% Zn, 80% Cu is harder than pure copper.

    **c.** The hardness of nickel is increased by adding particles of thorium oxide.

**10.2** Derive an expression for the shear stress $\tau$ needed to bow a dislocation line into a semicircle between small hard particles a distance $L$ apart.

**10.3 a.** A polycrystalline aluminum alloy contains a dispersion of hard particles of diameter $10^{-8}$ m and average center-to-center spacing of $6 \times 10^{-8}$ m measured in the slip planes. Estimate their contribution to the tensile yield strength, $\sigma_y$, of the alloy.

    **b.** The alloy is used for the compressor blades of a small turbine. Adiabatic heating raises the blade temperature to 150°C, and causes the particles to coarsen slowly. After 1000 hours they have grown to a diameter of $3 \times 10^{-8}$ m and are spaced $18 \times 10^{-8}$ m apart. Estimate the drop in yield strength. (The shear modulus of aluminum is 26 GN m$^{-2}$, and $b = 0.286$ nm.)

**Answers**

(a) 450 MN m$^{-2}$; (b) 300 MN m$^{-2}$

**10.4** The yield stress of a sample of brass with a large grain size was 20 MN m$^{-2}$. The yield stress of an otherwise identical sample with a grain size of 4 $\mu$m was 120 MN m$^{-2}$. Why did the yield stress increase in this way? What is the value of $\beta$ in Equation (10.5) for the brass?

**Answer**

0.2 MN m$^{-3/2}$

**10.5** Explain why a polycrystalline material does not usually have a sharp yield point. What is the ratio of $\tau / \tau_y$, where $\tau_y$ is the dislocation yield strength and $\tau$ is the shear yield strength of the complete polycrystal?

**10.6** Summarize the ways in which the dislocation yield strength can be maximized in a polycrystalline material.

## CHAPTER 11

# Continuum Aspects of Plastic Flow

## CONTENTS

## 11.1 INTRODUCTION

Plastic flow occurs by shear. Dislocations move when the shear stress on the slip plane exceeds *the dislocation yield strength* $\tau_y$ of a single crystal. If this is averaged over all grain orientations and slip planes, it can be related to the *tensile yield strength* $\sigma_y$ of a polycrystal by $\sigma_y = 3\tau_y$ (Chapter 10).

In solving problems of plasticity, however, it is more useful to define the *shear yield strength* $k$ of a polycrystal. It is equal to $\sigma_y/2$, and differs from $\tau_y$ because it is an average shear resistance over all orientations of slip plane. When a structure is loaded, the planes on which shear will occur can often be identified or guessed, and the collapse load calculated approximately by requiring that the stress exceed $k$ on these planes.

In this chapter, we show that $k = \sigma_y/2$, and use $k$ to relate the hardness to the yield strength of a solid. We then examine tensile instabilities which appear in the drawing of metals and polymers.

Engineering Materials I: An Introduction to Properties, Applications, and Design, Fourth Edition

## 11.2 THE ONSET OF YIELDING AND THE SHEAR YIELD STRENGTH, *k*

A tensile stress applied to a piece of material will create a shear stress at an angle to the tensile stress. Let us examine the stresses in more detail. Resolving forces in Figure 11.1 gives the shearing force as $F \sin \theta$. The area over which this force acts in shear is $A/\cos \theta$.

Thus the shear stress, $\tau$ is

$$\tau = \frac{F \sin\theta}{A/\cos\theta} = \frac{F}{A} \sin\theta \cos\theta = \sigma \sin\theta \cos\theta \qquad (11.1)$$

If we plot this against $\theta$ as in Figure 11.2 we find a maximum $\tau$ at $\theta = 45°$ to the tensile axis. This means that the *highest value of the shear stress is found at 45° to the tensile axis, and has a value of $\sigma/2$.*

Now, from what we have said in Chapters 9 and 10, if we are dealing with a single crystal, the crystal will *not* in fact slip on the 45° plane—it will slip on the nearest lattice plane to the 45° plane on which dislocations can glide (Figure 11.3). In a polycrystal, neighboring grains each yield on their nearest-to-45° slip planes. On a microscopic scale, slip occurs on a zigzag path;

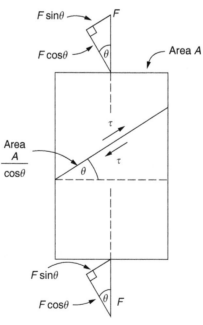

**FIGURE 11.1**
A tensile stress, *F/A*, produces a shear stress, $\tau$, on an inclined plane in the stressed material.

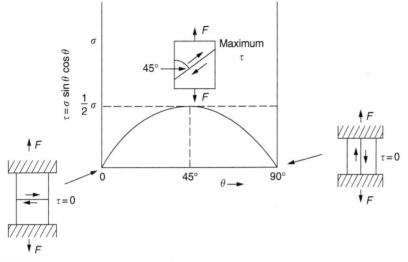

**FIGURE 11.2**

Shear stresses in a material have their maximum value on planes at 45° to the tensile axis.

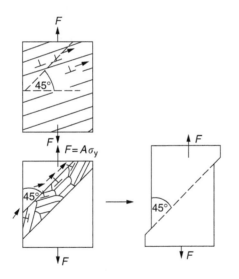

**FIGURE 11.3**

In a polycrystalline material the *average* slip path is at 45° to the tensile axis.

but the *average* slip path is at 45° to the tensile axis. The shear stress on this plane when yielding occurs is therefore $\tau = \sigma_y/2$, and we define this as the shear yield strength $k$:

$$k = \sigma_y/2 \qquad (11.2)$$

## 11.3 ANALYZING THE HARDNESS TEST

The concept of shear yielding—where we ignore the details of the grains in the polycrystal and treat the material as a *continuum*—is useful in many respects. For example, we can use it to calculate the loads that would make the material yield for all sorts of quite complicated geometries.

A good example is the problem of the *hardness indenter* that we referred to in the hardness test in Chapter 8. Then, we stated that the hardness

$$H = \frac{F}{A} = 3\sigma_y$$

We assume that the material does not work-harden so as the indenter is pushed into the material, the yield strength does not change. For simplicity, we consider a two-dimensional model. (A real indenter is three-dimensional, but the result is, for practical purposes, the same.)

As we press a flat indenter into the material, shear takes place on the 45° planes of maximum shear stress shown in Figure 11.4, at a value of shear stress equal to $k$. By equating the work done by the force $F$ as the indenter sinks a distance $u$ to the work done against $k$ on the shear planes, we get:

$$Fu = 2 \times \frac{Ak}{\sqrt{2}} \times u\sqrt{2} + 2 \times Ak \times u + 4 \times \frac{Ak}{\sqrt{2}} \times \frac{u}{\sqrt{2}}$$

This simplifies to

$$F = 6Ak$$

from which

$$\frac{F}{A} = 6k = 3\sigma_y$$

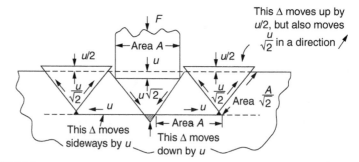

**FIGURE 11.4**

The plastic flow of material under a hardness indenter.

But $F/A = H$ so

$$H = 3\sigma_y \qquad (11.3)$$

(Strictly, shear occurs not just on the shear planes we have drawn, but on myriad 45° planes near the indenter. If our assumed geometry for slip is wrong it can be shown rigorously by a theorem called the *upper-bound* theorem that the value we get for $F$ at yield—the so-called "limit" load—is always on the high side.)

Similar treatments can be used for all sorts of two-dimensional problems: for calculating the plastic collapse load of structures of complex shape, and for analyzing metal-working processes like forging, rolling, and sheet drawing.

## 11.4 PLASTIC INSTABILITY: NECKING IN TENSILE LOADING

We now turn to the other end of the stress–strain curve and explain why, in tensile straining, materials eventually start to *neck*, a name for *plastic instability*. It means that flow becomes localized across one section of the specimen or component, as shown in Figure 11.5, and (if straining continues) the material fractures there. Plasticine necks readily; chewing gum is very resistant to necking.

We analyze the instability by noting that if a force $F$ is applied to the end of the specimen, then any section must carry this load. But is it capable of doing so? Suppose one section deforms a little more than the rest, as the figure shows. Its section is less, and the stress in it is therefore larger than elsewhere. If work-hardening has raised the yield strength enough, the reduced section can still

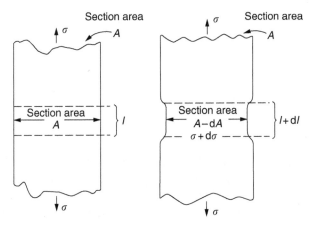

**FIGURE 11.5**
The formation of a neck in a bar of material that is being deformed plastically.

carry the force $F$; but if it has not, plastic flow will become localized at the neck and the specimen will fail there. Any section of the specimen can carry a force $A\sigma$, where $A$ is its area, and $\sigma$ its current strength. If $A\sigma$ increases with strain, the specimen is stable. If it decreases, it is unstable and will neck. The critical condition for the start of necking is that

$$A\sigma = F = \text{constant}$$

Then

$$Ad\sigma + \sigma dA = 0$$

or

$$\frac{d\sigma}{\sigma} = -\frac{dA}{A}$$

But volume is conserved during plastic flow, so

$$-\frac{dA}{A} = \frac{dl}{l} = d\varepsilon$$

(prove this by differentiating $Al = $ constant). So

$$\frac{d\sigma}{\sigma} = d\varepsilon$$

or

$$\frac{d\sigma}{d\varepsilon} = \sigma \qquad (11.4)$$

This equation is given in terms of true stress and true strain. As we saw in Chapter 8, tensile data are usually given in terms of nominal stress and strain. From Chapter 8:

$$\sigma = \sigma_n(1 + \varepsilon_n)$$
$$\varepsilon = \ln(1 + \varepsilon_n)$$

If these are differentiated and substituted into the necking equation we get

$$\frac{d\sigma_n}{d\varepsilon_n} = 0 \qquad (11.5)$$

In other words, on the point of instability, the *nominal* stress–strain curve is at its maximum as we know experimentally from Chapter 8.

To see what is going on physically, it is easier to return to our first condition. At low stress, if we make a little neck, the material in the neck will work-harden and will be able to carry the extra stress it has to stand because of its smaller area; load will therefore be continuous, and the material will be stable. At high stress, the *rate of work-hardening* is less as the true stress–true strain curve shows: that is, the slope of the $\sigma/\varepsilon$ curve is less. Eventually, we reach a point at which,

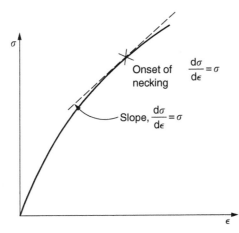

**FIGURE 11.6**
The condition for necking.

when we make a neck, the work-hardening is only *just* enough to stand the extra stress. This is the point of necking (Figure 11.6) with

$$\frac{d\sigma}{d\varepsilon} = \sigma$$

At still higher *true* stress $d\sigma/d\varepsilon$, the rate of work-hardening decreases further, becoming insufficient to maintain stability—the extra stress in the neck can no longer be accommodated by the work-hardening produced by making the neck, and the neck grows faster and faster, until final fracture takes place.

## Consequences of plastic instability

Plastic instability is very important in processes like deep drawing sheet metal to form car bodies, cans, and so on. Obviously we must ensure that the materials and press designs are chosen carefully to *avoid* instability.

Mild steel is a good material for deep drawing in the sense that it flows a great deal before necking starts. It can therefore be drawn very deeply without breaking (Figure 11.7).

Aluminum alloy is much less good (Figure 11.8)—it can only be drawn a little before instabilities form. Pure aluminum is not nearly as bad, but is much too *soft* to use for most applications.

Polythene shows a kind of necking that does *not* lead to fracture. Figure 11.9 shows its $\sigma_n/\varepsilon_n$ curve. At quite low stress

$$\frac{d\sigma_n}{d\varepsilon_n} \rightarrow 0$$

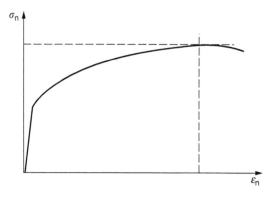

**FIGURE 11.7**
Mild steel can be drawn out a lot before it fails by necking.

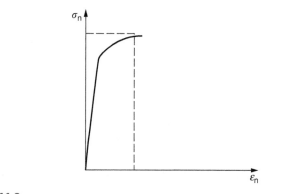

**FIGURE 11.8**
Aluminum alloy quickly necks when it is drawn out.

and necking begins. However, the neck never becomes *unstable*—it simply grows in length—because at high strain the material work-hardens considerably, and is able to support the increased stress at the reduced cross-section of the neck. This odd behavior is caused by the lining up of the polymer chains in the neck along the direction of the neck—and for this sort of reason *drawn* (i.e., "fully necked") polymers can be made to be very strong indeed—much stronger than the undrawn polymers.

Finally, mild steel can sometimes show an instability like that of polythene. If the steel is annealed, the stress–strain curve looks like that in Figure 11.10. A stable neck, called a Lüders Band, forms, and propagates (as it did in polythene) without causing fracture because the strong work-hardening of the later part of the stress–strain curve prevents this. Lüders Bands are a problem when sheet steel is pressed because they give lower precision and disfigure the pressing.

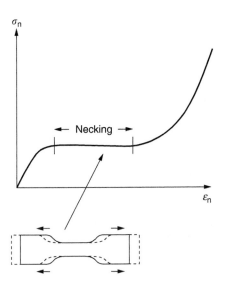

**FIGURE 11.9**
Polythene forms a *stable* neck when it is drawn out; drawn polythene is very strong.

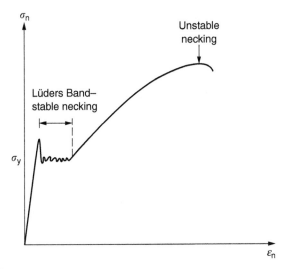

**FIGURE 11.10**
Mild steel often shows both stable and unstable necks.

## EXAMPLES

**11.1** By calculating the plastic work done in each process, determine whether the bolt passing through the plate will fail, when loaded in tension, by yielding of the shaft or shearing-off of the head. (Assume no work-hardening.)

**Answer**

The bolt will fail by shearing-off of the head.

**11.2** A metal bar of width $w$ is compressed between two hard anvils as shown in the following diagrams. The third dimension of the bar, $L$, is much greater than $w$. Plastic deformation takes place as a result of shearing along planes, defined by the dashed lines in the diagram on the next page, at a shear stress $k$. Find an upper bound for the load $F$ when (a) there is no friction between anvils and bar, and (b) there is sufficient friction to effectively weld the anvils to the bar. Show that the solution to case (b) satisfies the general formula

$$F \le 2wLk\left(1 + \frac{w}{4d}\right)$$

which defines upper bounds for all integral values of $w/2d$.

**Answers**

(a) $2wLk$; (b) $3wLk$

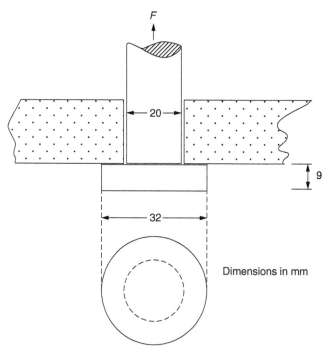

Dimensions in mm

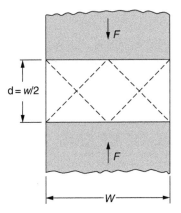

**11.3** A composite material used for rock-drilling bits consists of tungsten carbide cubes (each 20 μm in size) stuck together with a thin layer of cobalt. The material is required to withstand compressive stresses of 4000 MN m$^{-2}$ in service. Use the equation in Example 11.2 to estimate an upper limit for the thickness of the cobalt layer. You may assume that the compressive yield stress of tungsten carbide is well above 4000 MN m$^{-2}$, and that the cobalt yields in shear at $k = 175$ MN m$^{-2}$.

**Answer**

0.48 μm

**11.4 a.** Discuss the assumption that when a piece of metal is deformed at constant temperature, its volume is unchanged.

    **b.** A ductile metal wire of uniform cross-section is loaded in tension until it just begins to neck. Assuming that volume is conserved, derive a differential expression relating the *true* stress to the *true* strain at the onset of necking.

**11.5** The curve of true stress against true strain for the metal wire approximates to

$$\sigma = 350\varepsilon^{0.4} \text{ MN m}^{-2}$$

Estimate the tensile strength of the wire and the work required to take 1 m$^3$ of the wire to the point of necking.

**Answers**

163 MN m$^{-2}$; 69.3 MJ

**11.6 a.** If the *true* stress–*true* strain curve for a material is defined by
    $\sigma = A\varepsilon^n$, where $A$ and $n$ are constants, find the tensile strength $\sigma_{TS}$.

    **b.** For a nickel alloy, $n = 0.2$ and $A = 800$ MN m$^{-2}$. Evaluate the tensile strength of the alloy. Evaluate the true stress in an alloy specimen loaded to $\sigma_{TS}$.

**Answers**

(a) $\sigma_{TS} = An^n/e^n$; (b) $\sigma_{TS} = 475$ MN m$^{-2}$; $\sigma = 580$ MN m$^{-2}$

**11.7** The indentation hardness, $H$, is given by $H \approx 3\sigma_y$ where $\sigma_y$ is the true yield stress at a nominal plastic strain of 8%. If the true stress–strain curve of a material is given by

$$\sigma = A\varepsilon^n$$

and $n = 0.2$, calculate the tensile strength of a material for which the indentation hardness is 600 MN m$^{-2}$. You may assume that $\sigma_{TS} = An^n/\varepsilon^n$.

**Answer**

$\sigma_{TS} = 198$ MN m$^{-2}$

# PLASTIC BENDING OF BEAMS, TORSION OF SHAFTS, AND BUCKLING OF STRUTS

## Bending of beams

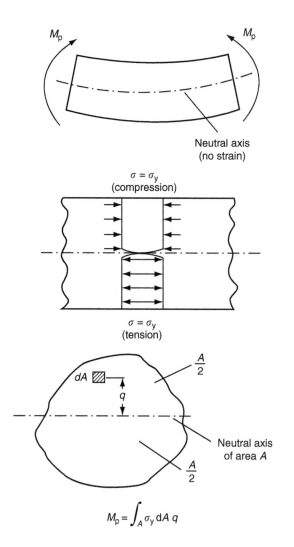

$$M_p = \int_A \sigma_y \, dA \, q$$

## Plastic moments

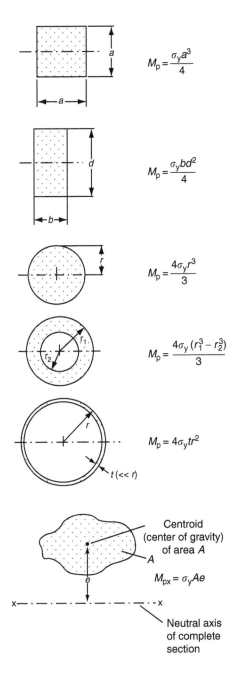

$$M_p = \frac{\sigma_y a^3}{4}$$

$$M_p = \frac{\sigma_y b d^2}{4}$$

$$M_p = \frac{4\sigma_y r^3}{3}$$

$$M_p = \frac{4\sigma_y (r_1^3 - r_2^3)}{3}$$

$$M_p = 4\sigma_y t r^2$$

Centroid (center of gravity) of area $A$

$$M_{px} = \sigma_y A e$$

Neutral axis of complete section

## Shearing torques

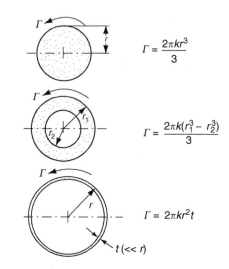

$$\Gamma = \frac{2\pi k r^3}{3}$$

$$\Gamma = \frac{2\pi k (r_1^3 - r_2^3)}{3}$$

$$\Gamma = 2\pi k r^2 t$$

$$t \, (<< r)$$

## Plastic buckling

The results given in Chapter 3 for the elastic buckling of struts can be applied to situations where the buckling is plastic provided. Young's modulus $E$ is replaced in the equations by the *tangent modulus* $E_t$ which is defined in the following diagram.

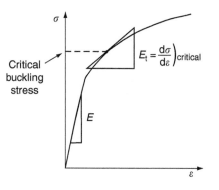

## CHAPTER 12

# Case Studies in Yield-Limited Design

## CONTENTS

## 12.1 INTRODUCTION

We now examine three applications of plasticity. The first (material selection for a spring) requires *no plasticity whatever*. The second (material selection for a pressure vessel) typifies plastic design of a large structure. It is unrealistic to expect no plasticity: there will always be some, at bolt holes, loading points, or changes of section. The important thing is that yielding should not spread entirely through any section of the structure—*plasticity must not become general*. Finally, we examine an instance (the rolling of metal strip) in which yielding is deliberately induced, to give *large-strain plasticity*.

## 12.2 CASE STUDY 1: ELASTIC DESIGN—MATERIALS FOR SPRINGS

Springs come in many shapes and have many purposes. One thinks of axial springs (a rubber band, for example), leaf springs, helical springs, spiral springs, torsion bars. Regardless of their shape or use, the best material for a spring of minimum volume is that with the greatest value of $\sigma_y^2/E$. Here $E$ is Young's modulus and $\sigma_y$ the failure strength of the material of the spring: its yield

Engineering Materials I: An Introduction to Properties, Applications, and Design, Fourth Edition

**Table 12.1** Materials for Springs

| Material | $E$ (GN m$^{-2}$) | $\sigma_y$ (MN m$^{-2}$) | $\sigma_y^2/E$ (MJ m$^{-3}$) | $\sigma_y/E$ |
|---|---|---|---|---|
| Brass (cold-rolled) |  | 638 | 3.38 | $5.32 \times 10^{-3}$ |
| Bronze (cold-rolled) |  | 640 | 3.41 | $5.33 \times 10^{-3}$ |
| Phosphor bronze | 120 | 770 | 4.94 | $6.43 \times 10^{-3}$ |
| Beryllium copper |  | 1380 | 15.9 | $1.1.5 \times 10^{-3}$ |
|  |  |  |  |  |
| Spring steel |  | 1300 | 8.45 | $6.5 \times 10^{-3}$ |
| Stainless steel (cold-rolled) | 200 | 1000 | 5.0 | $5.0 \times 10^{-3}$ |
| Nimonic (high-temperature spring) |  | 614 | 1.9 | $3.08 \times 10^{-3}$ |

strength if ductile, its fracture strength, or modulus of rupture if brittle. Some materials with high values of this quantity are listed in Table 12.1.

The argument, at its simplest, is as follows. The primary function of a spring is that of storing elastic energy and—when required—releasing it again. The elastic energy stored per unit volume in a block of material stressed uniformly to a stress $\sigma$ is:

$$U^{el} = \frac{\sigma^2}{2E}$$

It is this that we wish to maximize. The spring will be damaged if the stress $\sigma$ exceeds the yield stress or failure stress $\sigma_y$; the constraint is $\sigma \leq \sigma_y$. So the maximum energy density is

$$U^{el} = \frac{\sigma_y^2}{2E}$$

Torsion bars and leaf springs are less efficient than axial springs because some of the material is not fully loaded: the material at the neutral axis, for instance, is not loaded at all.

## The leaf spring

Even leaf springs can take many different forms, but all of them are basically elastic beams loaded in bending. For the loading shown in Figure 12.1, the beam bending results in Chapter 3 give

$$\delta = \frac{Fl^3}{4Ebt^3} \tag{12.1}$$

The elastic energy stored in the spring, per unit volume, is

$$U^{el} = \frac{1}{2}\frac{F\delta}{btl} = \frac{F^2 l^2}{8Eb^2 t^4} \tag{12.2}$$

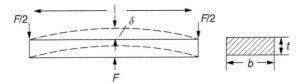

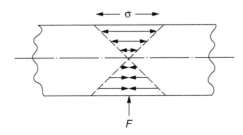

**FIGURE 12.1**

A leaf spring under load.

**FIGURE 12.2**

Stresses inside a leaf spring.

Figure 12.2 shows that the stress in the beam is zero along the neutral axis at its center, and is a maximum at the surface, at the midpoint of the beam (because the bending moment is biggest there). The beam bending results in Chapter 3 show that the maximum surface stress is given by

$$\sigma = \frac{Mc}{I} = \frac{F}{2} \times \frac{l}{2} \times \frac{t}{2} \times \frac{12}{bt^3} = \frac{3Fl}{2bt^2} \tag{12.3}$$

Now to be successful, a spring must not undergo a permanent set during use: it must always "spring" back. The condition for this is that the maximum stress must always be less than the yield stress:

$$\frac{3Fl}{2bt^2} < \sigma_y \tag{12.4}$$

Eliminating $t$ between this and Equation (12.2) gives

$$U^{el} = \frac{1}{18} \left( \frac{\sigma_y^2}{E} \right)$$

So if in service a spring has to undergo a given deflection $\delta$ under a force $F$, the ratio of $\sigma_y^2/E$ must be high enough to avoid a permanent set. This is why we have listed values of $\sigma_y^2/E$ in Table 12.1: the best springs are made of materials with high values of this quantity. For this reason spring materials are heavily strengthened (see Chapter 10): by solid solution strengthening plus work-hardening (cold-rolled, single-phase brass, and bronze), solid solution and precipitate strengthening (spring steel), and so on. Annealing any spring material removes the work-hardening, and may cause the precipitate to coarsen (increasing the particle spacing), reducing $\sigma_y$ and making the material useless as a spring.

## EXAMPLE

### Springs for a centrifugal clutch

Suppose you are asked to select a material for a spring with the following application. A spring-controlled clutch like that shown in Figure 12.3 is designed to transmit 20 horsepower at 800 rpm; the clutch is to begin to pick up load at 600 rpm. The blocks are lined with Ferodo or some other friction material. When properly adjusted, the maximum deflection of the springs is to be 6.35 mm (but the friction pads may wear, and larger deflections may occur; this is a standard problem with springs—they must often withstand extra deflections without losing their sets).

## Mechanics

The force on the spring is

$$F = Mr\omega^2 \tag{12.5}$$

where $M$ is the mass of the block, $r$ the distance of the center of gravity of the block from the center of rotation, and $\omega$ the angular velocity. The *net* force each block exerts on the clutch rim at full speed is

$$Mr(\omega_2^2 - \omega_1^2) \tag{12.6}$$

where $\omega_2$ and $\omega_1$ correspond to the angular velocities at 800 and 600 rpm (the *net* force must be zero for $\omega_2 = \omega_1$, at 600 rpm). The full power transmitted by

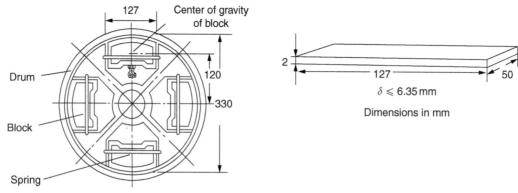

**FIGURE 12.3**
Leaf springs in a centrifugal clutch.

all four blocks is given by $4\mu_s Mr\,(\omega_2^2 - \omega_1^2) \times$ distance moved per second by inner rim of clutch at full speed, that is

$$\text{power} = 4\mu_s Mr(\omega_2^2 - \omega_1^2) \times \omega_2 r \qquad (12.7)$$

$\mu_s$ is the coefficient of static friction and $r$ is specified by the design (the clutch cannot be too big). Power, $\omega_2$, and $\omega_1$ are specified in Equation (12.7), so $M$ is also specified. The maximum force on the spring is determined by the design from $F = Mr\omega_1^2$. The requirement that this force deflect the beam by only 6.35 mm with the linings just in contact is what determines the thickness, $t$, of the spring via Equation (12.1) ($l$ and $b$ are fixed by the design).

## Metallic materials for the clutch springs

Given the spring dimensions ($t = 2$ mm, $b = 50$ mm, $l = 127$ mm) and given $\delta \le 6.35$ mm, all of which are specified by design, which material should we use? Eliminating $F$ between equations (12.1) and (12.4) gives

$$\frac{\sigma_y}{E} > \frac{6\delta t}{l^2} = \frac{6 \times 6.35 \times 2}{127 \times 127} = 4.7 \times 10^{-3} \qquad (12.8)$$

As well as seeking materials with high values of $\sigma_y^2/E$, we must also ensure that the material we choose meets the criterion of Equation (12.8).

Table 12.1 shows that spring steel, the cheapest material listed, is adequate for this purpose, but has a worryingly small safety factor to allow for wear of the linings. Only the expensive beryllium–copper alloy, of all the metals shown, would give a significant safety factor ($\sigma_y/E = 11.5 \times 10^{-3}$).

In many designs, the mechanical requirements are such that single springs of the type considered so far would yield even if made from beryllium copper. This commonly arises in the case of suspension springs for vehicles, and so on, where both large $\delta$ ("soft" suspensions) and large $F$ (good load-bearing capacity) are required. The solution can then be to use multi-leaf springs (Figure 12.4). $t$ can be made *small* to give *large* $\delta$ without yield according to

$$\left(\frac{\sigma_y}{E}\right) > \frac{6\delta t}{l^2} \qquad (12.9)$$

while the lost load-carrying capacity resulting from small $t$ can be made up by having several leaves combining to support the load.

## Nonmetallic materials

Finally, materials other than the metals originally listed earlier in Table 12.1 can make good springs. Glass, or fused silica, with $\sigma_y/E$ as large as $58 \times 10^{-3}$ is excellent, *provided* it operates under protected conditions where it cannot be scratched or suffer impact loading. Nylon is good—provided the forces are low—having $\sigma_y/E \approx 22 \times 10^{-3}$, and it is widely used in household appliances and children's toys (you probably brushed your teeth with little nylon springs

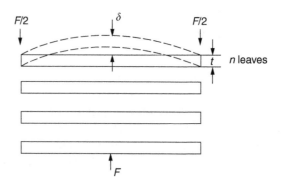

**FIGURE 12.4**
Multi-leaved springs (schematic).

this morning). Leaf springs for heavy trucks are now being made of CFRP: the value of $\sigma_y/E$ ($6 \times 10^{-3}$) is similar to that of spring steel, and the weight saving compensates for the higher cost. CFRP is always worth examining where an innovative use of materials might offer advantages.

## 12.3 CASE STUDY 2: PLASTIC DESIGN—MATERIALS FOR PRESSURE VESSELS

We now look at material selection for a pressure vessel able to contain a gas at pressure $p$, minimizing the *weight*. We seek a design that will not fail by plastic collapse (i.e., general yield). But we must be cautious: structures can also fail by *fast fracture*, by *fatigue*, and by *corrosion*. We shall discuss these later, but for now we assume that plastic collapse is our only problem.

The body of an aircraft, the casing of a solid rocket booster: these are examples of pressure vessels that must be as light as possible.

The stress in the vessel wall (Figure 12.5) is:

$$\sigma = \frac{pr}{2t} \tag{12.10}$$

$r$, the radius of the pressure vessel, is fixed by the design. For safety, $\sigma \leq \sigma_y/S$, where $S$ is the safety factor. The vessel mass is

$$M = 4\pi r^2 t \rho \tag{12.11}$$

giving

$$t = \frac{M}{4\pi r^2 \rho} \tag{12.12}$$

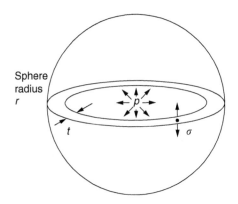

**FIGURE 12.5**
Thin-walled spherical pressure vessel.

**Table 12.2 Materials for Pressure Vessels**

| Material | $\sigma_y$ (MNm$^{-2}$) | $\rho$ (MG m$^{-3}$) | $\rho/\sigma_y \times 10^3$ |
|---|---|---|---|
| Reinforced concrete | 70 | 3.5 | 50 |
| Alloy steel (pressure-vessel steel) | 1000 | 7.8 | 7.8 |
| Mild steel | 220 | 7.8 | 36 |
| Aluminum alloy | 400 | 2.7 | 6.8 |
| CFRP quasi isotropic laminate | 200 | 1.5 | 7.5 |

Substituting for $t$ in Equation (12.10) we find that

$$\frac{\sigma_y}{S} \geq \frac{2\pi p r^3 \rho}{M} \tag{12.13}$$

From Equation (12.13) we have

$$M = S2\pi p r^3 \left(\frac{\rho}{\sigma_y}\right) \tag{12.14}$$

so for the lightest vessel we require the smallest value of $(\rho/\sigma_y)$. Table 12.2 gives values of $(\rho/\sigma_y)$ for candidate materials.

The lightest pressure vessels are made of CFRP, aluminum alloy, and pressure-vessel steel. Reinforced concrete or mild steel results in a very heavy vessel. Of course, CFRP is very expensive, which is why aluminum or high-strength steel is used for things like a aircraft fuselages and SRB casings.

## 12.4 CASE STUDY 3: LARGE-STRAIN PLASTICITY— METAL ROLLING

*Forging, sheet drawing,* and *rolling* are metal-forming processes in which the section of a billet or slab is reduced by compressive plastic deformation. When a slab is rolled (Figure 12.6) the section is reduced from $t_1$ to $t_2$ over a length $l$ as it passes through the rolls.

At first sight, it might appear that there would be no sliding (and thus no friction) between the slab and the rolls, since these move with the slab. But the metal is elongated in the rolling direction, so it speeds up as it passes through the rolls, and some slipping is inevitable. If the rolls are polished and lubricated (as they are for precision and cold-rolling) the frictional losses are small. We shall ignore them here (though all detailed treatments of rolling include them) and calculate the *rolling torque* for perfectly lubricated rolls.

From the geometry of Figure 12.6

$$l^2 + (r - x)^2 = r^2$$

or, if $x = \frac{1}{2}(t_1 - t_2)$ is small (as it almost always is),

$$l = \sqrt{r(t_1 - t_2)}$$

The rolling force $F$ must cause the metal to yield over the length $l$ and width $w$ (normal to Figure 12.6). Thus

$$F = \sigma_y wl$$

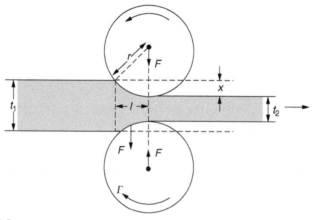

**FIGURE 12.6**
The rolling of metal sheet.

If the reaction on the rolls appears halfway along $l$ then the torque on each roll is

$$\Gamma = \frac{Fl}{2} = \frac{\sigma_y w l^2}{2}$$

giving

$$\Gamma = \frac{\sigma_y w r (t_1 - t_2)}{2} \qquad (12.16)$$

The torque required to drive the rolls increases with yield strength $\sigma_y$, so hot-rolling (when $\sigma_y$ is low—see Chapter 20) takes less power than cold-rolling. It obviously increases with the reduction in section $(t_1 - t_2)$. And it increases with roll diameter $2r$; this is one of the reasons why small-diameter rolls are used, often backed by two or more rolls of larger diameter (simply to stop them bending).

Rolling can be analyzed in much more detail to include important aspects, which we have ignored: friction, the elastic deformation of the rolls, and the constraint of two-dimensional strain imposed by the rolling geometry. But this case study gives an idea of why an understanding of plasticity, and the yield strength, is important in forming operations, both for metals and polymers.

## EXAMPLES

**12.1** Referring to Example 7.5, the sphere can fail by yield or compressive fracture at a pressure $p_f$ given by

$$p_f = 2\sigma_f \left(\frac{t}{r}\right)$$

where $\sigma_f$ is the yield stress or the compressive fracture stress as appropriate. The basic design requirement is that the pressure hull has the minimum possible mass compatible with surviving the design pressure.

By eliminating $t$ from the equations, show that the minimum mass of the hull is given by the expression

$$M_f = 2\pi r^3 p_f \left(\frac{\rho}{\sigma_f}\right)$$

Hence obtain a merit index to meet the design requirement for the failure mechanism. [You may assume that the surface area of the sphere is $4\pi r^2$.]

**Answer**

$\sigma_f / \rho$

**12.2** Consider the pressure hull of Examples 7.5 and 12.1. For each material listed in the following table, calculate the minimum mass and wall thickness of the pressure hull for both failure mechanisms (i.e., external-pressure buckling and yield/compressive fracture) at the design pressure.

| Material | E (GN m$^{-2}$) | $\sigma_f$ (M Nm$^{-2}$) | $\rho$ (kg m$^{-3}$) |
|---|---|---|---|
| Alumina | 390 | 5000 | 3900 |
| Glass | 70 | 2000 | 2600 |
| Alloy steel | 210 | 2000 | 7800 |
| Titanium alloy | 120 | 1200 | 4700 |
| Aluminum alloy | 70 | 500 | 2700 |

Hence determine the limiting failure mechanism for each material. [*Hint:* This is the failure mechanism which gives the larger of the two values of *t.*]

What is the optimum material for the pressure hull? What are the mass, wall thickness, and limiting failure mechanism of the optimum pressure hull?

**Answers**

| Material | $M_b$ (ton) | $t_b$ (mm) | $M_f$ (ton) | $t_f$ (mm) | Limiting Failure Mechanism |
|---|---|---|---|---|---|
| Alumina | 2.02 | 41 | 0.98 | 20 | Buckling |
| Glass | 3.18 | 97 | 1.63 | 50 | Buckling |
| Alloy steel | 5.51 | 56 | 4.90 | 50 | Buckling |
| Titanium alloy | 4.39 | 74 | 4.92 | 83 | Yielding |
| Aluminum alloy | 3.30 | 97 | 6.79 | 200 | Yielding |

The optimum material is alumina, with a mass of 2.02 ton, a wall thickness of 41 mm, and a limiting failure mechanism of external-pressure buckling.

12.3 The drawing on the next page shows a bolted flanged connection in a tubular supporting pier for an old bridge. You are required to determine the bending moment M needed to make the connection fail by yielding of the wrought-iron bolts. Because the data for the bridge are old, the units for the dimensions in the drawing are inches. The units for force and stress are tons and tons/square inch (tsi). Assume that the yield strength of wrought iron is 11 tsi. [*Hint:* Find the yield load for one bolt, then sum the moments generated by all the bolts about the axis *X–X.*]

**Answer**

126 ton foot (1 foot = 12 inches)

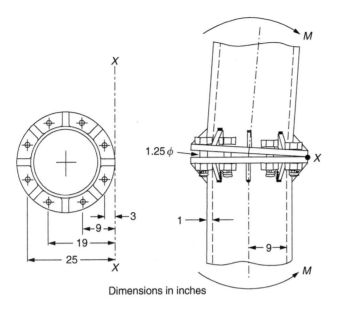

1.25 φ

3

9

19

25

X

X

M

M

1

9

Dimensions in inches

**12.4** The following diagram gives the dimensions of a steel bicycle chain. The chain is driven by a chain wheel which has pitch diameter of 190 mm. The chain wheel is connected to the pedals by a pair of cranks set at 180° in the usual way.

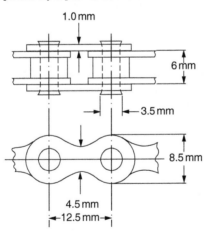

1.0 mm

6 mm

3.5 mm

8.5 mm

4.5 mm

12.5 mm

The center of each pedal is 170 mm away from the center of the chain wheel. If the cyclist weighs 90 kg, estimate the factor of safety of the chain. You may assume that a link would fail in simple tension at the position of minimum cross-sectional area and that a pin would fail in double shear. The yield strength of the steel is 1500 N m$^{-2}$.

**Answer**

8.5

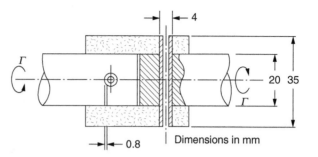

Dimensions in mm

**12.5** This diagram shows a coupling between two rotating shafts designed to transmit power from a low-speed hydraulic motor to a gearbox. The coupling sleeve was a sliding fit on the shafts and the torque was taken by the two Bissell pins as shown in the diagram. Owing to a malfunction in the gearbox, one of the pins sheared, disconnecting the drive. Assuming that the shear yield stress $k$ is 750 MN m$^{-2}$, estimate the failure torque.

**Answer**

12 kgf m

**12.6** The tubular casing of a solid rocket booster (SRB) for the space shuttle has a mean radius of 71.76" and a wall thickness of 0.479". The maximum working pressure it has to contain is 6.303 MN m$^{-2}$. What is the hoop stress at this pressure? The casing is made from D6AC steel with a yield strength of 1244 MN m$^{-2}$. What is the safety factor $S$ for the casing?

**Answers**

944 MN m$^{-2}$; 1.32

**12.7** The length and radius of the SRB casing are fixed by the design. Obtain an expression analogous to Equation (12.14) for the mass of the SRB casing. What is the value of $(\rho/\sigma_y)$ for the steel?

**Answers**

$$M = 1.32 \times 2\pi pr^2 \ell \left(\frac{\rho}{\sigma_y}\right); 6.3 \times 10^{-3}$$

**12.8** When considering composite materials for a spherical pressure vessel, why is it incorrect to use the strength of the unidirectional composite? Under what pressure loading conditions would it be appropriate to use the unidirectional strength for a *cylindrical* pressure vessel? Referring to Table 12.2, and assuming that the unidirectional strength is three times the strength of the quasi-isotropic laminate, what effect would this have on the choice of material to give the lightest pressure vessel?

**12.9** When a beam is subjected to a gradually increasing bending moment, the maximum bending stress—which is located at the surfaces furthest from the neutral axis—increases until it reaches the yield stress (when the moment is $M_y$). If the moment is increased further, a zone of plastic deformation spreads into the beam from these surfaces, until the whole of the cross section has yielded (when the moment is $M_p$). Calculate the value of $M_p/M_y$ for the following beam cross-sections: (a) rectangle, (b) solid round bar, (c) thin-walled tube.

**Answers**

(a) 1.5; (b) 1.70; (c) 1.27

**12.10** A metal strut is loaded in axial compression to produce a stress that is just a little less than the yield stress. At this load, there is a factor of safety of 5 on the critical load for elastic buckling (see the results for buckling of struts at the end of Chapter 3). The load is then increased until the metal yields in compression. The strut then buckles. Explain why.

# Fast Fracture, Brittle Fracture, and Toughness

## CHAPTER 13

# Fast Fracture and Toughness

## CONTENTS

## 13.1 INTRODUCTION

Sometimes, structures that were properly designed to avoid both excessive elastic deflection and plastic yielding fail in a catastrophic way by *fast fracture*. Common to these failures—of things such as welded ships, welded bridges, gas pipelines, and pressure vessels —is the presence of cracks, often the result of imperfect welding. Fast fracture is caused by the growth—at the speed of sound in the material—of existing cracks that suddenly became unstable. Why do they do this?

## 13.2 ENERGY CRITERION FOR FAST FRACTURE

If you blow up a balloon, energy is stored in it. There is the energy of the compressed gas in the balloon, and there is the elastic energy stored in the rubber membrane itself. As you increase the pressure, the total amount of elastic energy in the system increases.

If we then introduce a flaw into the system, by poking a pin into the inflated balloon, the balloon will explode, and all this energy will be released.

Engineering Materials I: An Introduction to Properties, Applications, and Design, Fourth Edition

The membrane fails by fast fracture, *even though well below its yield strength*. But if we introduce a flaw of the same dimensions into a system with *less* energy in it, as when we poke our pin into a *partially* inflated balloon, the flaw is stable and fast fracture does not occur. Finally, if we blow up the punctured balloon progressively, we eventually reach a pressure at which it suddenly bursts. In other words, we have arrived at a *critical* balloon *pressure* at which our pin-sized flaw is just unstable, and fast fracture *just* occurs. Why is this?

To make the flaw grow, say by 1 mm, we have to tear the rubber to create 1 mm of new crack surface, and this consumes energy: the tear energy of the rubber per unit area × the area of surface torn. If the work done by the gas pressure inside the balloon, plus the release of elastic energy from the membrane itself, is less than this energy the tearing simply cannot take place—it would infringe the laws of thermodynamics.

We can, of course, increase the energy in the system by blowing the balloon up a bit more. The crack or flaw will remain stable (i.e., it will not grow) until the system (balloon plus compressed gas) has stored in it enough energy that, if the crack advances, *more energy is released than is absorbed*. There is, then, a *critical pressure* for fast fracture of a pressure vessel containing a crack or flaw of a given *size*.

All sorts of accidents (e.g., the sudden collapsing of bridges, the sudden explosion of steam boilers) have occurred—and still do—due to this effect. In all cases, the critical stress—above which enough energy is available to provide the tearing energy needed to make the crack advance—was exceeded, taking the designer by surprise. But how do we calculate this critical stress?

From what we have said already, we can write down an energy balance that must be met if the crack is to advance, and fast fracture is to occur. Suppose a crack of length $a$ in a material of thickness $t$ advances by $\delta a$, then we require that: work done by loads ≥ change of elastic energy + energy absorbed at the crack tip, that is

$$\delta W \geq \delta U^{el} + G_c t \delta a \tag{13.1}$$

where $G_c$ is the energy absorbed per unit area of *crack* (*not* unit area of new surface), and $t\delta a$ is the crack area.

$G_c$ is a material property—it is the energy absorbed in making a unit area of crack, and we call it the *toughness* (or, sometimes, the "critical strain energy release rate"). Its units are energy m$^{-2}$ or J m$^{-2}$. A high toughness means that it is hard to make a crack propagate (as in copper, for which $G_c \approx 10^6$ J m$^{-2}$). Glass, on the other hand, cracks very easily; $G_c$ for glass is only $\approx 10$ J m$^{-2}$.

This same quantity $G_c$ measures the strength of adhesives. You can measure it for the adhesive used on sticky tape (e.g., Sellotape) by hanging a weight on a

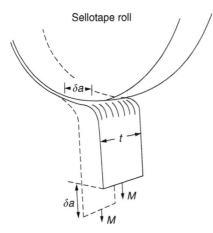

Sellotape roll

**FIGURE 13.1**
How to determine $G_c$ for Sellotape adhesive.

partly peeled length while supporting the roll so that it can freely rotate (hang it on a pencil) as shown in Figure 13.1. Increase the load to the value $M$ that just causes rapid peeling (= fast fracture). For this geometry, the quantity $\delta U^{el}$ is small compared to the work done by $M$ (the tape has comparatively little "give") and it can be neglected. Then, from our energy formula,

$$\delta W = G_c t \delta a$$

for fast fracture. In our case,

$$Mg\delta a = G_c t \delta a$$
$$Mg = G_c t$$

and therefore,

$$G_c = \frac{Mg}{t}$$

Typically, $t = 0.02$ m, $M = 0.15$ kg, and $g \approx 10$ m s$^{-2}$, giving

$$G_c \approx 75 \, \text{J} \, \text{m}^{-2}$$

This is a reasonable value for adhesives, and a value bracketed by the values of $G_c$ for many polymers.

Naturally, in most cases, we cannot neglect $\delta U^{el}$, and must derive more general relationships. Let us first consider a cracked plate of material loaded so that the displacements at the boundary of the plate are fixed. This is a common mode of loading a material—it occurs frequently in welds between large pieces of steel, for example—and is one that allows us to calculate $\delta U^{el}$ quite easily.

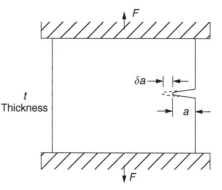

**FIGURE 13.2**

Fast fracture in a fixed plate.

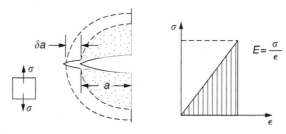

**FIGURE 13.3**

The release of stored strain energy as a crack grows.

## Fast fracture at fixed displacements

The plate shown in Figure 13.2 is clamped under tension so that its upper and lower ends are fixed. Since the ends cannot move, the forces acting on them can do no work, and $\delta W = 0$. Accordingly, our energy formula gives, for the onset of fast fracture,

$$-\delta U^{\text{el}} = G_c t \delta a \tag{13.2}$$

Now, as the crack grows into the plate, it allows the material of the plate to *relax*, so that it becomes less highly stressed, and *loses* elastic energy. $\delta U^{\text{el}}$ is thus *negative*, so that $-\delta U^{\text{el}}$ is *positive*, as it must be since $G_c$ is defined positive. We can estimate $\delta U^{\text{el}}$ in the way shown in Figure 13.3.

Let us examine a small cube of material of unit volume inside our plate. Due to the load $F$ this cube is subjected to a stress $\sigma$, producing a strain $\varepsilon$. Each unit cube therefore has strain energy $U^{\text{el}}$ of $\frac{1}{2} \sigma\varepsilon$, or $\sigma^2/2E$. If we now introduce a crack of length $a$, we can consider that the material in the dotted region relaxes (to zero stress) so as to lose all its strain energy. The energy change is shown in the following equation.

$$U^{\text{el}} = -\frac{\sigma^2}{2E}\frac{\pi a^2 t}{2}$$

As the crack spreads by length $\delta a$, we can calculate the appropriate $\delta U^{\text{el}}$ as

$$\delta U^{\text{el}} = \frac{dU^{\text{el}}}{da}\delta a = \frac{\sigma^2}{2E}\frac{2\pi at}{2}\delta a$$

The critical condition (Equation (13.2)) then gives

$$\frac{\sigma^2 \pi a}{2E} = G_c$$

at onset of fast fracture.

Actually, our assumption about the way in which the plate material relaxes is obviously rather crude, and a rigorous mathematical solution of the elastic stresses and strains indicates that our estimate of $\delta U^{\text{el}}$ is too low by exactly a factor of 2. Thus, correctly, we have

$$\frac{\sigma^2 \pi a}{E} = G_c$$

which reduces to

$$\sigma\sqrt{\pi a} = \sqrt{EG_c} \qquad (13.3)$$

at fast fracture.

## Fast fracture at fixed loads

Another, obviously very common way of loading a plate of material, or any other component for that matter, is simply to hang weights on it (fixed loads) (Figure 13.4). Here the situation is a little more complicated than it was in the case of fixed displacements. As the crack grows, the plate becomes less *stiff*, and relaxes so that the applied forces move and do work. $\delta W$ is therefore finite and positive. However, $\delta U^{\text{el}}$ is now positive also (it turns out that some of $\delta W$ goes into increasing the strain energy of the plate) and our final result for fast fracture is in fact found to be unchanged.

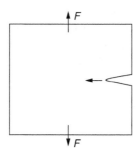

**FIGURE 13.4**
Fast fracture of a dead-loaded plate.

## The fast-fracture condition

Let us now return to our condition for the onset of fast fracture, knowing it to be general for engineering structures

$$\sigma\sqrt{\pi a} = \sqrt{EG_c}$$

The left side of our equation says that *fast fracture will occur when, in a material subjected to a stress $\sigma$, a crack reaches some critical size a: or, alternatively, when material containing cracks of size a is subjected to some critical stress $\sigma$.* The right side of our result *depends on material properties only*; $E$ is obviously a material constant, and $G_c$, the energy required to generate a unit area of crack, again must depend only on the basic properties of our material. Thus, the important point about the equation is that *the critical combination of stress and crack length at which fast fracture commences is a material constant.*

The term $\sigma\sqrt{\pi a}$ crops up so frequently in discussing fast fracture that it is usually abbreviated to a single symbol, $K$, having units MN m$^{-3/2}$; it is called the *stress intensity factor*. Fast fracture therefore occurs when

$$K = K_c$$

where $K_c(= \sqrt{EG_c})$ is the *critical* stress intensity factor, more usually called the *fracture toughness*.

To summarize:

$G_c = $ *toughness* (sometimes, critical strain energy release rate); usual units: kJ m$^{-2}$

$K_c = \sqrt{EG_c} = $ *fracture toughness* (sometimes: critical stress intensity factor); usual units: MN m$^{-3/2}$

$K = \sigma\sqrt{\pi a} = $ stress intensity factor; usual units: MN m$^{-3/2}$

Fast fracture occurs when $K = K_c$.

## 13.3 DATA FOR $G_C$ AND $K_C$

$K_c$ can be determined experimentally for any material by inserting a crack of known length $a$ into a piece of the material and loading until fast fracture occurs. $G_c$ can be derived from the data for $K_c$ using the relation $K_c = \sqrt{EG_c}$. Figures 13.5 and 13.6 and Table 13.1 show experimental data for $K_c$ and $G_c$ for a wide range of metals, polymers, ceramics, and composites.

The values of $K_c$ and $G_c$ range considerably, from the least tough materials such as ice and ceramics, to the toughest such as ductile metals; polymers have intermediate toughness, $G_c$, but low fracture toughness, $K_c$ (because their *moduli* are low). However, reinforcing polymers to make *composites* produces materials having good fracture toughnesses. Finally, although most metals are tough at or

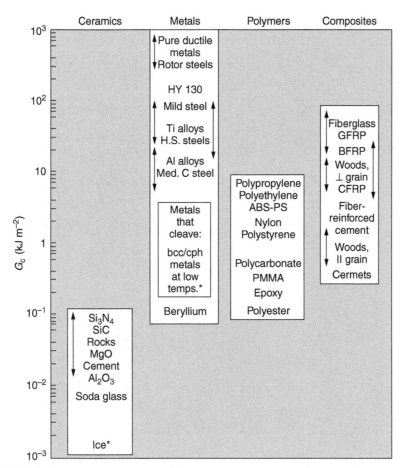

**FIGURE 13.5**
Toughness, $G_c$ (values at room temperature unless starred).

above room temperature, when many (e.g., b.c.c. metals such as steels, or c.p.h. metals) are cooled sufficiently, they become quite brittle, as the data show.

Obviously these figures for toughness and fracture toughness are extremely important—ignorance of such data has led, and can continue to lead, to engineering disasters of the sort we mentioned at the beginning of this chapter. But just how do these large variations between various materials arise? Why *is* glass so brittle and annealed copper so tough? We shall explain why in Chapter 14.

## A note on the stress intensity factor, *K*
In Section 13.2 we showed that
$$K = \sigma\sqrt{\pi a} = \sqrt{EG_c}$$

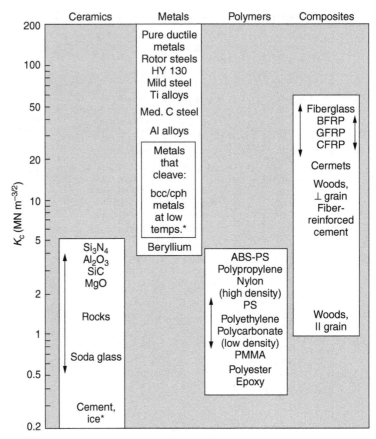

**FIGURE 13.6**
Fracture toughness, $K_c$ (values at room temperature unless starred).

at onset of fast fracture. Strictly speaking, this result is valid only for a crack through the center of a wide plate of material. In practice, the problems we encounter seldom satisfy this geometry, and a numerical correction to $\sigma\sqrt{\pi a}$ is required to get the strain energy calculation right. In general we write:

$$K = Y\sigma\sqrt{\pi a}$$

where $Y$ is the numerical correction factor. Values of $Y$ are given at the end of this chapter. However, provided the crack length $a$ is small compared to the width of the plate $W$, it is usually safe to assume that $Y \approx 1$.

**Table 13.1** Toughness, $G_c$, and Fracture Toughness, $K_c$

| Material | $G_c$ (kJ m$^{-2}$) | $K_c$ (MN m$^{-3/2}$) |
|---|---|---|
| Pure ductile metals (e.g., Cu, Ni, Ag, Al) | 100–1000 | 100–350 |
| Rotor steels (A533; Discalloy) | 220–240 | 204–214 |
| Pressure-vessel steels (HY130) | 150 | 170 |
| High-strength steels (HSS) | 15–118 | 50–154 |
| Mild steel | 100 | 140 |
| Titanium alloys (Ti6Al4V) | 26–114 | 55–115 |
| GFRPs | 10–100 | 20–60 |
| Fiberglass (glassfiber epoxy) | 40–100 | 42–60 |
| Aluminum alloys (high strength–low strength) | 8–30 | 23–45 |
| CFRPs | 5–30 | 32–45 |
| Common woods, crack $\perp$ to grain | 8–20 | 11–13 |
| Boron-fiber epoxy | 17 | 46 |
| Medium-carbon steel | 13 | 51 |
| Polypropylene | 8 | 3 |
| Polyethylene (low density) | 6–7 | 1 |
| Polyethylene (high density) | 6–7 | 2 |
| ABS Polystyrene | 5 | 4 |
| Nylon | 2–4 | 3 |
| Steel-reinforced cement | 0.2–4 | 10–15 |
| Cast iron | 0.2–3 | 6–20 |
| Polystyrene | 2 | 2 |
| Common woods, crack \|\| to grain | 0.5–2 | 0.5–1 |
| Polycarbonate | 0.4–1 | 1.0–2.6 |
| Cobalt/tungsten carbide cermets | 0.3–0.5 | 14–16 |
| PMMA | 0.3–0.4 | 0.9–1.4 |
| Epoxy | 0.1–0.3 | 0.3–0.5 |
| Granite (Westerly granite) | 0.1 | 3 |
| Polyester | 0.1 | 0.5 |
| Silicon nitride, $Si_3N_4$ | 0.1 | 4–5 |
| Beryllium | 0.08 | 4 |
| Silicon carbide SiC | 0.05 | 3 |
| Magnesia, MgO | 0.04 | 3 |
| Cement/concrete, unreinforced | 0.03 | 0.2 |
| Calcite (marble, limestone) | 0.02 | 0.9 |
| Alumina, $Al_2O_3$ | 0.02 | 3–5 |
| Shale (oilshale) | 0.02 | 0.6 |
| Soda glass | 0.01 | 0.7–0.8 |
| Electrical porcelain | 0.01 | 1 |
| Ice | 0.003 | 0.2* |

*Values at room temperature unless starred.*

# EXAMPLES

**13.1** Two wooden beams are butt-jointed using an epoxy adhesive as shown in the following diagram. The adhesive was stirred before application, entraining air bubbles which, under pressure in forming the joint, deform to flat, penny-shaped discs of diameter $2a = 2$ mm. If the beam has the dimensions shown, and epoxy has a fracture toughness of 0.5 MN m$^{-3/2}$, calculate the maximum load $F$ that the beam can support. Assume $K = Y\sigma\sqrt{\pi a}$ for the disc-shaped bubbles, where $Y = 0.64$.

**Answer**

4.65 kN

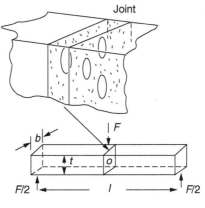

$t = b = 0.1\,\text{m};\ l = 2\,\text{m}$

**13.2** A large thick plate of steel is examined by X-ray methods, and found to contain no detectable cracks. The equipment can detect a single edge-crack of depth $a = 5$ mm or greater. The steel has a fracture toughness $K_c$ of 40 MN m$^{-3/2}$ m and a yield strength of 500 MN m$^{-2}$. Assuming that the plate contains cracks on the limit of detection, determine whether the plate will undergo general yield or will fail by fast fracture before general yielding occurs. What is the stress at which fast fracture would occur? ($Y = 1.12$ for the edge crack.)

**Answer**

Failure by fast fracture at 285 MN m$^{-2}$

**13.3** The fuselage of a passenger aircraft can be considered to be an internally pressurized thin walled tube of diameter 7 m and wall thickness 3 mm. It is made from aluminum alloy plate with a fracture toughness $K_c$ of 100 MN m$^{-3/2}$. At cruising altitude, the internal gauge pressure is 0.06 MN m$^{-2}$. Multiple fatigue cracks initiated at a horizontal row of rivet holes, and linked to form a single long axial through thickness crack in the fuselage. Estimate the critical length at which this crack will run, resulting in the break-up of the fuselage.

**Answer**

0.65 m

**13.4** A large furnace flue operating at 440°C was made from a low-alloy steel. After 2 years in service, specimens were removed from the flue and fracture toughness tests were carried out at room temperature. The average value of $K_c$ was only about 30 MN m$^{-3/2}$ compared to a value when new of about 80. The loss in toughness was due to temper embrittlement caused by the impurity phosphorus.

   The skin of the flue was made from plate 10 mm thick. Owing to the self-weight of the flue the plate has to withstand primary membrane stresses of up to 60 MN m$^{-2}$. Estimate the length of through-thickness crack that will lead to the fast fracture of the flue when the plant is shut down.

**Answer**

160 mm

**13.5** A ceramic bidet failed catastrophically in normal use, causing serious lacerations to the person concerned. The fracture initiated at a pre-existing crack of length $a = 10$ mm located at a circular flush hole. Strain gauge tests on an identical bidet gave live load stress $= 1.0$ MN m$^{-2}$ (tensile) and residual stress (from shrinkage during manufacture) $= 4$ MN m$^{-2}$ (tensile). Specimen tests gave $K_c = 1.3$ MN m$^{-3/2}$. Given that the hole introduces a local stress concentration factor of 3, account for the failure using fracture mechanics. Assume that $Y = 1.2$.

**13.6** A cylindrical pressure vessel in an ammonia plant was 7 m long, had an internal diameter of 1 m and had a wall thickness of 62 mm. The operating hoop stress was 285 MN m$^{-2}$.

   After 16 years in service the vessel exploded into a large number of fragments, some of which were hurled a distance of 1 km. Semicircular "thumbnail" cracks, typically 6 mm deep, were found at the inner surface of the vessel; they had initiated at the edges of a series of fillet welds used to attach internal fittings to the vessel wall. The gas in the vessel contained 58% hydrogen and the cracks were blamed on hydrogen cracking.

   Tests on samples of the steel gave a value for $K_c$ of about 40 MN m$^{-3/2}$. Account for the failure using fracture mechanics. You should assume that there is a residual tensile stress of 100 MN m$^{-2}$ at the welds in addition to the hoop stress.

## Y VALUES

**Case 1**

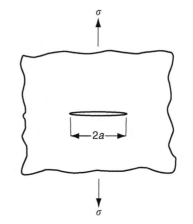

$Y = 1$

**Case 2**

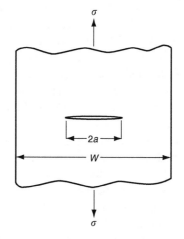

$$Y = \left\{ \cos \left( \frac{\pi a}{W} \right) \right\}^{-1/2}$$

*Note*: When $W \gg a$, $Y = 1$ (Case 1).

Examples: When $W = 4a$, $Y = 1.20$; when $W = 3a$, $Y = 1.43$.

**Case 3**

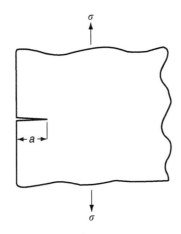

$Y = 1.12$

This situation is like one half of Case 1. The factor of 1.12 is added to compensate for introducing a free surface.

**Case 4**

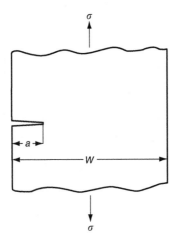

| Y | a/W |
|---|-----|
| 1.12 | 0 (Case 3) |
| 1.37 | 0.2 |
| 2.11 | 0.4 |
| 2.83 | 0.5 |

**Case 5**

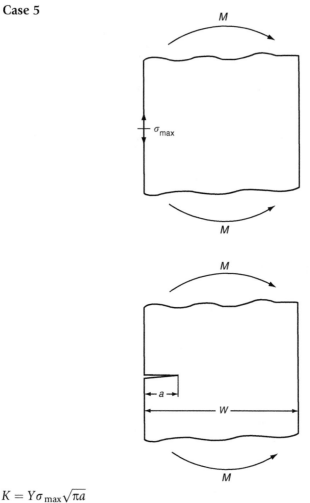

$$K = Y\sigma_{max}\sqrt{\pi a}$$

| Y | a/W |
|------|-----|
| 1.00 | 0 |
| 1.06 | 0.2 |
| 1.32 | 0.4 |
| 1.62 | 0.5 |
| 2.10 | 0.6 |

**Case 6**

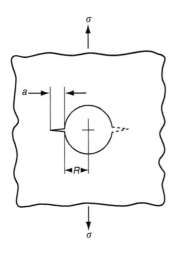

| Y (one crack) | Y (two cracks) | a/R |
|---|---|---|
| 3.36 | 3.36 | 0 |
| 2.73 | 2.73 | 0.1 |
| 2.30 | 2.41 | 0.2 |
| 1.86 | 1.96 | 0.4 |
| 1.64 | 1.71 | 0.6 |
| 1.47 | 1.58 | 0.8 |
| 1.37 | 1.45 | 1.0 |
| 1.18 | 1.29 | 1.5 |
| 0.71 | 1.00 (Case 1) | ∞ |

Note that, for a round hole in uniaxial tension, the stress concentration factor is 3. For $a/R = 0$ we have a Case 3 crack embedded in a local stress field of $3\sigma$. Thus $Y = 3 \times 1.12 = 3.36$ as shown in the table.

Case 7

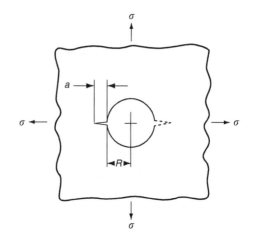

| Y (one crack) | Y (two cracks) | a/R |
|---|---|---|
| 2.24 | 2.24 | 0 |
| 1.98 | 1.98 | 0.1 |
| 1.82 | 1.83 | 0.2 |
| 1.58 | 1.61 | 0.4 |
| 1.42 | 1.52 | 0.6 |
| 1.32 | 1.43 | 0.8 |
| 1.22 | 1.38 | 1.0 |
| 1.06 | 1.26 | 1.5 |
| 0.71 | 1.00 | ∞ |

Note that, for a round hole in equi-biaxial tension, the stress concentration factor is 2. For $a/R = 0$ we have a Case 3 crack embedded in a local stress field of $2\sigma$. Thus $Y = 2 \times 1.12 = 2.24$.

Case 8

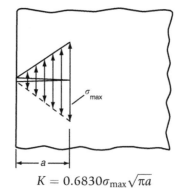

$$K = 0.6830\sigma_{max}\sqrt{\pi a}$$

**Case 9**

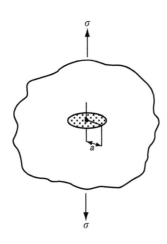

$Y = 0.64$

**Case 10**

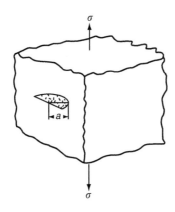

$Y = 1.12 \times 0.64$

## *K* CONVERSIONS

Values of *K* can be written differently, or can use different units. This can be very confusing. Conversions are given next.

$1\ \text{MN m}^{-3/2} = 1\ \text{MPa} \sqrt{m} = 31.6\ \text{N mm}^{-3/2} = 31.6\ \text{MPa} \sqrt{mm} = 0.907\ \text{ksi} \sqrt{in}$ (kilopounds per sq. in. $\sqrt{in}$)

# Micromechanisms of Fast Fracture

## 14.1 INTRODUCTION

In Chapter 13 we showed that, if a material contains a crack and is sufficiently stressed, the crack becomes unstable and grows—at up to the speed of sound in the material—to cause catastrophically rapid fracture, or *fast fracture* at a stress less than the yield stress. We were able to quantify this phenomenon and obtained a relationship for the onset of fast fracture

$$Y\sigma\sqrt{\pi a} = K_c$$

or, in more succinct notation,

$$K = K_c \text{ for fast fracture.}$$

It is helpful to compare this with other, similar, "failure" criteria:

$$\sigma = \sigma_y \text{ for yielding}$$
$$\sigma = \sigma_{TS} \text{ for tensile failure}$$

Engineering Materials I: An Introduction to Properties, Applications, and Design, Fourth Edition

The left side of each equation describes the *loading conditions*; the right side is a *material property*. When the left side (which increases with load) equals the right side (which is fixed), failure occurs.

Some materials, such as glass, have low $K_c$, and crack easily; ductile metals have high $K_c$ and are very resistant to fast-fracture; polymers have intermediate $K_c$, but can be made tougher by making them into composites; and (finally) many metals, when cold, become brittle—that is, $K_c$ decreases with temperature. How can we explain these important observations?

## 14.2 MECHANISMS OF CRACK PROPAGATION 1: DUCTILE TEARING

Let us first of all look at what happens when we load a cracked piece of a *ductile* metal—in other words, a metal that can flow readily to give large plastic deformations (e.g., pure copper; or mild steel at, or above, room temperature). If we load the material sufficiently, we can get fracture to take place starting from the crack. If you examine the surfaces of the metal after it has fractured (Figure 14.1) you see that the fracture surface is extremely rough, indicating that a great deal of plastic work has taken place.

Let us explain this observation. Whenever a crack is present in a material, the stress close to the crack, $\sigma_{local}$, is greater than the average stress $\sigma$ applied to the

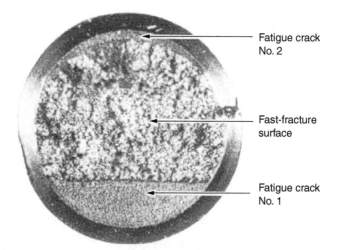

**FIGURE 14.1**

Before it broke, this steel bolt held a seat onto its mounting at Milan airport. Whenever someone sat down, the lower part of the cross-section went into tension, causing a crack to grow there by *metal fatigue* (Chapter 17; crack No. 1). When someone got up again, the upper part went into tension, causing fatigue crack No. 2 to grow. Eventually the bolt failed by fast fracture from the larger of the two fatigue cracks. The victim was able to escape with the fractured bolt!

piece of material; the crack has the effect of *concentrating* the stress. Mathematical analysis shows that the local stress ahead of a *sharp* crack in an elastic material is

$$\sigma_{local} = \sigma + \sigma\sqrt{\frac{a}{2r}} \qquad (14.1)$$

The closer one approaches to the tip of the crack, the higher the local stress becomes, until at some distance $r_y$ from the tip of the crack the stress reaches the yield stress, $\sigma_y$ of the material, and plastic flow occurs (Figure 14.2). The distance $r_y$ is easily calculated by setting $\sigma_{local} = \sigma_y$ in Equation (14.1). Assuming $r_y$ to be small compared to the crack length, $a$, the result is

$$
\begin{aligned}
r_y &= \frac{\sigma^2 a}{2\sigma_y^2} \\
&= \frac{K^2}{2\pi\sigma_y^2}
\end{aligned}
\qquad (14.2)
$$

The crack propagates when $K$ is equal to $K_c$; the width of the *plastic zone, $r_y$,* is then given by Equation (14.2) with $K$ replaced by $K_c$. Note that the zone of plasticity shrinks rapidly as $\sigma_y$ increases: cracks in soft metals have a large plastic zone; cracks in hard ceramics have a small zone, or none at all.

Even when nominally pure, most metals contain tiny *inclusions* (or particles) of chemical compounds formed by reaction between the metal and impurity atoms. Within the plastic zone, plastic flow takes place around these inclusions, leading to elongated cavities, as shown in Figure 14.2. As plastic flow progresses, these cavities link up, and the crack advances by means of this *ductile*

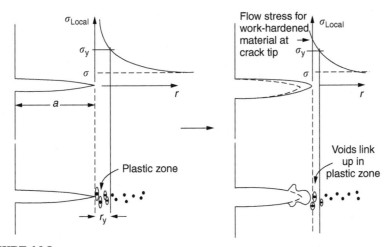

**FIGURE 14.2**
Crack propagation by ductile tearing.

*tearing*. The plastic flow at the crack tip naturally turns our initially sharp crack into a *blunt* crack, and it turns out from the stress mathematics that this *crack blunting* decreases $\sigma_{local}$ so that, at the crack tip itself, $\sigma_{local}$ is just sufficient to keep on plastically deforming the work-hardened material there, as the figure shows.

The important thing about crack growth by ductile tearing is that *it consumes a lot of energy by plastic flow*; the bigger the plastic zone, the more energy is absorbed. High energy absorption means that $G_c$ is high, and so is $K_c$. This is why ductile metals are so tough. Other materials, too, owe their toughness to this behavior—plasticine is one, and some polymers also exhibit toughening by processes similar to ductile tearing.

## 14.3 MECHANISMS OF CRACK PROPAGATION 2: CLEAVAGE

If you now examine the fracture surface of something like a ceramic, or a glass, you see a very different state of affairs. Instead of a very rough surface, indicating massive local plastic deformation, you see a rather featureless, flat surface suggesting little or no plastic deformation. How is it that cracks in ceramics or glasses can spread without plastic flow taking place? Well, the local stress ahead of the crack tip, given by our formula

$$\sigma_{local} = \sigma + \sigma\sqrt{\frac{a}{2r}}$$

can clearly approach very high values very near to the crack tip *provided that blunting of our sharp crack tip does not occur.*

As we showed in Chapter 8, ceramics and glasses have very high yield strengths, and thus very little plastic deformation takes place at crack tips in these materials. Even allowing for a small degree of crack blunting, the local stress at the crack tip is still in excess of the ideal strength and is thus large enough to literally break apart the interatomic bonds there; the crack then spreads between a pair of atomic planes giving rise to an atomically flat surface by *cleavage* (Figure 14.3). The energy required simply to break the interatomic bonds is *much* less than that absorbed by ductile tearing in a tough material, and this is why materials like ceramics and glasses are so brittle (see Figure 14.4). It is also why some steels become brittle and fail, like glass, at low temperatures—as we shall now explain.

At low temperatures metals having b.c.c. and c.p.h. structures become brittle and fail by cleavage, even though they may be tough at or above room temperature. In fact, only those metals with an f.c.c. structure (e.g., copper, lead, aluminum) remain unaffected by temperature in this way. In metals not having an

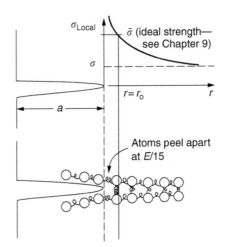

**FIGURE 14.3**
Crack propagation by cleavage.

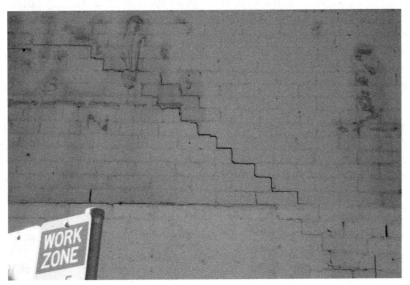

**FIGURE 14.4**
Brittle cracking of cement, Vista Street, Mosman, NSW, Australia. – 33 49 30.60 S 151 14 25,72 E

f.c.c. structure, the motion of dislocations is assisted by the *thermal agitation* of the atoms (we shall talk in more detail about *thermally activated* processes in Chapter 21). At lower temperatures this thermal agitation is less, and the dislocations cannot move as easily as they can at room temperature in response to a stress—the intrinsic lattice resistance (Chapter 10) increases. The result is that the yield strength rises, and the plastic zone at the crack tip shrinks until it becomes so small that the fracture mechanism changes from ductile tearing to cleavage. This effect is called the *ductile-to-brittle* transition; for steels it can be as high as $\approx 0°C$, depending on the composition of the steel; steel structures, such as ships, bridges, and oil rigs, are much more likely to fail in winter than in summer.

A somewhat similar thing happens in many polymers at the *glass–rubber transition* that we mentioned in Chapter 6. Below the transition these polymers are much more brittle than above it, as you can easily demonstrate by cooling a piece of rubber or polyethylene in liquid nitrogen. (Many other polymers, such as epoxy resins, have low $K_c$ values at *all* temperatures simply because they are heavily cross-linked at all temperatures by *covalent* bonds and the material does not flow at the crack tip to cause blunting.)

## 14.4 COMPOSITES, INCLUDING WOOD

As Figures 13.5 and 13.6 show, composites are tougher than ordinary polymers. The low toughness of materials, such as epoxy resins or polyester resins, can be enormously increased by reinforcing them with carbon fiber or glass fiber. But why is it that putting a second, equally (or more) brittle material such as graphite or glass into a brittle polymer makes a tough composite? The reason is the *fibers act as crack stoppers* (Figure 14.5).

The sequence in the figure shows what happens when a crack runs through the brittle matrix toward a fiber. As the crack reaches the fiber, the stress field just ahead of the crack separates the matrix from the fiber over a small region (a process called *debonding*) and the crack is blunted so much that its motion is *arrested*. Naturally, this only works if the crack is running normal to the fibers: wood is very tough across the grain, but can be split easily (meaning that $K_c$ is low) along it.

One of the reasons why fiber composites are so useful in engineering design—in addition to their high *stiffnesses* that we talked about in Chapter 6—is their high *toughness* produced in this way. Of course, there are other ways of making

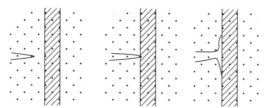

**FIGURE 14.5**
Crack stopping in composites.

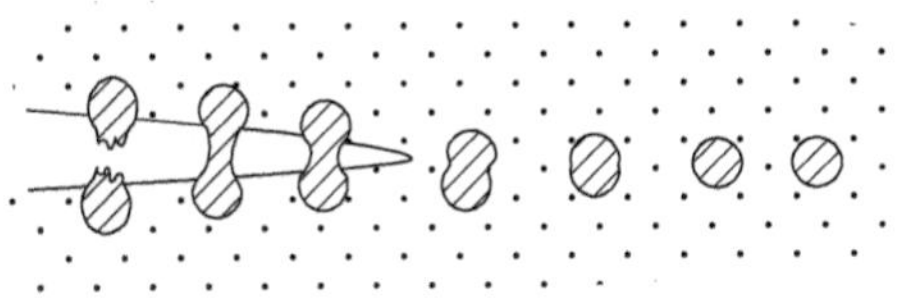

**FIGURE 14.6**
Rubber-toughened polymers.

polymers tough. The addition of small particles ("fillers") of various sorts to polymers can modify their properties considerably.

Rubber-toughened polymers (e.g., ABS), for example, derive their toughness from the small rubber particles they contain. A crack intersects and stretches them as shown in Figure 14.6. The particles act as little springs, clamping the crack shut, and thereby increasing the load needed to make it propagate.

## 14.5 AVOIDING BRITTLE ALLOYS

Let us finally return to the toughnesses of metals and alloys, as these are by far the most important class of materials for highly stressed applications. Even at, or above, room temperature, when nearly all common pure metals are tough, alloying of these metals with other metals or elements (e.g., with carbon to produce steels) can reduce the toughness. This is because alloying increases the resistance to dislocation motion (Chapter 10), raising the yield strength and causing the plastic zone to shrink. A more marked decrease in toughness can occur if enough impurities are added to make precipitates of chemical *compounds* formed between the metal and the impurities.

These compounds can often be very brittle and, if they are present in the shape of extended plates (e.g., sigma-phase in stainless steel; graphite in cast iron), cracks can spread along the plates, leading to brittle fracture. Finally, heat-treatments of alloys such as steels can produce different *crystal structures* having great hardness (but also therefore great brittleness because crack blunting cannot occur).

A good example of such a material is high-carbon steel after quenching into water from bright red heat: it becomes as brittle as glass. Proper heat treatment, following suppliers' specifications, is essential if materials are to have the properties you want.

## WORKED EXAMPLE

Underground tunnels in deep coal mines are usually supported by mine arches, made from curved lengths of steel I-beam bolted together on site. At the mine arch factory, straight lengths of I-beam stock were stored outside in the open, and were brought into the factory for bending to curve, punching bolt holes, and cutting to length. Fractures were common when the bending process was carried out in cold weather (steel temperature $\leq 0°C$). This was disruptive and expensive—the bending process had to be stopped, then the broken I-beam had to be removed from the bending machine, and scrapped.

The following photographs show a typical fracture, from the edge of a punched hole, in the tensile part of the bending stress field. The fracture surface had the characteristics of a substantially brittle (cleavage) fracture:

- Essentially flat, with clear "chevron" markings pointing back to the fracture origin
- Perpendicular to the direction of tensile stress
- No significant plastic deformation-sharp edges, and no thinning down of the web

Once the first fracture occurred, further bending of the I-beam produced a tensile stress on the opposite side of the hole, which caused a second fracture. However, this fracture was a ductile shear fracture, with a fracture surface at 45° to the web surface (angle of maximum shear stress).

These brittle fractures are an excellent example of the ductile-to-brittle transition in b.c.c. steels. The change from a ductile to a cleavage fracture mechanism was caused by a combination of:

- Low temperature
- Considerable work-hardening at and near the bore of the hole, from the punching operation

Both make it harder for plastic flow to occur, and the cleavage stress is now less than the yield stress.

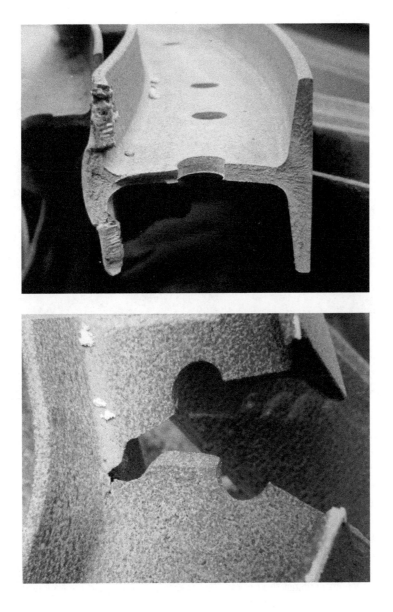

## EXAMPLES

**14.1** Account for the following observations.
    **a.** Ductile metals have high toughnesses whereas ceramics, glasses, and rigid polymers have low toughnesses.
    **b.** Aligned fiber composites are much tougher when the crack propagates perpendicular to the fibers than parallel to them.

**14.2** The following photograph shows a small part of a ductile tearing fracture surface taken in the scanning electron microscope (SEM). The material shown is aluminum alloy. Comment on the features you observe, and relate them to the description in the text.

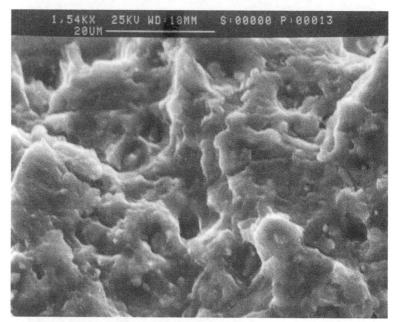

**14.3** The photograph on the next page shows a small part of a cleavage fracture surface taken in the scanning electron microscope (SEM). The material is low-alloy steel. Comment on the features you observe, and relate them to the description in the text. The white marker is 100 $\mu$m long.

**14.4** A student residence has balconies which are accessed from public areas in the building. Each balcony in flanked by a pair of brick walls between which is secured a balustrade fabricated from wooden components. Unfortunately, while a number of people were gathered on a second-floor balcony, the balustrade gave way. As a consequence, people fell to the ground below and were injured. The failure was caused by the propagation of preexisting cracks in the balustrade while people were leaning against the top rail. The cracks had developed over a long period of time as a result of exposure to the weather.

The diagrams on the next two pages show the design of the balustrades and a typical preexisting crack in a balustrade which has yet to fail (see photograph on page 216).

The numbers on the drawings are identified as follows: 1, end post (31-inch long); 2, baluster; 3, top rail (105-inch long); 4, top stringer; 5, bottom stringer; 6, kick board; 7, rebate groove; 8, blind mortise; 9, tenon; 10, 2-inch nail (head at outer face of baluster); 11, 2-inch nail (head at inner face of baluster); 12, galvanized steel fixing bracket cemented into brick wall; 13, steel coach screw (1/4-inch diameter).

Sketch the made of failure, and account for it in terms of the fracture toughness of wood (numbers are not required).

You are called to a meeting with the architect. Explain to him or her the errors in the design, and how you think the balustrade should have been designed.

**14.5** Give examples from your own experience of the many *advantages* of wood having a low fracture toughness along the grain.

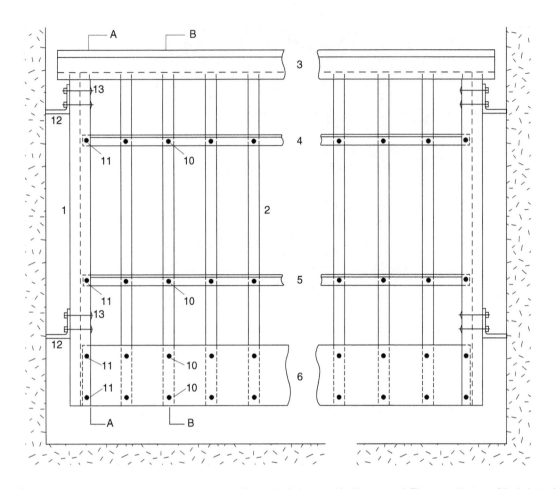

Elevation of the balustrade as seen by an observer standing on the balcony and looking outward. The top and bottom of the balustrade are respectively 39 and 6 inches above the balcony floor. There is a total of 24 balusters along the length of the balustrade (not all shown).

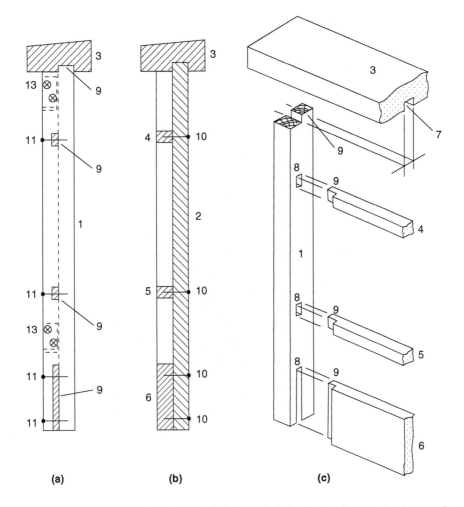

(a) Cross section drawn on A A looking at the left end of the balustrade. (b) Cross-section drawn on B B looking at the left end of the balustrade. (c) Exploded view showing joints.

An 18-inch crack in another balustrade.

## CHAPTER 15

# Probabilistic Fracture
# of Brittle Materials

## CHAPTER CONTENTS

## 15.1 INTRODUCTION

We saw in Chapter 13 that the fracture toughness $K_c$ of ceramics and rigid polymers was very low compared to that of metals and composites. Cement and ice have the lowest $K_c$ at $\approx 0.2$ MN m$^{-3/2}$. Traditional manufactured ceramics (brick, pottery, china, porcelain) and natural stone or rock are better, at $\approx 0.5$–2 MN m$^{-3/2}$. But even engineering ceramics such as silicon nitride, alumina, and silicon carbide only reach $\approx 4$ MN m$^{-3/2}$. Rigid polymers (thermoplastics below the glass transition temperature, or heavily cross-linked thermosets such as epoxy) have $K_c \approx 0.5$–4 MN m$^{-3/2}$. This low fracture toughness makes ceramics and rigid polymers very vulnerable to the presence of crack-like defects. They are *defect-sensitive* materials, liable to fail by fast fracture from defects well before they can yield.

Unfortunately, many manufactured ceramics contain cracks and flaws left by the production process (e.g., the voids left between the particles from which the ceramic was fabricated). The defects are worse in cement, because of the rather crude nature of the mixing and setting process. Ice usually contains small

Engineering Materials I: An Introduction to Properties, Applications, and Design, Fourth Edition

bubbles of trapped air (and, in the case of sea ice, concentrated brine). Molded polymer components often contain small voids. And all but the hardest brittle materials accumulate additional defects when they are handled or exposed to an abrasive environment.

The design strength of a brittle material in tension is therefore determined by its low fracture toughness in combination with the lengths of the crack-like defects it contains. If the longest microcrack in a given sample has length $2a_m$, then the tensile strength is simply given by

$$\sigma_{TS} \approx \frac{K_c}{\sqrt{\pi a_m}} \tag{15.1}$$

from the fast fracture equation. Some engineering ceramics have tensile strengths about half that of steel—around 200 MN m$^{-2}$. Taking a typical fracture toughness of 2 MN m$^{-3/2}$, the largest microcrack has a size of 60 µm, which is of the same order as the original particle size. Pottery, brick, and stone generally have tensile strengths that are much lower than this—around 20 MN m$^{-2}$. This indicates defects of the order of 2 mm for a typical fracture toughness of 1 MN m$^{-3/2}$. The tensile strength of cement and concrete is even lower— 2 MN m$^{-2}$ in large sections—implying the presence of at least one crack 6 mm or more in length for a fracture toughness of 0.2 MN m$^{-3/2}$.

## 15.2 THE STATISTICS OF STRENGTH

The chalk with which I write on the blackboard when I teach is a brittle solid. Some sticks of chalk are weaker than others. On average, I find (to my slight irritation), that about 3 out of 10 sticks break as soon as I start to write with them; the other 7 survive. The failure probability, $P_f$, for this chalk, loaded in bending under my (standard) writing load, is 3/10, that is,

$$P_f = 0.3$$

When you write on a blackboard with chalk, you are not unduly inconvenienced if 3 pieces in 10 break while you are using it; but if 1 in 2 broke, you might seek an alternative supplier. So the failure probability, $P_f$, of 0.3 is acceptable (just barely). If the component were a ceramic cutting tool, a failure probability of 1 in 100 ($P_f = 10^{-2}$) might be acceptable, because a tool is easily replaced. But if it were the glass container of a cafetiere, the failure of which could cause injury, one might aim for a $P_f$ of $10^{-4}$.

When using a brittle solid under load, it is not possible to be certain that a component will not fail. But if an acceptable risk (the failure probability) can be assigned to the function filled by the component, then it is possible to design so that this acceptable risk is met. This chapter explains why ceramics have this dispersion of strength; and shows how to design components so they have a given probability of survival. The method is an interesting one, with application beyond ceramics to the malfunctioning of any complex system in which the breakdown of one component will cause the entire system to fail.

Chalk is a porous ceramic. It has a fracture toughness of 0.9 MN m$^{-3/2}$ and, being poorly consolidated, is full of cracks and angular holes. The average tensile strength of a piece of chalk is 15 MN m$^{-2}$, implying an average length for the longest crack of about 1 mm (calculated from Equation (15.1)). But the chalk itself contains a distribution of crack lengths. Two nominally identical pieces of chalk can have tensile strengths that differ greatly—by a factor of 2 or more. This is because one was cut so that, by chance, all the cracks in it are small, whereas the other was cut so that it includes one of the longer flaws of the distribution.

Figure 15.1 illustrates this: if the block of chalk is cut into pieces, piece A will be weaker than piece B because it contains a larger flaw. It is inherent in the strength of brittle materials that there will be a statistical variation in strength. There is no single "tensile strength" but there is a certain, definable, *probability* that a given sample will have a given strength.

The distribution of crack lengths has other consequences. A large sample will fail at a lower stress than a small one, on average, because it is more likely that it will contain one of the larger flaws (Figure 15.1). So there is a *volume dependence* of the strength. For the same reason, a brittle rod is stronger in bending

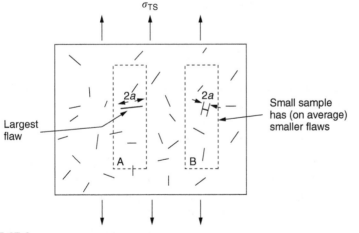

**FIGURE 15.1**
If small samples are cut from a large block of a brittle material, they will show a dispersion of strengths because of the dispersion of flaw sizes. The average strength of the small samples is greater than that of the large sample.

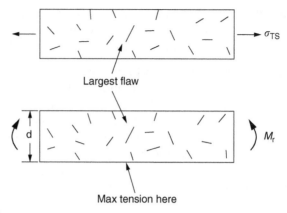

Largest flaw

Max tension here

**FIGURE 15.2**
Brittle materials appear to be stronger in bending than in tension because the largest flaw may not be near the surface.

than in simple tension: in tension the entire sample carries the tensile stress, while in bending only a thin layer close to one surface (and thus a relatively smaller volume) carries the peak tensile stress (Figure 15.2).

## 15.3 THE WEIBULL DISTRIBUTION

The Swedish engineer, Weibull, invented the following way of handling the statistics of strength. He defined the *survival probability* $P_s(V_0)$ as the fraction of identical samples, each of volume $V_0$, which survive loading to a tensile stress $\sigma$. He then proposed that

$$P_s(V_0) = \exp\left\{-\left(\frac{\sigma}{\sigma_0}\right)^m\right\} \tag{15.2}$$

where $\sigma_0$ and $m$ are constants. This equation is plotted in Figure 15.3(a). When $\sigma = 0$ all the samples survive, of course, and $P_s(V_0) = 1$. As $\sigma$ increases, more and more samples fail, and $P_s(V_0)$ decreases. Large stresses cause virtually all the samples to break, so $P_s(V_0) \rightarrow 0$ and $\sigma \rightarrow \infty$.

If we set $\sigma = \sigma_0$ in Equation (15.2) we find that $P_s(V_0) = 1/e \, (= 0.37)$. So $\sigma_0$ is simply the tensile stress that allows 37 % of the samples to survive. The constant $m$ tells us how rapidly the strength falls as we approach $\sigma_0$ (see Figure 15.3(b)). It is called the *Weibull modulus*. The lower $m$, the greater the *variability* of strength. For ordinary chalk, $m$ is about 5, and the variability is great. Brick, pottery, and cement are like this too.

The engineering ceramics (e.g., SiC, $Al_2O_3$, and $Si_3N_4$) have values of $m$ of about 10; for these, the strength varies much less. Even steel shows some

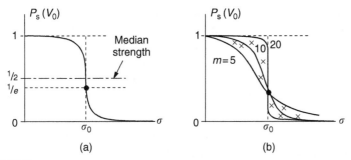

**FIGURE 15.3**

(a) The Weibull distribution. (b) Relation between the modulus $m$ and the spread of strength.

variation in strength, but it is small: it can be described by a Weibull modulus of about 100. Figure 15.3(b) shows that, for $m$ greater than about 20, a material can be treated as having a single, well-defined failure stress.

So much for the stress dependence of $P_s$. But what of its volume dependence? We have already seen that the probability of one sample surviving a stress $\sigma$ is $P_s(V_0)$. The probability that a batch of $n$ such samples all survive the stress is just $\{P_s(V_0)\}^n$. If these $n$ samples were stuck together to give a single sample of volume $V = nV_0$ then its survival probability would still be $\{P_s(V_0)\}^n$. So

$$P_s(V) = \{P_s(V_0)\}^n = \{P_s(V_0)\}^{V/V_0}$$

This is equivalent to

$$\ln P_s(V) = \frac{V}{V_0} \ln P_s(V_0)$$

or

$$P_s(V) = \exp\left\{\frac{V}{V_0} \ln P_s(V_0)\right\} \tag{15.3}$$

The Weibull distribution (Equation (15.2)) can be rewritten as

$$\ln P_s(V_0) = -\left(\frac{\sigma}{\sigma_0}\right)^m$$

If we insert this result into equation (15.3) we get

$$P_s(V) = \exp\left\{-\frac{V}{V_0}\left(\frac{\sigma}{\sigma_0}\right)^m\right\} \tag{15.4}$$

or

$$\ln P_s(V) = -\frac{V}{V_0}\left(\frac{\sigma}{\sigma_0}\right)^m$$

This, then, is our final design equation. It shows how the survival probability depends on both the stress $\sigma$ and the volume $V$ of the component. In using it, the first step is to fix on an acceptable failure probability, $P_f$: 0.3 for chalk, $10^{-2}$ for the cutting tool, $10^{-4}$ for the cafetiere. The survival probability is then given by $P_s = 1 - P_f$. (It is useful to remember that, for small $P_f$, $\ln P_s = \ln (1 - P_f) \approx - P_f$.) We can then substitute suitable values of $\sigma_0$, $m$, and $V/V_0$ into the equation to calculate the design stress.

Note that Equation (15.4) assumes that the component is subjected to a *uniform* tensile stress $\sigma$. In many applications, $\sigma$ is not constant, but instead varies with position throughout the component. Then, we rewrite Equation (15.4) as

$$P_s(V) = \exp\left\{-\frac{1}{\sigma_0^m V_0}\int_V \sigma^m dV\right\} \tag{15.5}$$

## 15.4 THE MODULUS OF RUPTURE

In Chapter 8 we saw that properties such as the yield and tensile strengths could be measured easily using a long cylindrical specimen loaded in simple tension. But it is difficult to perform tensile tests on brittle materials—the specimens tend to break where they are gripped by the testing machine. This is because the local contact stresses exceed the fracture strength, and premature failure occurs at the grips.

It is much easier to measure the force required to break a beam in bending (Figure 15.4). The maximum tensile stress in the surface of the beam when it breaks is called the modulus of rupture, $\sigma_r$; for an elastic beam it is related to the maximum moment in the beam, $M_r$, by

$$\sigma_r = \frac{6M_r}{bd^2} = \frac{3\,Fl}{2\,bd^2} \tag{15.6}$$

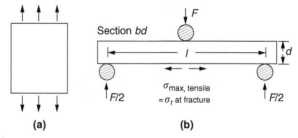

**FIGURE 15.4**
Tests that measure the fracture strengths of brittle materials. (a) The tensile test measures the tensile strength, $\sigma_{TS}$ at fracture. (b) The bend test measures the modulus of rupture, $\sigma_r$.

You might think that $\sigma_r$, should be equal to the tensile strength $\sigma_{TS}$. But it is actually larger than the tensile strength. There are two reasons for this. The first is that half of the beam—the half above the neutral axis—is subjected to compressive stresses. Flaws here will close up, and will be very unlikely to propagate to failure. The second is that the peak tensile stress is located in the lower surface of the beam, vertically below the loading point. Everywhere else, the tensile stress is lower—in many cases much lower. So only a small volume is subjected to high tensile stress.

The tensile strength may be found from the modulus of rupture using the following equation:

$$\sigma_{TS} = \frac{\sigma_r}{\{2(m+1)^2\}^{1/m}}$$

This is derived using Equation (15.5), together with the (known) stress state in the beam. For example, if $m = 10$, $\sigma_{TS} = \sigma_r/1.73$. If $m = 5$, $\sigma_{TS} = \sigma_r/2.35$.

## WORKED EXAMPLE

How do we find $\sigma_0$, $m$, and the median strength from test data? The table gives results from a set of six separate bend tests (numbered 1–6). Listed for each specimen is the maximum tensile stress at mid-span when fracture occurred—the modulus of rupture, calculated using Equation 15.6. Note that the initiating flaw was not always located at the position of maximum tensile stress.

The table ranks the test results in order of increasing modulus of rupture ($j = 1$ to 6). Then $P_s$ can be found from the standard statistical result

$$P_s = 1 - \frac{j - 0.375}{n + 0.25} = 1 - \frac{j - 0.375}{6.25}$$

Values for $P_s$ are listed in this table.

| j | Specimen Number | $\sigma_r$ (MN m$^{-2}$) | $P_s$ | ln $\sigma$ | ln ln (1/$P_s$) |
|---|---|---|---|---|---|
| 1 | 4 | 25 | 0.900 | 3.22 | −2.25 |
| 2 | 2 | 34.6 | 0.740 | 3.54 | −1.20 |
| 3 | 3 | 38.5 | 0.580 | 3.65 | −0.61 |
| 4 | 1 | 43 | 0.420 | 3.76 | −0.14 |
| 5 | 5 | 49 | 0.260 | 3.90 | +0.30 |
| 6 | 6 | 55 | 0.100 | 4.00 | +0.83 |

The trick for analyzing the data is as follows. We start with the basic Weibull equation—Equation (15.2)—for specimens all having the same volume $V_0$. Taking natural logs on each side of Equation (15.2), we get

$$\ln\left\{\frac{1}{P_s(V_0)}\right\} = \left(\frac{\sigma}{\sigma_0}\right)^m$$

Taking natural logs on each side again gives

$$\ln\left\{\ln\left(\frac{1}{P_s(V_0)}\right)\right\} = m\ \ln\sigma - m\ \ln\sigma_0$$

Note that the last term in this equation is a constant. The preceding table lists values for $\ln\ln(1/P_s)$ and $\ln\sigma$, and these are plotted in the following diagram. The slope of the best-fit line is 4 (note that the scales of the vertical and horizontal log axes are not the same!). Therefore, $m = 4$. We can also easily see that $\sigma_0$ and the median strength are 46 and 41 MN m$^{-2}$. This type of graph is called a Weibull plot.

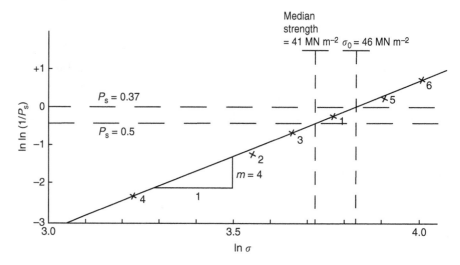

## EXAMPLES

**15.1** $Al_2O_3$ has a fracture toughness $K_c$ of about 3 MN m$^{-3/2}$. A batch of $Al_2O_3$ samples is found to contain surface flaws about 30 μm deep. Estimate the tensile strength of the samples. Assume that $Y = 1{,}12$.

**Answer**

276 MN m$^{-2}$

**15.2** Modulus-of-rupture tests are carried out using the arrangement shown in Figure 15.4. The specimens break at a load $F$ of about 330 N. Find the modulus of rupture, given that $l = 50$ mm and $b = d = 5$ mm.

**Answer**

198 MN m$^{-2}$

**15.3** To test the strength of a ceramic, cylindrical specimens of length 25 mm and diameter 5 mm are put into axial tension. The tensile stress $\sigma$ which causes 50% of the specimens to break is 120 MN m$^{-2}$. Cylindrical ceramic components of length 50 mm and diameter 11 mm are required to withstand an axial tensile stress $\sigma^1$ with a survival probability of 99%. Given that $m = 5$, use Equation (15.4) to determine $\sigma^1$.

**Answer**

32.6 MN m$^{-2}$

**15.4** Modulus-of-rupture tests were carried out on samples of silicon carbide using the three-point bend test geometry shown in Figure 15.4. The samples were 100 mm long and had a 10 mm by 10 mm square cross section. The median value of the modulus of rupture was 400 MN m$^{-2}$. Tensile tests were also carried out using samples of identical material and dimensions, but loaded in tension along their lengths. The median value of the tensile strength was only 230 MN m$^{-2}$. Account in a qualitative way for the difference between the two measures of strength.

**Answer**

In the tensile test, the whole volume of the sample is subjected to a tensile stress of 230 MN m$^{-2}$. In the bend test, only the lower half of the sample is subjected to a tensile stress. Furthermore, the average value of this tensile stress is considerably less than the peak value of 400 MN m$^{-2}$ (which is only reached at the underside of the sample beneath the central loading point). The probability of finding a fracture-initiating defect in the small volume subjected to the highest stresses is small.

**15.5** Modulus-of-rupture tests were done on samples of ceramic with dimensions $l = 100$ mm and $b = d = 10$ mm. The median value of $\sigma_r$ (i.e., $\sigma_r$ for $P_s = 0.5$) was 300 MN m$^{-2}$. The ceramic is to be used for components with dimensions $l = 50$ mm, $b = d = 5$ mm loaded in simple tension along their length. Calculate the tensile stress $\sigma$ that will give a probability of failure, $P_f$ of $10^{-6}$. Assume that $m = 10$. Note that, for $m = 10$, $\sigma_{TS} = \sigma_r/1.73$.

**Answer**

55.7 MN m$^{-2}$

**15.6** The following diagram is a schematic of a stalactite, a cone-shaped mineral deposit hanging down ward from the roof of a cave. Its failure due to self weight loading is to be modeled using Weibull statistics. The geometry of the stalactite is idealized as a cone of length $L$ and semiangle $\alpha$. The cone angle is assumed small so that the base radius equals $\alpha L$. The stalactite density is $\rho$.

 a. Show that the variation of tensile stress $\sigma$ with height $x$ is given by $\sigma = 1/3 \, \rho g x$. You may assume that the volume of a cone is given by $(\pi/3) \times (\text{base radius})^2 \times \text{height}$

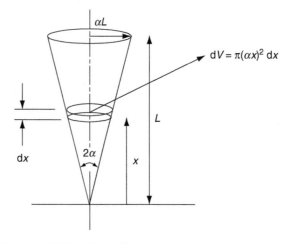

b.  Use Equation (15.5) for Weibull statistics with a varying stress to show that the probability of survival $P_s(L)$ for a stalactite of length $L$ is given by

$$P_s(L) = \exp\left\{-\left(\frac{\rho g}{3\sigma_0}\right)^m \frac{\pi \alpha^2 L^{m+3}}{(m+3)V_0}\right\}$$

Explain why there is a dependency on cone angle $\alpha$, even though the stress variation up the stalactite is independent of $\alpha$.

## CHAPTER 16

# Case Studies in Fracture

## CONTENTS

## 16.1 INTRODUCTION

In this chapter we look at three real situations where failure occurred because of the catastrophic growth of a crack by fast fracture: a steel ammonia tank which exploded because of weld cracks; a perspex pressure window which exploded during hydrostatic testing; and a polyurethane foam jacket on a liquid methane tank which cracked during cooling.

## 16.2 CASE STUDY 1: FAST FRACTURE OF AN AMMONIA TANK

Figure 16.1 shows part of a steel tank which came from a road tank vehicle. The tank consisted of a cylindrical shell about 6 m long. A hemispherical cap was welded to each end of the shell with a circumferential weld. The tank was used to transport liquid ammonia. In order to contain the liquid ammonia the pressure had to be equal to the saturation pressure (the pressure at which a mixture

Engineering Materials I: An Introduction to Properties, Applications, and Design, Fourth Edition

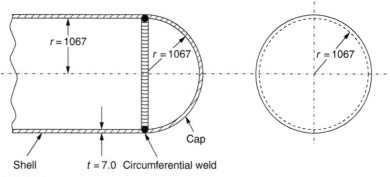

**FIGURE 16.1**

The weld between the shell and the end cap of the pressure vessel (dimensions in mm).

of liquid and vapor is in equilibrium). The saturation pressure increases rapidly with temperature: at 20°C the absolute pressure is 8.57 bar; at 50°C it is 20.33 bar. The *gauge* pressure at 50°C is 19.33 bar, or 1.9MN m$^{-2}$. Because of this the tank had to function as a pressure vessel. The maximum operating pressure was 2.07MN m$^{-2}$ gauge. This allowed the tank to be used safely to 50°C, above the maximum temperature expected in even a hot climate.

While liquid was being unloaded from the tank a fast fracture occurred in one of the circumferential welds and the cap was blown off the end of the shell. In order to decant the liquid the space above the liquid had been pressurized with ammonia gas using a compressor. The normal operating pressure of the compressor was 1.83MN m$^{-2}$; the maximum pressure (set by a safety valve) was 2.07MN m$^{-2}$. One can imagine the effect on nearby people of this explosive discharge of a large volume of highly toxic vapor.

### Details of the failure

The geometry of the failure is shown in Figure 16.2. The initial crack, 2.5 mm deep, had formed in the heat-affected zone between the shell and the circumferential weld. The defect went some way around the circumference of the vessel. The cracking was intergranular, and had occurred by a process called stress corrosion cracking (see Chapter 26). The final fast fracture occurred by transgranular cleavage (see Chapter 14). This indicates that the heat-affected zone must have had a very low fracture toughness. In this case study we predict the critical crack size for fast fracture using the fast fracture equation.

### Material properties

The tank was made from high-strength low-alloy steel with a yield strength of 712 MN m$^{-2}$ and a fracture toughness of 80 MN m$^{-3/2}$. The heat from the welding process had altered the structure of the steel in the heat-affected

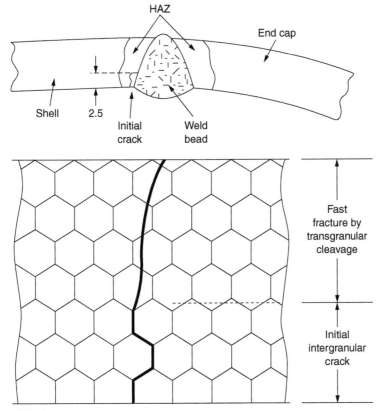

**FIGURE 16.2**
The geometry of the failure (dimensions in mm).

zone to give a much greater yield strength (940 MN m$^{-2}$) but a much lower fracture toughness (39 MN m$^{-3/2}$).

## Calculation of critical stress for fast fracture

The longitudinal stress $\sigma$ in the wall of a cylindrical pressure vessel containing gas at pressure $p$ is given by

$$\sigma = \frac{pr}{2t}$$

provided that the wall is thin ($t \ll r$). $p = 1.83$ MN m$^{-2}$, $r = 1067$ mm, and $t = 7$ mm, so $\sigma = 140$ MN m$^{-2}$. The fast fracture equation is

$$Y\sigma\sqrt{\pi a} = K_c$$

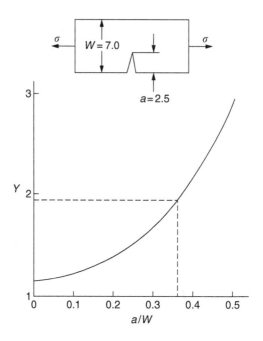

**FIGURE 16.3**
Y value for the crack (dimensions in mm).

Figure 16.3 shows that $Y = 1.92$ for our crack. The critical stress for fast fracture is given by

$$\sigma = \frac{K_c}{Y\sqrt{\pi a}} = \frac{39}{1.92\sqrt{\pi \times 0.0025}} = 229 \text{ MN m}^{-2}$$

The critical stress is 64% greater than the longitudinal stress. However, the change in section from a cylinder to a sphere produces something akin to a stress concentration; when this is taken into account the failure is accurately predicted.

## Conclusions

This case study provides a good example of the consequences of having an in-adequate fracture toughness. However, even if the heat-affected zone had a high toughness, the crack would have continued to grow through the wall of the tank by stress-corrosion cracking until fast fracture occurred. The critical crack size would have been greater, but failure would have occurred eventually. The only way of avoiding failures of this type is to prevent stress corrosion cracking in the first place.

# 16.3 CASE STUDY 2: EXPLOSION OF A PERSPEX PRESSURE WINDOW DURING HYDROSTATIC TESTING

Figure 16.4 shows the general arrangement drawing for an experimental rig, which is designed for studying the propagation of buckling in externally pressurized tubes. A long open-ended tubular specimen is placed on the horizontal axis of the rig with the ends emerging through pressure seals. The rig is partially filled with water and the space above the water is filled with nitrogen. The nitrogen is pressurized until buckling propagates along the length of the specimen.

The volumes of water and nitrogen in the rig can be adjusted to give stable buckling propagation. Halfway along the rig is a flanged perspex connector which allows the propagation of the buckle to be observed directly using a high-speed camera. Unfortunately, this perspex window exploded during the first hydrostatic test at a pressure of only one-half of the specified hydrostatic test pressure.

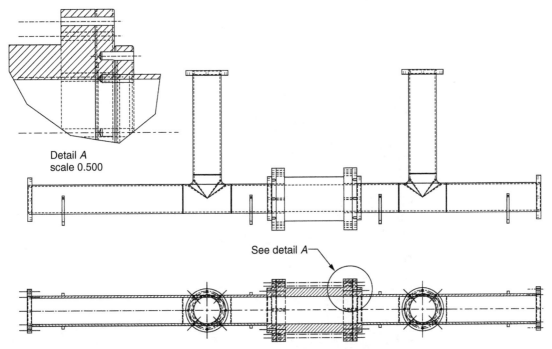

Detail *A*
scale 0.500

See detail *A*

**FIGURE 16.4**
General arrangement drawing of the experimental rig.

## Design data

Relevant design data for the perspex connector are given as follows.

- Internal diameter of cylindrical portion $2B = 154$ mm
- External diameter of cylindrical portion $2A = 255$ mm
- Forming process: casting
- Tensile strength, $\sigma_f \approx 62$ MN m$^{-2}$ (minimum), 77 MN m$^{-2}$ (average)
- Fracture toughness $K_c \approx 0.8$ to $1.75$ MN m$^{-3/2}$
- Working pressure $= 7$ MN m$^{-2}$ gauge
- Hydraulic test pressure $= 8.6$ MN m$^{-2}$ gauge
- Failure pressure $= 4.8$ MN m$^{-2}$ gauge

## Failure analysis

Figure 16.5 is a photograph of the perspex connector taken after the explosion. Detailed visual inspection of the fracture surface indicated that the fracture initiated as a hoop stress tensile failure in the cylindrical portion and subsequently propagated toward each flange.

The hoop stress $\sigma$ in the cylindrical portion can be calculated from the standard result for thick-walled tubes

$$\sigma = p\left(\frac{B}{r}\right)^2 \frac{A^2 + r^2}{A^2 - B^2}$$

**FIGURE 16.5**
Photograph of the perspex connector taken after the explosion.

where $p$ is the internal gauge pressure and $r$ is the radius at which the stress is calculated. The hoop stress is a maximum at the bore of the tube, with $r = B$ and

$$\sigma = p \frac{A^2 + B^2}{A^2 - B^2} = 2.13\, p$$

The hoop stress is a minimum at the external surface of the tube, with $r = A$ and

$$\sigma = p \frac{2B^2}{A^2 - B^2} = 1.13\, p$$

We can see that the most probable site for failure initiation is the bore of the tube: the hoop stress here is calculated to be 10 MN m$^{-2}$ at the failure pressure. This is only *one-sixth* of the minimum tensile strength. Using the fast fracture equation

$$K_c = Y\, \sigma \sqrt{\pi a}, \; Y \approx 1$$

with a fracture toughness of 1 MN m$^{-3/2}$ and a hoop stress of 10 MN m$^{-2}$ gives a critical defect size $a$ of 3.2 mm at the failure pressure. At the operating pressure the critical defect size would only be 1.5 mm. A defect of this size would be difficult to find under production conditions in such a large volume. In addition, it would be easy to introduce longitudinal scratches in the bore of the connector during routine handling and use.

## Conclusions

The most probable explanation for the failure is that a critical defect was present in the wall of the perspex connector. The connector was a standard item manufactured for flow visualization in pressurized systems. The designers had clearly used a stress-based rather than a fracture-mechanics based approach with entirely predictable consequences.

## 16.4 CASE STUDY 3: CRACKING OF A FOAM JACKET ON A LIQUID METHANE TANK

Figure 16.6 is a schematic half-section through a tank used for storing liquid methane at atmospheric pressure. Because methane boils at −162°C, the tank is made from an aluminum alloy in order to avoid any risk of brittle failure. Even so, it is necessary to have a second line of defense should the tank spring a leak. This is achieved by placing the tank into a mild-steel jacket, and inserting a layer of thermal insulation into the space between the two. The jacket is thereby protected from the cooling effect of the methane, and the temperature of the steel is kept above the ductile-to-brittle transition temperature. But what happens if the tank does spring a leak? If the insulation is porous (e.g., fiberglass matting) then the liquid methane will flow through the insulation to the wall of the jacket and will boil off.

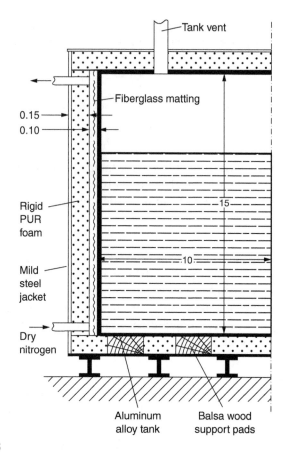

**FIGURE 16.6**

Schematic half-section through a typical liquid methane storage tank using closed-cell polyurethane foam for thermal insulation/secondary containment (typical dimensions in m).

As a result the jacket will cool down to −162°C and may fail by brittle fracture. To avoid this possibility the inner wall of the jacket is coated with a layer of closed-cell foam made from rigid polyurethane (PUR). The tank is then lowered into the jacket and the assembly gap is filled with fiberglass matting. The theory is that if the tank leaks, the flow of methane will be arrested by the closed-cell structure of the PUR and the jacket will be protected.

This system has been used in ships designed for the bulk carriage of liquid methane. In such applications, the mild-steel jacket surrounding the tank is the hull of the ship itself. It is vital that the PUR provides effective containment. However, incidents have occurred where the PUR has cracked, compromising the integrity of the hull. In one reported instance, three brand-new bulk carriers had to be written off as a result of multiple cracking of the PUR foam layer.

## Thermal stresses in the foam

Under normal operating conditions, the temperature of the foam decreases linearly with distance through the layer, as shown in Figure 16.7. The foam wants to contract as it gets cold, but is prevented from doing so by the rigid steel wall of the jacket to which it is stuck. The temperature differential $\Delta T$ generates a biaxial tensile stress $\sigma$ in the plane of the layer which is given by

$$\sigma = \frac{\alpha \Delta T E}{(1 - v)}$$

$\alpha$, $E$, and $v$ are the coefficient of thermal expansion, Young's modulus, and Poisson's ratio of the foam. Figure 16.7 also shows the variation of stress with distance through the layer. The thermal stress is a maximum at the inner surface of the PUR layer.

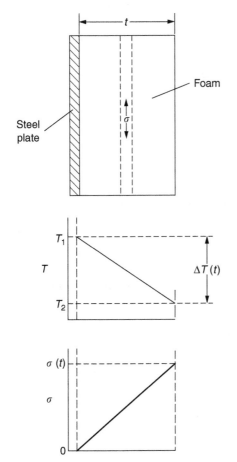

**FIGURE 16.7**
Temperature and thermal stress in the foam layer.

We saw in Chapter 6 that polymers behave as elastic-brittle solids provided they are colder than the glass transition temperature, $T_G$. For PUR, $T_G \approx 100°C$, or 373 K. Presumably the foam failed by brittle cracking when the maximum thermal stress reached the fracture stress of the foam. To check this hypothesis we first list the relevant data for a typical PUR foam used in cryogenic applications.

- Thermal expansion coefficient $\alpha \approx 10^{-4}°C^{-1}$
- Poisson's ratio $v \approx 0.3$
- Young's modulus $E \approx 34$ MN m$^{-2}$ at $-100°C$
- Fracture stress $\sigma_f \approx 1.4$ MN m$^{-2}$ at $-100°C$

Referring to Figure 16.7, $T_1 \approx 0°C$, $T_2 \approx -100°C$, and $\Delta T(t) \approx 100°C$. For a temperature differential of this magnitude, $\sigma \approx 0.5$ MN m$^{-2}$. However, this value for the maximum stress is substantially less than the fracture stress of approximately 1.4 MN m$^{-2}$ expected at $-100°C$. This means that, on a simple basis, the foam should not have fractured in service. In order to understand why the foam did in fact break it is necessary to analyze the problem using Weibull statistics.

Tensile tests were carried out on foam samples at $-100°C$. The stressed volume of each sample $V_0$ was $\approx 5 \times 10^{-5}$ m$^3$. The test data gave a median fracture strength ($P_s = 0.5$) of 1.4 MPa, and a Weibull modulus $m$ of 8.

The Weibull equation for the tensile tests is

$$P_s(V_0) = 0.5 = \exp\left\{-\left(\frac{\sigma_f}{\sigma_0}\right)^m\right\} \tag{16.1}$$

The Weibull equation for the foam layer is

$$P_s(V) = 0.5 = \exp\left\{-\frac{1}{\sigma_0{}^m V_0}\int_V \sigma^m dV\right\} \tag{16.2}$$

Referring to Figure 16.8, we set up a representative volume element

$$dV = a^2\, dx$$

"containing" a thermal stress

$$\sigma = \sigma_{max}\left(\frac{x}{t}\right)$$

Using these two results to substitute for $\sigma$ and $dV$ in Equation 16.2, we get

$$P_s(V) = 0.5 = \exp\left\{-\frac{1}{\sigma_0^m V_0}\int_0^t \sigma_{max}^m \left(\frac{x}{t}\right)^m a^2\, dx\right\} \tag{16.3}$$

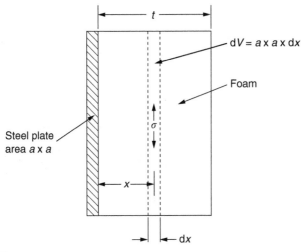

**FIGURE 16.8**
Volume element with constant tensile stress in the foam jacket.

Combining Equations (16.1) and (16.3) we get

$$\left(\frac{\sigma_f}{\sigma_0}\right)^m = \left(\frac{\sigma_{max}}{\sigma_0}\right)^m \frac{a^2}{V_0 t^m} \int_0^t x^m dx = \frac{1}{(m+1)}\left(\frac{\sigma_{max}}{\sigma_0}\right)^m \frac{a^2 t}{V_0}$$

This gives

$$a^2 t = (m+1)\left(\frac{\sigma_f}{\sigma_{max}}\right)^m V_0$$

and

$$a^2 t = 9\left(\frac{1.4}{0.5}\right)^8 5 \times 10^{-5} \text{ m}^3 = 1.70 \text{ m}^3 = 3.4 \times 3.4 \times 0.15 \text{ m}^3$$

This means that there is a 50% chance that cracks will occur in the foam jacket at intervals of approximately 3.4 m.

## Conclusions

Considerable financial loss has resulted from the cracking of PUR foam in liquid methane insulation/containment systems. This has been caused by a combination of the very low fracture toughness of PUR foam at low temperature and the likelihood that the method of applying the foam will introduce defects exceeding the critical size for crack propagation. An elementary knowledge of Weibull Statistics would have suggested that the use of plain (unreinforced) PUR in the present situation was fundamentally unsound.

# WORKED EXAMPLE

The photographs and diagrams that follow show how the Tay Bridge was built, and how it collapsed. This single-track railway bridge was erected over the Firth of Tay, at Dundee, Scotland. It entered service in May 1878, but was blown over

View of the bridge from the north bank of the firth, looking south. *(Courtesy of University of St Andrews.)*

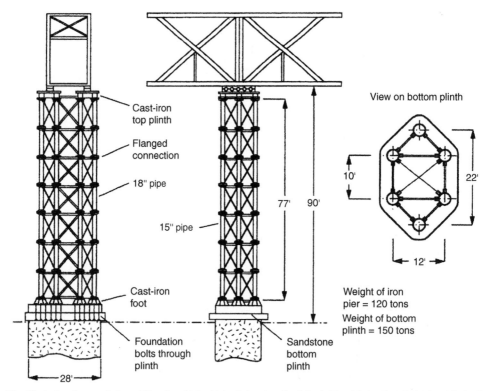

Cast-iron top plinth

Flanged connection

18" pipe

15" pipe

Cast-iron foot

Foundation bolts through plinth

Sandstone bottom plinth

View on bottom plinth

10'     22'

12'

77'    90'

Weight of iron pier = 120 tons

Weight of bottom plinth = 150 tons

28'

Simplified elevations and plan of the piers in the high girders length of the bridge (dimensions in feet and inches).

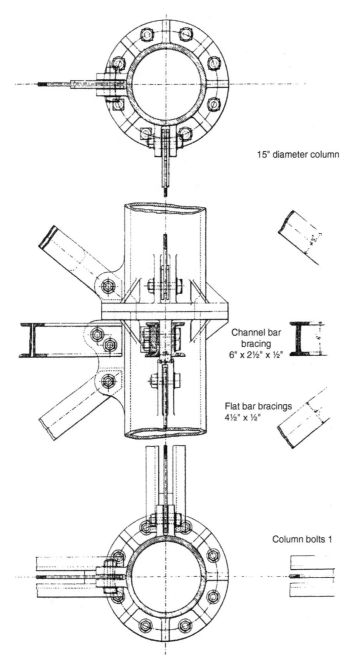

15" diameter column

Channel bar
bracing
6" x 2½" x ½"

Flat bar bracings
4½" x ½"

Column bolts 1

Details of the flanges and lugs of the cast-iron columns (dimensions in inches).

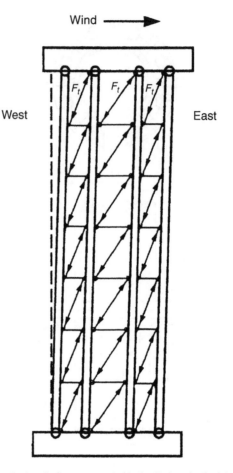

Wind ⟶

West

East

$F_t$ $F_t$ $F_t$

Schematic of a pier showing the tensile forces generated in the tie-bars by the lateral wind loading on the high girders plus train.

on 28 December 1879 by a storm force wind from the west, with a train that was carrying 75 people inside the high girders (the 13 central navigation spans of the bridge). The failure occurred because the sideways wind loading on the bridge girders and train put too much tension into the diagonal tie bars (wind braces), and broke the cast iron lugs to which they were attached.

The bridge was subsequently demolished, and replaced with a massive double-track bridge. There is a great aerial view of the replacement bridge (casting reflections on the water), with the stumps of the old bridge alongside, on Google Earth at 56 26 20.00 N 2 59 18.70 W (see also *http://taybridgedisaster.co.uk/index/index*).

Fallen piers as seen from the south bank of the firth. For the purposes of the investigation, the piers between 28 and 41 were renumbered 1 to 12. *(Courtesy of University of St. Andrews.)*

Tests done after the disaster found that the cast iron lugs failed when the tension in the tie bars reached about 24 tons. The fracture area of a lug is about 20 sq. in. Tests on specimens taken from the cast-iron columns gave a tensile strength of 9 tons/sq. in. This indicates that the lug should have been able to carry a load of 180 tons applied by a tie bar. There is a factor of 7.5 times between this ultimate load of 180 tons and the actual measured breaking load of only 24 tons! Yet there was no indication that the cast iron was defective in any way.

Pin and lug connections are important in the aerospace industry for attaching major fuselage components together (e.g., keel splice joints). They have been analyzed by linear elastic finite element stress analysis (FEA). Stress concentrations as high as 7.5 have been found from FEA. So it seems reasonable to think that the lugs broke because of local stress hot spots of the order of 9 tons/sq. in. next to the bolt holes.

If the lugs had been made from a ductile material, like mild steel, they would not have fractured. The steel would have yielded at the stress concentration, and the local stress would have been truncated at the yield stress (which is much less than the tensile strength). However, in a brittle material, fracture intervenes before yield can take place, so the stress concentration predicted by FEA is still valid.

We saw an example of this in Chapter 15, Section 15.4. The modulus of rupture test is used for brittle materials because specimens tend to break in the grips in a

Broken lug on cast-iron column. *(Courtesy of Dundee Central Library.)*

tensile testing machine. This is because the local contact stresses exceed the fracture strength—and because fracture occurs before yield, there can be no truncation of the local contact stresses. The conclusion is that ductile materials are much more forgiving than brittle materials—essentially for the same reason that ductile materials are much tougher than brittle materials (see Chapter 14).

## EXAMPLES

**16.1** A cylindrical pressure vessel is rolled up from flat steel plate 10 mm thick, and the edges are welded together to produce a longitudinal welded seam. Calculate the stress intensity factor $K$ at a hoop stress of 100 MN m$^{-2}$ for two possible types of weld defect:

  **a.** A crack running along the whole length of the weld, and extending from the inner surface of the vessel to a depth of 5 mm

  **b.** A through-thickness crack of length $2a = 40$ mm

Which is the more dangerous crack, and why? [*Hint:* Make sure you use the correct values for $Y$.]

**Answers**

(a) 37.6 MN m$^{-3/2}$; (b) 25.1 MN m$^{-3/2}$; crack (a), because it has the larger $K$, and also because it does not reveal itself by creating a leak.

**16.2** With reference to Case Study 2, why was perspex an unsuitable material for the pressure window?

**16.3** With reference to Case Study 3, how would you redesign the insulation system to prevent cracking in a new installation?

**16.4** The following diagram shows a schematic of one of the hydraulic rams used on 28 November 1857 in a failed attempt to get the 22,500-ton ship Great Eastern off the builder's slipway. *http://en.wikipedia.org/wiki/SS_Great_Eastern*
51 29 1562 N 0 01 15.38 W (Isle of Dogs, London, England)

The cylinder of the ram (made from gray cast iron) split open under the internal water pressure. Estimate the tensile strength of the cast iron, given the following data. Water pressure inside cylinder $= 83$ MN m$^{-2}$. External radius of cylinder $= 279$ mm. Internal radius of cylinder $= 127$ mm. Comment on the advisability of using gray cast iron for this application.

**Answer**

126 MN m$^{-2}$

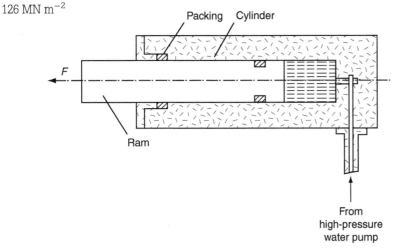

Packing   Cylinder

$F$

Ram

From
high-pressure
water pump

# Fatigue Failure

# Fatigue Failure

## CONTENTS

## 17.1 INTRODUCTION

In Chapters 13 and 14, we examined the conditions under which a crack was stable, and would not grow, and the condition

$$K = K_c$$

under which it would propagate catastrophically by fast fracture. If we know the maximum size of crack in the structure we can then choose a working load at which fast fracture will not occur.

But cracks can form, and grow slowly, at loads lower than this, if the load is cycled. The process of slow crack growth—*fatigue*—is the subject of this chapter. When the clip of your pen breaks, when the pedals fall off your bicycle, when the handle of the refrigerator comes away in your hand, it is usually fatigue which is responsible.

Engineering Materials I: An Introduction to Properties, Applications, and Design, Fourth Edition

## 17.2 FATIGUE OF UNCRACKED COMPONENTS

Fatigue tests are done by subjecting specimens of the material to a cyclically varying load or displacement (Figure 17.1). For simplicity, the time variation of stress and strain is shown as a sine wave, but modern testing machines are capable of applying almost any chosen load or displacement spectrum. The specimens are subjected to increasing numbers of fatigue cycles $N$, until they finally crack (when $N = N_f$—the number of cycles to failure).

Figure 17.2 shows how the cyclic stress and cyclic strain are coupled in a linear elastic specimen. The stress and strain are linearly related through Young's modulus, and the stress range $\Delta\sigma$ produces (or is produced by) an elastic strain range $\Delta\varepsilon^{el}$ (equal to $\Delta\sigma/E$).

Figure 17.3 shows how the cyclic stress and cyclic strain are coupled in a specimen that is cycled outside the elastic limit. There is no longer a simple relationship between stress and strain. Instead, they are related by a *stress–strain loop*, which must be determined from tests on the material. In addition, the shape of the loop changes with the number of cycles, and it can take several hundred

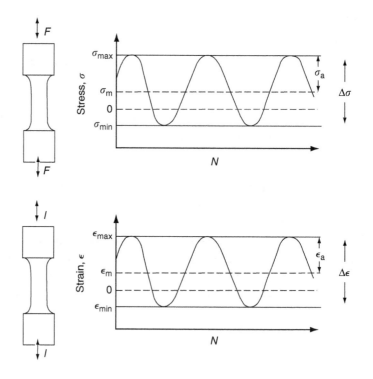

**FIGURE 17.1**

Fatigue testing.

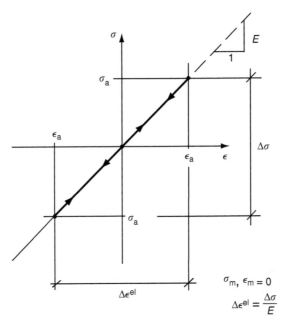

**FIGURE 17.2**
Coupling of cyclic stress and cyclic strain for a linear elastic specimen.

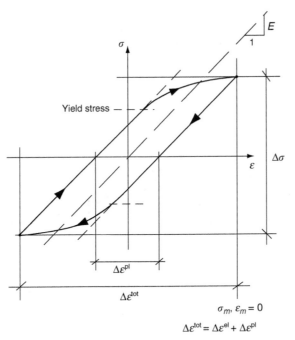

**FIGURE 17.3**
Coupling of cyclic stress and cyclic strain for a linear elastic/yielding specimen.

(or even thousand) cycles before the shape of the loop stabilizes. The stress range $\Delta\sigma$ produces (or is produced by) a *total* strain range $\Delta\varepsilon^{tot}$—which is the sum of the *elastic* strain range $\Delta\varepsilon^{el}$ and the *plastic* strain range $\Delta\varepsilon^{pl}$.

The best way to correlate fatigue test data is on a log-log plot of the total strain amplitude $\Delta\varepsilon^{tot}/2$ versus the number of *reversals* to failure $2N_f$ (there are two reversals of load or displacement in each complete cycle). Figure 17.4 shows the shape of the curve on which the data points typically fall (although there is usually a lot of experimental scatter on either side of the curve). It is useful to know that this curve is the sum of two linear relationships on the log–log plot: (a) between the *elastic* strain amplitude and $2N_f$, and (b) between the *plastic* strain amplitude and $2N_f$. It can be approximated mathematically as:

$$\frac{\Delta\varepsilon^{tot}}{2} \approx \frac{\sigma_f'(2N_f)^b}{E} + \varepsilon_f'(2N_f)^c \tag{17.1}$$

$b$ and $c$ are constants determined by fitting the test data—they are generally in the range −0.05 to −0.12 for $b$, and −0.5 to −0.7 for $c$. $\sigma_f'$ and $\varepsilon_f'$ are the *true* fracture stress and *true* fracture strain (derived from a standard tensile test on the material).

The data in Figure 17.4 can be divided into two régimes:

- Low-cycle fatigue (less than about $10^4$ cycles; plastic strain > elastic strain)
- High-cycle fatigue (more than about $10^4$ cycles; elastic strain > plastic strain)

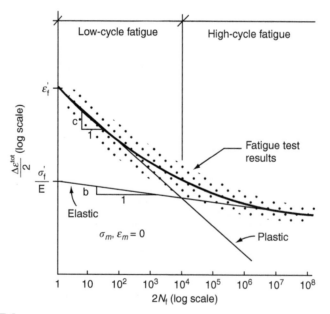

**FIGURE 17.4**
Relation between total strain amplitude and fatigue life.

Until the 1950s, most fatigue studies were concerned with high-cycle fatigue (HCF), since engineering components subjected to cyclic loadings (e.g., railway axles, engine crankshafts, bicycle frames) were designed to keep the maximum stress below the elastic limit. Because of this, it is still common practice to plot HCF data on a log–log plot of the *stress* amplitude $\Delta\sigma/2$ versus $2N_f$—see Figure 17.5. The test data can then be approximated as:

$$\frac{\Delta\sigma}{2} \approx \sigma_f'(2N_f)^b \tag{17.2}$$

So far, we have only considered test data obtained with zero mean stress ($\sigma_m = 0$). However, in many design situations, there will be a tensile mean stress ($\sigma_m > 0$). Intuitively, we would expect that the component would be more prone to fatigue if it were subjected to a large mean stress in addition to having to cope with repeated cycles of stress. The test data confirm this—to keep the fatigue life the same when the mean stress is increased from 0 to some large tensile value, the stress (or strain) cycles must be reduced in amplitude to compensate.

In terms of the strain approach to fatigue, the test data can be approximated as follows:

$$\frac{\Delta\varepsilon^{tot}}{2} \approx \frac{(\sigma_f' - \sigma_m)(2N_f)^b}{E} + \varepsilon_f'(2N_f)^c \tag{17.3}$$

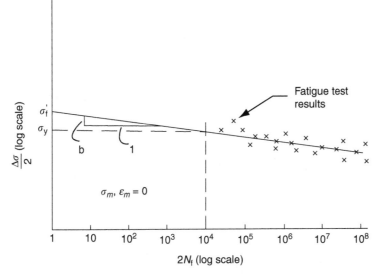

**FIGURE 17.5**
Relation between stress amplitude and fatigue life in high-cycle fatigue.

The equation clearly shows that, to keep the fatigue life the same, the strain amplitude must be decreased to compensate for a mean stress.

In terms of the stress approach to fatigue, the test data can be approximated as:

$$\frac{\Delta\sigma_{\sigma m}}{2} \approx \frac{\Delta\sigma\left(\sigma_f' - \sigma_m\right)}{2\sigma_f'} \tag{17.4}$$

$\Delta\sigma/2$ is the stress amplitude for failure after a given number of cycles with zero mean stress, and $\Delta\sigma_{\sigma m}/2$ is the stress amplitude for failure after the same number of cycles but with a mean stress. So if, for example, the mean stress is half the fracture stress, then the applied stress amplitude must be halved to keep the fatigue life the same.

## 17.3 FATIGUE OF CRACKED COMPONENTS

Large structures—particularly welded structures such as bridges, ships, oil rigs, and nuclear pressure vessels—always contain cracks. All we can be sure of is that the initial length of these cracks is less than a given length—the length we can reasonably detect when we check or examine the structure. To assess the safe life of the structure we need to know how long (for how many cycles) the structure can last before one of these cracks grows to a length at which it propagates catastrophically.

Data on fatigue crack propagation are gathered by cyclically loading specimens containing a sharp crack like that shown in Figure 17.6. We define

$$\Delta K = K_{max} - K_{min} = Y\Delta\sigma\sqrt{\pi a}$$

The cyclic stress intensity $\Delta K$ increases with time (at constant load) because the crack grows in tension. It is found that the crack growth per cycle, $da/dN$, increases with $\Delta K$ in the way shown in Figure 17.7.

In the steady-state régime, the crack-growth rate is described by

$$\frac{da}{dN} = A(\Delta K)^m \tag{17.5}$$

where $A$ and $m$ are material constants. Obviously, if $a_0$ (the initial crack length) is given, and the final crack length ($a_f$) at which the crack becomes unstable and runs rapidly is known or can be calculated, then the safe number of cycles can be estimated by integrating the equation

$$N_f = \int_0^{N_f} dN = \int_{a_0}^{a_f} \frac{da}{A(\Delta K)^m} \tag{17.6}$$

remembering that $\Delta K = Y\Delta\sigma\sqrt{\pi a}$. Case Study 3 of Chapter 19 gives a worked example of this method of estimating fatigue life.

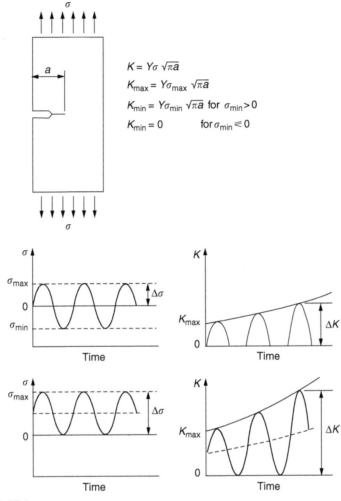

**FIGURE 17.6**
Fatigue-crack growth in precracked components.

## 17.4 FATIGUE MECHANISMS

Cracks grow in the way shown in Figure 17.8. In a pure metal or polymer (left diagram), the tensile stress produces a plastic zone (Chapter 14) which makes the crack tip stretch open by the amount $\delta$, creating new surface there. As the stress is removed the crack closes and the new surface folds forward extending the crack (roughly by $\delta$). On the next cycle the same thing happens again, and the crack inches forward, roughly at $da/dN \approx \delta$. Note that the crack cannot grow when the stress is compressive because the crack faces come into contact and carry the load (crack closure).

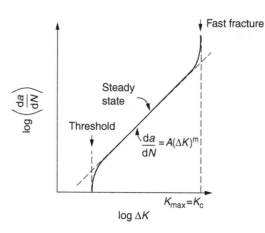

**FIGURE 17.7**
Fatigue crack-growth rates for precracked material.

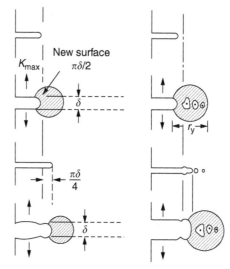

**FIGURE 17.8**
How fatigue cracks grow.

We mentioned in Chapter 14 that real engineering alloys always have little inclusions in them. Then (see right diagram of Figure 17.8), within the plastic zone, holes form and link with each other, and with the crack tip. The crack now advances a little faster than before, aided by the holes.

In *precracked structures* these processes determine the fatigue life. In uncracked components subject to *low-cycle fatigue*, the general plasticity quickly roughens the surface, and a crack forms there, propagating first along a slip path ("Stage 1" crack) and then, by the mechanism we have described, normal to the tensile axis (Figure 17.9).

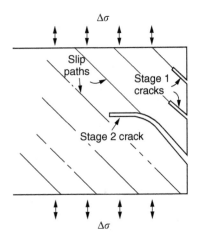

**FIGURE 17.9**
How cracks form in low-cycle fatigue. Once formed, they grow as shown in Figures 17.10 through 12.

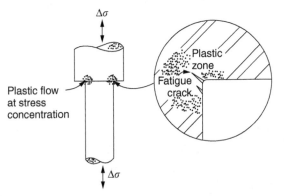

**FIGURE 17.10**
How cracks form in high-cycle fatigue.

High-cycle fatigue is different. When the stress is below general yield, almost all of the life is taken up in initiating a crack. Although there is no *general* plasticity, there is *local* plasticity wherever a notch or scratch or change of section concentrates stress. A crack ultimately initiates in the zone of one of these stress concentrations (Figure 17.10) and propagates, slowly at first, and then faster, until the component fails. For this reason, sudden changes of section or scratches are very dangerous in high-cycle fatigue, often reducing the fatigue life by orders of magnitude.

Figure 17.11 shows the crack surface of a steel plate (40 mm thick) which failed by fatigue. The shape of the fatigue crack at any given time is indicated by "beach marks". Figure 17.12 shows a fatigue crack in the deck of a ship which had started near to an access hole. The hole creates a stress concentration.

**FIGURE 17.11**
Fatigue crack surface in a steel plate. The arrows show the direction of crack growth.

**FIGURE 17.12**
Fatigue crack in the deck of a ship.

# WORKED EXAMPLE

You have just been told that some copper water-cooling plates have begun failing in a furnace. The suspected cause is low-cycle fatigue, caused by thermal expansion movements between the cooling plates and the furnace shell. To make a preliminary assessment, you urgently need the strain–life plot for annealed deoxidized copper—the material of the plates. But where do you start?

Remember that Figure 17.4 shows the typical form of the strain–life plot. Looking at the left side of the plot, you can see that it should at least be possible to find standard tensile testing data for copper, which would give you values for the true fracture stress and strain. There are plenty of handbooks and/or websites that can give you tensile test data, but of course this is always listed as nominal stress and strain, so needs to be converted into true stress and strain. This is OK, because you already know about the equations that do this conversion. You also need a value for Young's modulus, but again that is easily available (and will be the same whatever the grade of copper). You easily locate the necessary data for annealed copper: $\sigma_{TS} = 216$ MN m$^{-2}$, $\varepsilon_f = 48\% \rightarrow 0.48$, $E = 130$ GN m$^{-2}$. Then:

$$\varepsilon_f' = \ln(1 + \varepsilon_f) = \ln(1 + 0.48) = \ln(1.48) = 0.39$$

$$\sigma_f' = \sigma_{TS}(1 + \varepsilon_f) = 216(1 + 0.48) = 320 \text{ MN m}^{-2}, \quad \frac{\sigma_f'}{E} = \frac{320 \text{ MN m}^{-2}}{130 \text{ GN m}^{-2}}$$

$$= 0.0025$$

So you already have two critical points to put on your plot—we have marked them on the diagram.

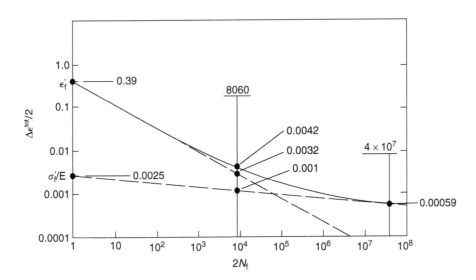

We could now do with something at the right side of the plot, to define the elastic fatigue line. There are lots of data available for HCF—in fact, much more than there are for LCF. But almost always, handbooks and/or websites list just one value—the stress amplitude for failure after typically $10^8$ cycles. In the case of annealed copper, a quick handbook search comes up with a stress of $\pm76$ MN m$^{-2}$ after $2 \times 10^7$ cycles ($4 \times 10^7$) reversals. The strain amplitude corresponding to the stress amplitude of 76 MN m$^{-2}$ is:

$$\frac{\Delta\varepsilon^{el}}{2} = \frac{\Delta\sigma}{2E} = \frac{76 \text{ MN m}^{-2}}{130 \text{ GN m}^{-2}} = 0.00059$$

Marking this point on the diagram fixes the elastic fatigue line.

Finally, we need some points in the middle of the plot, to define the plastic line. Then we can just add the plastic line and the elastic line to get the overall strain–life plot (but remember that the vertical axis is not a linear scale!). Fortunately, a quick trawl through some standard textbooks comes up with a drawing of the cyclic stress–strain loop for annealed copper under a fixed strain range of 0.0084 (Hertzberg). The stabilized stress range is 252 MN m$^{-2}$, and the number of reversals to failure is 8060. The total strain amplitude is then 0.0042. The elastic part of the total strain amplitude is:

$$\frac{\Delta\varepsilon^{el}}{2} = \frac{\Delta\sigma}{2E} = \frac{252 \text{ MN m}^{-2}}{2 \times 130 \text{ GN m}^{-2}} = 0.001$$

The plastic part of the total strain amplitude is then $0.0042 - 0.001 = 0.0032$. The three strain amplitudes are marked on the preceding diagram, and they allow the overall strain–life plot to be drawn. This is sufficient for your preliminary assessment. However, predictive design should always rely on actual fatigue test data—and plenty of it, so that the variability of the data (mean and standard deviation) can be established.

## EXAMPLES

17.1 The copper water-cooling plates in the Worked Example failed at welded joints which connected the plates to the inlet and outlet water pipes. During thermal cycling of the furnace, the pipes were bent back and forth, which resulted in cyclic straining of the welds. The strain amplitude was estimated to have been 0.01. Using the strain–life plot in the diagram, estimate the number of cycles required to cause the welded joints to fail. Your estimate of the strain amplitude could be in error by a factor of 2. What would be the failure life if the strain amplitude were as large as 0.02?

**Answers**

Approximately 1000 and 300 reversals (500 and 150 cycles).

**17.2** An electric iron for smoothing clothes was being used when there was a loud bang and a flash of flame from the electric flex next to the iron. An inspection showed that the failure had occurred at the point where the flex entered a polymer tube which projected about 70 mm from the body of the iron. There was a break in the live wire and the ends of the wire showed signs of fusion. The fuse in the electric plug was intact. Explain the failure. Relevant data are given in the table that follows.

| | |
|---|---|
| Rating of appliance | 1.2 kW |
| Power supply | 250 V, 50 Hz AC |
| Fuse rating of plug | 13 A |
| Flex | 3 conductors (live, neutral, and earth) each rated 13 A |
| Individual conductor | 23 strands of copper wire in a polymer sheath; each strand 0.18 mm in diameter |
| Age of iron | 14 years |
| Estimated number of movements of iron | $10^6$ |

**17.3** A large component is made of a steel for which $K_c = 54$ MN m$^{-3/2}$. Nondestructive testing by ultrasonic methods shows that the component contains long surface cracks up to $a = 2$ mm deep. Laboratory tests show that the crack-growth rate under cyclic loading is given by

$$\frac{da}{dN} = A(\Delta K)^3$$

where $A = 6 \times 10^{-12}$ m (MN m$^{-3/2}$)$^{-3}$. The component is subjected to an alternating stress of range

$$\Delta \sigma = 180 \text{ MN m}^{-2}$$

about a mean tensile stress of $\Delta\sigma/2$. Given that $\Delta K = Y\Delta\sigma\sqrt{\pi a}$, estimate the number of cycles to failure.

**Answer**

$1.2 \times 10^5$ cycles

**17.4** The preceding photograph shows the fracture surfaces of two broken tools from a pneumatic drill. The circular fracture surface is 35 mm in diameter, and the "rectangular" fracture surface measures 24 mm × 39 mm. The shape of the fatigue crack at any given time is indicated by the beach marks, which are clearly visible on the fatigue part of the fracture surface. Indicate the following features on each fracture surface:

**a**   The point where the fatigue crack initiated

**b**   The position of the crack just before the final fast fracture event

The fatigue crack traverses much more of the cross section in the circular tool than in the rectangular tool. What does this tell you about the maximum stress in the fatigue cycle?

**17.5** An aluminum alloy for an airframe component was tested in the laboratory under an applied stress which varied sinusoidally with time about a mean stress of zero. The alloy failed under a stress range, $\Delta\sigma$, of 280 MN m$^{-2}$ after $10^5$ cycles; under a range of 200 MN m$^{-2}$, the alloy failed after $10^7$ cycles. The fatigue behavior of the alloy can be represented by

$$\Delta\sigma = C(N_f)^b$$

where $b$ and $C$ are constants. Find the number of cycles to failure, $N_f$, for a component subjected to a stress range of 150 MN m$^{-2}$.

**Answer**

$5.2 \times 10^8$ cycles

**17.6** The following photograph shows a small part of a fatigue fracture surface taken in the scanning electron microscope (SEM). The material is stainless steel. The position of the crack at the end of each stress cycle is indicated by parallel lines, or "striations." By taking measurements from the photograph, estimate the striation spacing.

**Answer**

0.5 μm

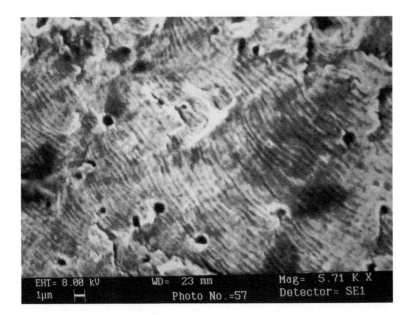

**17.7** After 3 years in service, a steel boiler started to leak steam from a position below the water line. The leak was traced to a fine crack 20 mm long in the wall of the boiler barrel alongside the entry position of the water feedpipe. An internal inspection showed severe cracking of the boiler plate. It was concluded that the cracks had initiated at the inner surface of the barrel and had then propagated through the wall. A baffle had been fitted in the water space a short distance in from the feed pipe in order to deflect the cold feed water sideways along the boiler barrel. Indicate the likely mechanism of failure and support it with a simple calculation. Supporting data is as follows: water temperature of boiler = 180°C; feed water temperature = 10°C; coefficient of thermal expansion of steel = 12 × $10^{-6}$ °C$^{-1}$; yield strength of boiler plate = 250 MN m$^{-2}$.

# Fatigue Design

## CONTENTS

## 18.1 INTRODUCTION

In this chapter we look at a number of aspects of fatigue that are relevant to designing structures or components against fatigue failure in service. We give data for the fatigue strengths of metals and alloys (useful for designing mechanical components) and for welded joints (important in large structures such as bridges and oil rigs).

We look at the problems of stress concentrations produced by abrupt changes in cross-section (e.g., shoulders or holes). We see how fatigue strength can be improved by better surface finish, better component geometry, and compressive residual surface stress. In addition, we look at how the preloading of bolts is essential in bolted connections such as car engine big-end bearings.

Engineering Materials I: An Introduction to Properties, Applications, and Design, Fourth Edition

## 18.2 FATIGUE DATA FOR UNCRACKED COMPONENTS

Table 18.1 gives high-cycle fatigue data for uncracked specimens tested about zero mean stress. The data are for specimens with an excellent surface finish tested in clean dry air. Fatigue strengths can be considerably less than these if the surface finish is poor, or if the environment is corrosive.

Obviously, if we have a real component with an excellent surface finish in clean dry air, then if it is to survive $10^8$ cycles of constant-amplitude fatigue loading about zero mean stress, the stress amplitude $\Delta\sigma/2$ in service must be less than that given in Table 18.1 by a suitable safety factor. If the mean stress is not zero, then Equation (17.4) can be used to calculate the fatigue strength under conditions of non-zero mean stress. In the absence of specific data, it is useful to know that $\Delta\sigma/2 = C\sigma_{TS}$. The value of the constant $C$ is typically 0.3 to 0.5 depending on the material.

## 18.3 STRESS CONCENTRATIONS

Any abrupt change in the cross section of a loaded component causes the local stress to increase above that of the background stress. The ratio of the maximum local stress to the background stress is called the stress concentration factor, or SCF for short. Figure 18.1 gives details of the SCF for two common changes in section—a hole in an axially loaded plate and a shouldered shaft in bending.

The hole gives an SCF of 3. The SCF of the shaft is critically dependent on the ratio of the fillet radius $r$ to the minor shaft diameter $d$—to minimize the SCF,

| Table 18.1 Approximate Fatigue Strengths of Metals and Alloys | |
| --- | --- |
| **Metal or Alloy** | **Stress Amplitude $\Delta\sigma/2$ for Failure after $10^8$ Cycles (Zero Mean Stress) (MN m$^{-2}$)** |
| Aluminum | 35–60 |
| Aluminum alloys | 50–170 |
| Copper | 60–120 |
| Copper alloys | 100–300 |
| Magnesium alloys | 50–100 |
| Nickel | 230–340 |
| Nickel alloys | 230–620 |
| Steels | 170–500 |
| Titanium | 180–250 |
| Titanium alloys | 250–600 |

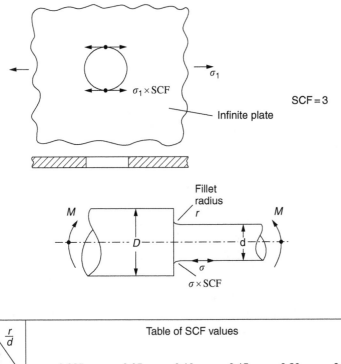

**FIGURE 18.1**
Typical stress concentration factors.

| $\frac{r}{d}$ | Table of SCF values | | | | | |
|---|---|---|---|---|---|---|
| $\frac{D}{d}$ | 0.025 | 0.05 | 0.10 | 0.15 | 0.20 | 0.30 |
| 1.02 | 1.90 | 1.64 | 1.43 | 1.34 | 1.24 | 1.20 |
| 1.05 | 2.13 | 1.79 | 1.54 | 1.40 | 1.31 | 1.23 |
| 1.10 | 2.25 | 1.86 | 1.59 | 1.43 | 1.37 | 1.26 |
| 1.50 | 2.59 | 2.06 | 1.67 | 1.50 | 1.40 | 1.29 |
| 3.00 | 2.85 | 2.30 | 1.80 | 1.58 | 1.43 | 1.32 |

$r/d$ should be maximized. Obviously, fatigue failure will occur preferentially at sites of local stress concentration. If a component has an SCF then it is the maximum local stress which must be kept below the material fatigue strength, and not the background stress.

## 18.4 THE NOTCH SENSITIVITY FACTOR

Taking the shouldered shaft as an example, we can see that as the $r/d$ ratio decreases toward zero (a sharp corner) the SCF should increase toward infinity. This implies that any component with a sharp corner, or notch, will always fail by fatigue no matter how low the background stress! Clearly, this is not correct, because there are many components with sharp corners which are used success-fully in fatigue loading (although this is very bad practice).

In fatigue terminology, we define an effective stress concentration factor, $SCF_{eff}$ such that $SCF_{eff} < SCF$. The two are related by the equation

$$SCF_{eff} = S(SCF - 1) + 1 \qquad (18.1)$$

where $S$, the notch sensitivity factor, lies between 0 and 1. If the material is fully notch sensitive, $S = 1$ and $SCF_{eff} = SCF$. If the material is not notch sensitive, $S = 0$ and $SCF_{eff} = 1$.

Figure 18.2 shows that $S$ increases with increasing $\sigma_{TS}$ and fillet radius $r$. We would expect $S$ to increase with $\sigma_{TS}$. As we saw in Chapter 14 for sharp cracks, material at the fillet radius can yield in response to the local stress, and this will limit the maximum local stress to the yield stress. In general, increasing $\sigma_{TS}$ increases $\sigma_y$. In turn, this increases the maximum local stress which can be sustained before yielding limits the stress, and helps keep $SCF_{eff} \approx SCF$.

The decrease in $S$ with decreasing $r$ has a different origin. As Figure 18.2 shows, as $r$ tends toward zero, $S$ also tends toward zero for all values of $\sigma_{TS}$. This is be-cause a sharp notch produces a small process zone (the zone in which the fatigue crack initiates) and this makes it harder for a fatigue crack to grow. We saw

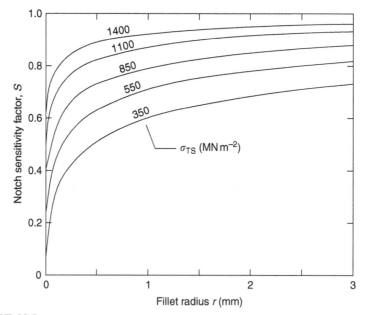

**FIGURE 18.2**
Effect of tensile strength and fillet radius on notch sensitivity factor.

in Chapter 16 that the tensile strength of a brittle component increases as the volume decreases. This size effect also applies to the formation of fatigue cracks—the smaller the process zone, the larger the fatigue strength of the component.

The notch sensitivity curves in Figure 18.2 have an interesting implication for designing components with small fillet radii. One would think that increasing the tensile strength (and hence the fatigue strength) of the material would increase the fatigue strength of the component. However, this is largely offset by the increase in notch sensitivity, which increases the value of the effective SCF by Equation (18.1). Fortunately, as we shall see later, there are other ways of increasing the fatigue strength of notched components.

## 18.5 FATIGUE DATA FOR WELDED JOINTS

Welding is the preferred method for joining structural steels and aluminum alloys in many applications. The world is awash with welds subjected to fatigue loading—bridges, oil rigs, ships, boats, chemical plants, and so on. Because welded joints are so important (and because they have some special features) there is a large amount of data in constructional standards for weld fatigue strength.

Figure 18.3 shows how the various types of welded joints can be categorized into standard weld classes. Figure 18.4 gives the fatigue strengths of the classes for structural steel. The 97.7% survival lines are used for design purposes, and the 50% lines for analyzing welds which actually failed. It is important to note that the vertical axis of the fatigue lines in Figure 18.4 is the full stress range $\Delta\sigma$, and not the $\Delta\sigma/2$ conventionally used for high-cycle fatigue data (see Table 18.1). From that table, we can see that $\Delta\sigma$ for steel ($10^8$ cycles) is at least $2 \times 170 = 340$ MN m$^{-2}$. The $\Delta\sigma$ for a class G weld ($10^8$ cycles) is only 20 MN m$^{-2}$. This huge difference is due mainly to three special features of the weld—the large SCF, the rough surface finish, and the presence of small crack-like defects produced by the welding process.

It is important to note that the fatigue strength of welds does not depend on the value of the mean stress in the fatigue cycle. Equation 17.4 should not be used for welds. This makes life much easier for the designer—the data in Figure 18.4 work for any mean stress, and the input required is simply the stress range. This major difference from conventional fatigue data is again due to a special feature of welded joints. Welds contain tensile residual stresses that are usually equal to the yield stress (these residual stresses are produced when the weld cools and contracts after the weld bead has been deposited). Whatever the applied stress cycle, the actual stress cycle in the weld itself always has a maximum stress of $\sigma_y$ and a minimum stress of $\sigma_y - \Delta\sigma$.

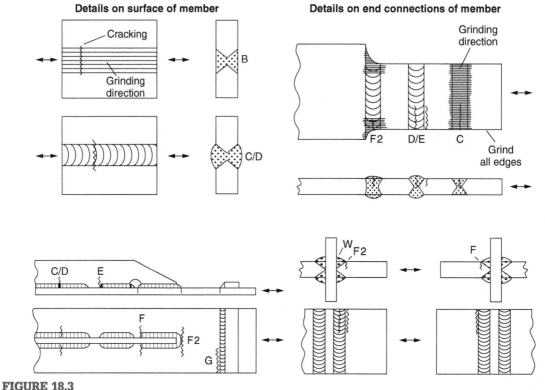

**FIGURE 18.3**
Standard weld classes.

## 18.6 FATIGUE IMPROVEMENT TECHNIQUES

We have already seen that the fatigue strength of a component can be increased by minimizing stress concentration factors, and having a good surface finish (a rough surface is, after all, just a collection of small stress concentrations). However, it is not always possible to remove SCFs completely. A good example is a screw thread, or the junction between the shank and the head of a bolt, which cannot be removed without destroying the functionality of the component.

The answer here is to introduce a residual compressive stress into the region of potential crack initiation. This can be done using thread rolling (for screw threads), roller peening (for fillet radii on bolts or shafts), hole expansion (for pre-drilled holes), and shot peening (for relatively flat surfaces). The compressive stress makes it more difficult for fatigue cracks to grow away from the initiation sites in the surface.

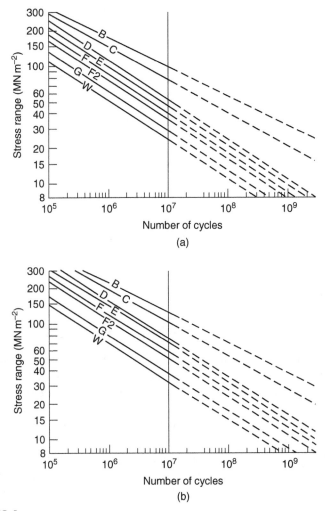

**FIGURE 18.4**

Fatigue strengths of the standard weld classes for structural steel. (a) Curves for 97.7% survival; (b) Curves for 50% survival.

Figure 18.5 shows how the fatigue strengths of welds can be improved. The first step is to improve the class of weld, if this is possible. By having a full penetration weld, the very poor class W weld is eliminated, and the class of the connection is raised to class F. Further improvements are possible by grinding the weld bead to improve surface finish, reduce SCFs, and remove welding defects. Finally, shot peening can be used to put the surface into residual compression.

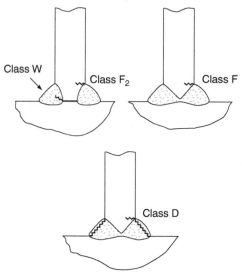

**FIGURE 18.5**
Improving the fatigue strength of a typical welded connection.

## 18.7 DESIGNING OUT FATIGUE CYCLES

In some applications, the fatigue strength of the component cannot easily be made large enough to avoid failure under the applied loading. But there may be design-based solutions, which involve reducing or even eliminating the stress range that the loading cycle produces in the component. A good example is the design of bolted connections in the bearing housings of automotive crankshafts and con-rod big ends. Looking at Figure 18.6, it is easy to see that if the bolts are left slightly slack on assembly, the whole of the applied loading is taken by the two bolts (there is nothing else to take a tensile load). The load in each bolt therefore cycles from 0 to $P$ to 0 with each cycle of applied loading.

The situation is quite different if the bolts are torqued-up to produce a large tension (or preload) in the bolts at assembly. The situation can be modeled very clearly as shown in Figure 18.7. Here, the bolt is represented by a rubber band, assembled with a tension $T$. There is an equal and opposite compressive force at the interface between the two halves of the housing. As the connection is loaded, nothing obvious happens, until the applied load $P$ reaches $T$. Then, the two halves of the housing begin to separate. Provided $P < T$, the bolt sees no variation in stress at all.

The variations in the applied load are provided by variations in the compressive force at the interface between the two halves of the housing. Of course, in a real connection, the bolt does not behave in such a springy way as a rubber band. But provided the bolts are long and have a small diameter, it is possible to make

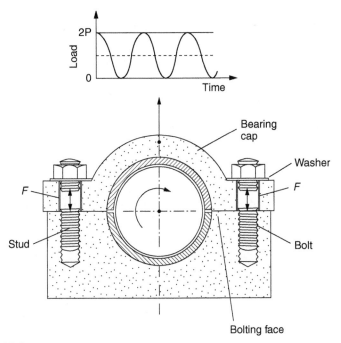

**FIGURE 18.6**
Typical bearing housing, with studs or bolts used to secure the bearing cap.

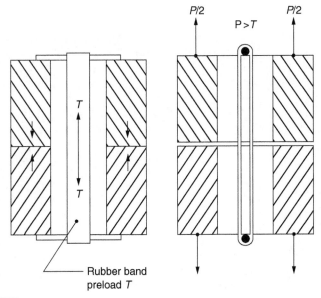

**FIGURE 18.7**
Model of a bolted connection, with assembly preload $T$.

them springy enough that they are shielded from most of the variations in the applied loading cycle. In critical cases, the bolts may be "waisted"—except where the bolt must be threaded to take the nut, it is machined to a smooth diameter which is less than the core diameter of the thread. This increases the "give" of the bolt, and also removes unnecessary threads, which are a potential source of fatigue cracking. Incidentally, this is one reason why bolts for fatigue critical applications are made from high tensile steel—they must not break in tension under the high preload even if they have a small diameter.

## WORKED EXAMPLE

The photographs that follow show traffic lights in Manhattan (Upper West Side, Broadway at W 80th and W 77th). These are everyday sights in New York City and are ignored by all the pedestrians. But we are different. We are engineers, and we take delight in seeing interesting things in even the most ordinary surroundings—things that non-engineers are completely unaware of.

Look at the lamp unit hanging from its mast. How is it attached to the mast? There is a double clevis arrangement, with bolts to allow the clevises to rotate. This can be seen in the zoom view, but is quite obvious from street level. Why bother to do this? Surely it would be cheaper and quicker just to weld the top of the lamp unit straight on to the end of the mast. But what is the loading? Self weight, yes—but variable loadings?

40 47 04.22 N 73 58 45.56 W

40 46 56.85 N 73 58 50.45 W

Of course, wind loadings—remember just how fast the wind can whistle down the avenues between the high-rise buildings. If it is windy today, maybe you can see the lamp units swinging below their masts. If you let them swing, you will *design out* the fatigue stresses! Then, there is no need to determine the fatigue loadings (expensive), or choose a class of weld with a large enough fatigue strength (many engineers don't know how to do this), or make sure that the weld is well made every time (not easy with low-tech fabrications). So, by designing out the fatigue problem, you save a lot of money—and you avoid a significant chance that, sooner or later, a lamp unit will fall off and kill someone.

And you also make it easy to remove the lamp unit for servicing. Is that all? No—what about wear on the bolts? You need to check this periodically on typical units, and if necessary put in place a bolt renewal protocol. In fact, all structures should be inspected from time to time in case faults have developed that the designer did not envisage.

## EXAMPLES

**18.1** Indicate briefly how the following affect fatigue life:
    **a.** A good surface finish
    **b.** The presence of a rivet hole
    **c.** A significant mean tensile stress
    **d.** A corrosive atmosphere

**18.2** The following diagram shows the crank-pin end of the connecting rod on a large-scale miniature steam locomotive. The locomotive weighs about 900 kg and is designed for hauling passengers around a country park. In the full-size prototype the connecting rod and big-end were forged from a single billet of steel. However, to save cost in building the miniature version, it is intended to weld the two parts together with a full-penetration double-sided weld, grinding the surface flush to hide the joint. Do you think that this design solution will have the required fatigue properties? Design and operational data are given in the following table.

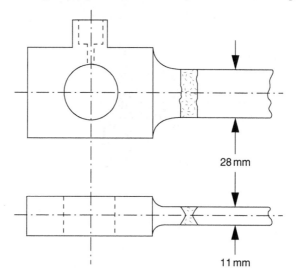

28 mm

11 mm

| Diameter of double-acting cylinder | 90 mm |
| Diameter of driving wheel | 235 mm |
| Steam pressure inside cylinder at point of admission | 7-bar gauge |
| Estimated annual distance traveled | 6000 km |
| Design life | 20 years minimum |

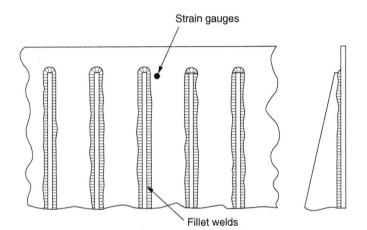

Strain gauges

Fillet welds

**18.3** Vibrating screens are widely used in the mining industry for sizing, feeding, and washing crushed mineral particles. A typical screen consists of a box fabricated from structural steel plate which contains a mesh screen. During operation the box is shaken backward and forward at a frequency of up to 20 Hz. Although the major parts of the box are often fixed together using bolts or rivets, individual subassemblies such as a side of the box frequently have welded joints, particularly where frame stiffeners or gussets are added. Owing to the inertial forces generated by the rapid shaking, the stresses in the unit can be significant and the fatigue design of the welded connections has to be considered carefully.

The side of one box consists of a relatively thin plate which is stiffened with triangular gussets, as shown in the preceding diagram. Strain gauges attached to the plate during operation show that the maximum principal stress range in the plate near end of the gusset is 8 MN m$^{-2}$.

  **a.** Given that the screen is expected to work for 12 hours per day and 6 days per week, estimate the time that it will take for there to be a 50% chance of a crack forming in the plate at the end of each gusset.

  **b.** What would be the time for a 2.3% chance of cracking?

**Answers**

a. 11 years; b. 4 years

**18.4** The diagrams that follow show the mechanical linkage in a pipe organ. The organist presses down on the pedal with his or her foot, which in turn pulls the vertical tracker down, rotates the roller, and pulls the horizontal tracker sideways. While the mechanism is being actuated, the tension in the trackers rises from zero to approximately 1 kgf. The roller arms were made from aluminum alloy rod 4.75 mm in diameter. The end of the rod was turned down to a diameter of 4 mm to fit the hole in the roller, and was riveted over to hold it in place. The turning had been done with a sharp-ended lathe tool, which produced a sharp corner. A fatigue crack had initiated at this sharp corner, and led to the failure of the mechanism after only 2 years in service.

    **a.** Why did the fatigue crack form at the sharp corner?

    **b.** Where around the sharp corner did the fatigue crack initiate?

    **c.** How would you modify the design to increase the fatigue life?

    **d.** What other material would you choose to increase the fatigue life?

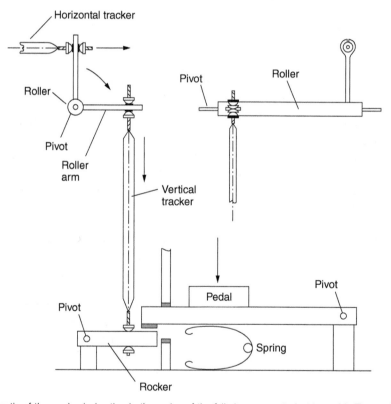

Schematic of the mechanical action in the region of the failed components (not to scale). The vertical tracker and the rocker are both made from light wood, and their weight can be neglected in the loading calculations

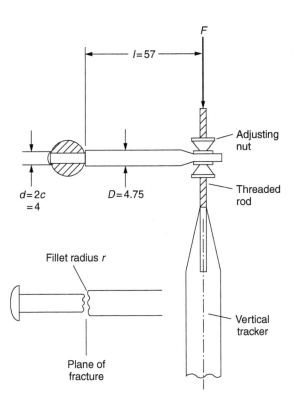

Fatigue fractures occurred at a sharp change of section in the roller arms. To scale—dimensions in mm.

The maximum bending stress in the reduced section of the roller arm is given by the standard equation

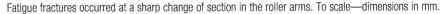

$$\sigma_{max} = \frac{4Fl}{\pi c^3}$$

Using this equation, calculate (a) the stress amplitude and (b) the mean stress of the stress cycle. Given that the true fracture stress of the alloy is 250 MN m$^{-2}$, estimate (c) the stress amplitude that would produce the same fatigue life under conditions of zero mean stress.

**Answers**

(a) 45 MN m$^{-2}$; (b) 45 MN m$^{-2}$ (tensile); (c) 55 MN m$^{-2}$

**18.5** Referring to Figures 18.3 and 18.4, the fatigue stress range for welded joints is the stress range in the load carrying member, except for class W, where it is the stress range in the weld metal, defined as $\Delta F/2gL$.

For a "T" joint, as shown in the diagram, calculate the ratio of plate thickness to leg length, $t/a$, which produces the same stress range in both plate and weld throat.

Estimate the $t/a$ ratio that allows both class W and class F2 cracking to occur at the same time.

**Answers**

√2; 0.92

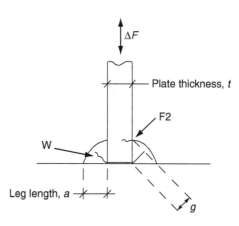

Applied load range = $\Delta F$
Effective throat dimension = $g$
Length of weld runs = $L$

**18.6** Cambridge (England) is flat, peak traffic can cause gridlock, and public transport is inadequate. Many people find a bicycle is the best mode of transport. The following photographs show the bicycle that I (DRHJ) use to travel to Christ's College most days. It is 30 years old and has experienced a large number of fatigue loading cycles. Two years ago, the bicycle suddenly started to wobble, and I had to stop in a hurry. One of the steel tubes in the main frame had snapped right across. The photograph also shows how I fixed the bicycle with stuff from my junk box. Initially, I just wrapped a split tube of softened copper around the frame tube to bridge across the break, and compressed it onto the frame tube with screw clips. However, the ends of the frame tube soon began to pull out of the sleeve, verifying that the fatigue loading had a large tensile component. I then added the steel stay, which cured the problem.

The fatigue crack had started from a 3 mm air hole, which had been drilled through the tube wall at original manufacture. Why did the fatigue crack initiate at the hole? Where around the circumference of the hole would the fatigue crack have initiated? Estimate the margin of safety (in terms of loading) against fatigue cracking if the hole had not been there at all.

**18.7** The following diagram shows the output pipe from a reciprocating slurry pump. With each pump stroke, the pressure in the pipe cycles from zero to a large positive pressure and back to zero. The pipe goes through a right-angled bend, which means that the flanged connection "sees" the full pressure loading. How would you specify and fit the bolts to minimize the risk that they will fail by fatigue? In spite of your best efforts, bolts start failing by fatigue. What simple design modification could you try in order to increase the fatigue life? (Ideally, your modification should not involve dismantling the flanged joint, so the pump can be kept operating throughout the repairs).

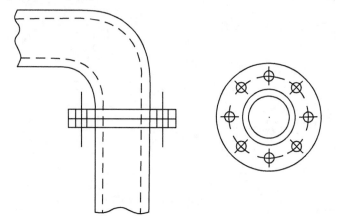

18.8 The next diagrams show a pulley block used to deploy an underwater platform from an offshore exploration rig. There are two modes of use: (a) when lowering or raising the platform off the side of the rig, (b) when slewing the platform over the deck of the rig. After about 200 deployments, the bolt broke and the platform was lost overboard. Where do you think the fracture occurred? Would you categorize the fracture as low-cycle or high-cycle fatigue? Support your answer by an approximate calculation. The bending moment at which a circular cross-section starts to yield is given by the standard equation

$$M_{\text{elastic}} = \frac{\pi \sigma_y c^3}{4}$$

The steel of the bolt had a yield strength of 540 MN m$^{-2}$. How would you modify the design to ensure that the connection between the lifting eye and the pulley block is less likely to fail by fatigue?

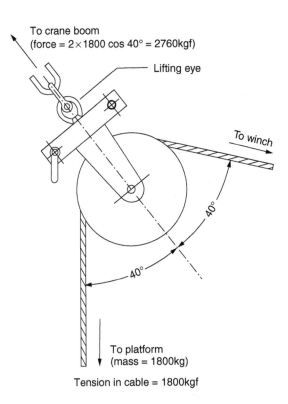

To crane boom
(force = 2×1800 cos 40° = 2760kgf)

Lifting eye

To winch

40°

40°

To platform
(mass = 1800kg)

Tension in cable = 1800kgf

(a) Schematic of pulley block when lowering or raising. The crane exerts a straight pull along the axis of the lifting eye.

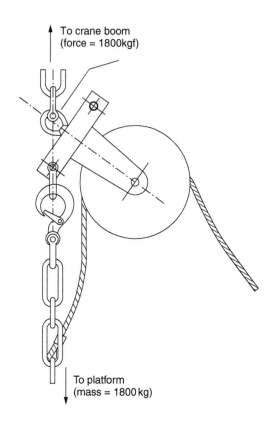

To crane boom
(force = 1800kgf)

To platform
(mass = 1800 kg)

(b) Schematic of pulley block when slewing. The crane now exerts an oblique pull on the lifting eye.

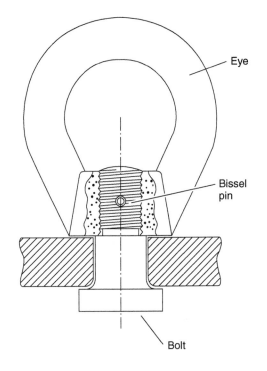

Eye

Bissel pin

Bolt

(c) Details of lifting eye.

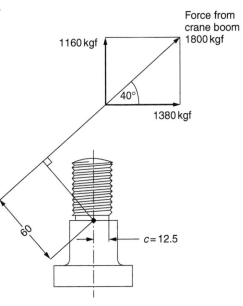

Force from crane boom 1800 kgf

1160 kgf

1380 kgf

40°

60

c = 12.5

(d) Forces acting on the bolt when slewing.

## CHAPTER 19

# Case Studies in Fatigue Failure

## CONTENTS

## 19.1 CASE STUDY 1: THE COMET AIR DISASTERS

Figure 19.1 is a drawing of the Comet 1 aircraft. This aircraft was designed and built by de Havilland, and put into service with BOAC in May 1952. It was powered by four Ghost turbo-jet engines, and had a cruising altitude of 35,000 ft. It was intended for long-distance high-speed flights on passenger and freight routes with a maximum all-up weight of 49 tons. By comparison, existing civil aircraft used turbo-prop engines and had an altitude ceiling of about 17,000 ft, so the Comet was the most advanced civil aircraft of its time.

Unfortunately, a disturbing crash occurred on 10 January 1954. Comet G-ALYP left Rome airport at 0931 bound for London. The crew kept in contact with control and at 0950 reported that they were over Orbetello. The flight plan indicated that the plane should have climbed to 26,000 ft. A message from the Comet was broken off at 0951 in mid-sentence. At the same time, eyewitnesses on Elba saw debris fall into the sea to the south of the island. None of the 29 passengers and six crew survived.

On 8 April 1954, Comet G-ALYY left Rome at 1832 bound for Cairo. At 1857 the crew reported that they were alongside Naples, and were approaching the cruising altitude of 35,000 ft. At 1905 the crew radioed Cairo to give their

Engineering Materials I: An Introduction to Properties, Applications, and Design, Fourth Edition

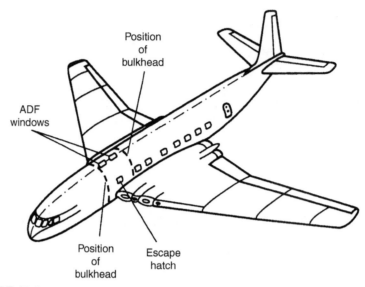

**FIGURE 19.1**
The Comet I aircraft.

estimated arrival time. This was the last transmission from the aircraft, and neither Rome nor Cairo was able to make contact again. Some light wreckage was recovered from the sea the next day. There was now mounting evidence that something was wrong with the structural integrity of the aircraft, so BOAC grounded the whole fleet.

The first priority was to recover as much of the wreckage as possible. Unfortunately, G-ALYY had gone down in water 1000 m deep, and there was no realistic prospect of recovering any wreckage from such depths with the undersea technology of the time. G-ALYP had sunk in water 180 m deep, and recovery was feasible if difficult. By August 1954 the recovery team had raised 70% of the structure, 80% of the engines, and 50% of the equipment.

The wreckage of the fuselage was reconstructed and the failure was traced to a crack that had started near the corner of a window in the cabin roof. The window was one of a pair that were positioned one behind the other just aft of the leading wing spar. The windows housed the antennae for the automatic direction finding (ADF) system. As Figure 19.2 shows, the crack had started near the rear starboard corner of the rear ADF window, and had then run backward along the top of the cabin parallel to the axis of the fuselage. Circumferential cracks had developed from the main longitudinal crack and as a result whole areas of the skin had peeled away from the structure.

A second longitudinal crack then formed at the forward port corner of the rear window, and ran into the forward ADF window. Finally, two more cracks

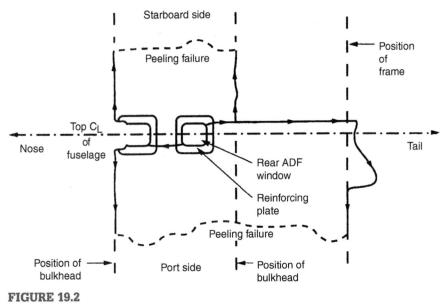

**FIGURE 19.2**
The failure on Comet G-ALYP.

nucleated at the forward corners of the forward window and developed into a pair of circumferential fractures. As one might expect, the circumferential cracks all followed the line of a transverse frame member. Figure 19.3 is a close-up of the rear starboard corner of the rear window showing where the first crack probably originated. Fatigue markings were found on the fracture surface at this location: they had probably started at the edge of a countersunk hole, which had been drilled through the skin to take a fastener.

Passengers cannot be carried at high altitude unless the cabin is pressurized. The Comet cabin was pressurized in service to 0.57 bar gauge, 50% more than the pressure in other civil aircraft at the time. During flight, the fuselage functions as a pressure vessel, and is subjected both to an axial and a circumferential tensile stress as a result. With each flight the fuselage experiences a single cycle of pressure loading. Comets G-ALYP and G-ALYY had flown 1290 and 900 flights respectively. The conclusion was that this small number of cycles had generated a fatigue crack, which in time had become long enough to cause fast fracture.

The skin of the fuselage was made from aluminum alloy sheet, with yield and tensile strengths of approximately 350 and 450 MN m$^{-2}$. Approximate hand calculations by de Havilland (at normal cabin pressure) gave an average stress of about 195 MN m$^{-2}$ near the corner of a typical window. The rules of the International Civil Aviation Organisation (ICAO) contained two major requirements for pressurized cabins: there was to be no deformation of the structure when the cabin was pressurized to 33% above normal; and the maximum

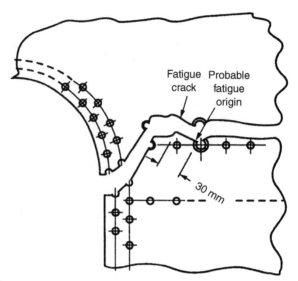

**FIGURE 19.3**

Close-up showing the origin of the failure on Comet G-ALYP.

stress in the structure at a pressure 100% above normal was to be less than the tensile strength.

De Havilland decided to go one better—it increased the "design" pressure from 100 to 150% above normal, and decided to pressure test a section of the fuselage to 100% above normal. At this test pressure, the maximum stress near the windows should have gone up to 390 MN m$^{-2}$ if the stress calculations were to be believed. This was still less than the tensile strength, and de Havilland was sure that the cabin would stand up to this increased test pressure without excessive deformation. In the event, the test section withstood two pressure tests to 100% above normal without any problems, showing that there was an ample margin of safety on static pressure loading.

The ICAO rules did not consider the possibility of fatigue caused by repeated cycles of pressure loading. But there was a growing awareness that this was a potential problem. As a result, de Havilland decided to determine the likely fatigue life of the fuselage. Between July and September 1953 (more than a year after the aircraft had entered service!), the test section was subjected to a continuous fatigue test, which involved cycling the pressure between zero and the normal operating value. After 18,000 cycles, the section failed from a fatigue crack, which had initiated at a small defect in the skin next to the corner of a window. The design life of the Comet was 10,000 flights, so de Havilland considered that it had demonstrated a reasonable margin of safety against fatigue from pressure cycling.

The air accident investigators decided to do a full-scale simulation test on a Comet aircraft, in order to confirm that the failure had indeed been caused by fatigue from repeated cycles of cabin pressure. The aircraft chosen was G-ALYU, which had already made 1230 flights. The aim was to cycle the cabin pressure between zero and the normal operating value until it failed by fast fracture. This could not be done using air, because the energy released when the cabin broke up would be equivalent to a 225 kg bomb. It was decided instead to pressurize the cabin using water. In order to balance the weight of water in the cabin the whole of the fuselage was put into a large tank and immersed in water at atmospheric pressure. Finally, the required pressure differential was generated by pumping a small volume of water into the fuselage.

The cabin failed after 1830 cycles in the tank, giving a total of $1230 + 1830 = 3060$ "flights." The crack started at an escape hatch, which was positioned in the port side of the cabin just forward of the wing. As shown in Figure 19.4, the fatigue crack initiated at the bottom corner of the hatch. Once the crack had reached the critical length it ran backward along the axis of the fuselage. The crack was diverted in a circumferential direction at the site of a transverse bulkhead. A second crack then nucleated at the forward bottom corner of the hatch. This crack ran forward along the axis of the fuselage until it was diverted into a circumferential crack at another transverse bulkhead. The panel of skin detached by the cracks was pushed out of the fuselage by a few centimetres, relieving the pressure inside the cabin and arresting the failure process. If the cabin had been filled with air, the panel would have blown right out, and the whole fuselage would have exploded.

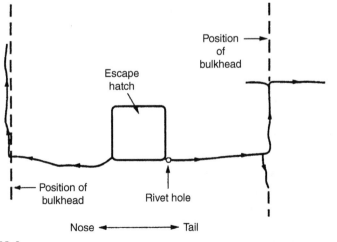

**FIGURE 19.4**
The failure on Comet G-ALYU produced by the simulation test.

The fuselage was repaired and strain gauges were stuck on to the skin immediately next to the edge of a window. The cabin was taken up to normal operating pressure and the stresses were found from the strain gauge readings. The maximum stress appeared at the corner of the window, and had a value of 297 MN m$^{-2}$. The background stress at the location of the critical bolt hole would obviously have been much less, since it was located some distance away from the edge of the window. However, the bolt hole would have acted as a stress raiser, and this could have increased the local stress by as much as a factor of 3 above the background stress.

The designers should have been aware that introducing bolt holes near the windows was bound to lead to high local stresses, and that under repeated cycles of cabin pressurization these bolt holes were likely to be preferential sites for fatigue crack initiation. In addition, the investigators were able to point to two crucial fallacies in the de Havilland figure for the fatigue life. The first was purely statistical. Data for low-cycle fatigue were already known to show a large amount of scatter. The probability that a single Comet would fail by fatigue could only be found if both the mean and the standard deviation of the test data were known. This information could not have been obtained from one solitary fatigue test.

The second was that the test section had been pressurized to 100% above normal before the fatigue test, whereas the production aircraft themselves had only been proof tested to 33% above normal. When the test section was taken to 100% above normal, the bolt holes could have yielded in tension. In this case, when the structure was let down to atmosphere again, the elastic spring-back would have put the holes into compression. When, later on, the test section was taken up to normal operating pressure during the fatigue test, the metal around the holes would have experienced a reduced tensile stress and consequently an increased fatigue life.

The final irony is that the doublers and other stiffeners around the window openings were not meant to be attached to the skin with bolts and rivets at all, but instead with glued joints. However, the design was revised at a late stage because it was felt that it would be difficult and expensive to make glued joints in these geometrically complex locations (although glued joints were extensively used elsewhere in the airframe—yet another pioneering innovation, and one much used in modern aircraft).

In conclusion, although the enquiry established the technical reasons for the failures, in reality the most important factors were probably that de Havilland tried to introduce too many innovations at once, and as a medium-sized company overstretched its design department. It is no surprise that civil aircraft construction today is dominated by huge companies or consortia (most notably Boeing in the United States and Airbus in Europe) because of the sheer range and complexity of design, manufacturing, and maintenance capability needed to support modern civil aircraft fleets. Even so, air crashes caused by the failure of structures or components continue to occur (although nowadays many are due to faulty maintenance procedures rather than design errors, and the numbers are much smaller in relation to the size of aircraft fleets than they used to be).

## 19.2 CASE STUDY 2: THE ESCHEDE RAILWAY DISASTER

On 5 June 1998, a high-speed train traveling from Munich to Hamburg (Germany) derailed near the village of Eschede (see Figure 19.5). The speed of the train was approximately 250 km h$^{-1}$. The train consisted of two power cars (one at each end) plus 12 carriages. The steel tire of one of the wheels suffered a fatigue fracture, came away from the wheel, and jammed under the floor of the carriage. The broken tire then became stuck in a set of points (or switch), and switched them from the running line to the junction line. The broken tire came from the rear bogie of the leading carriage, so although the front power car was still on the running line, the rear of the leading carriage was diverted on to the junction line, and it and the remainder of the train were derailed as a result.

**FIGURE 19.5**
Photograph of the wreckage of the high-speed train after the crash. *(Courtesy of Elsevier © 2005.)*

By a very rare coincidence, the derailment occurred just before an over-bridge. The derailed part of the train (still traveling at high speed) demolished one of the supporting piers of the bridge, which promptly collapsed on to the train. This led to the almost total destruction of much of the train. One hundred people were killed in the crash, and more than 100 injured. Criminal proceedings for negligent homicide were commenced against the operators of the railway and the manufacturer of the wheel. The subsequent technical investigations involved a total of 13 experts from five different countries.

Of the 112 wheels in the train, 36 were of the monobloc type (consisting of a one-piece steel wheel with an integral tire). However, the remaining 76 were rubber-sprung wheels (see Figure 19.6) in which rubber blocks were fitted between the tire and the wheel center to help damp out the vibrations generated by wheel–rail contact. This type of wheel had been introduced into the high-speed trains in Germany in 1991, although similar wheels had been used successfully in commuter trains and trams for some time previously.

The fracture surface of the broken tire is shown in Figure 19.7. It has classic fatigue beach marks, which show that the fatigue crack initiated at the bore of the tire near the position of maximum thickness. The area proportion of final fast fracture is small, indicating a relatively low maximum stress.

The stress geometry is shown in Figure 19.8. The most important load on the wheel is the vertical load Q at the wheel–rail interface. This is produced by the

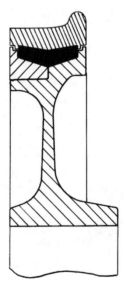

**FIGURE 19.6**
Construction of the rubber-sprung wheel.

**FIGURE 19.7**
Fatigue fracture surface from the broken tire. *(Courtesy of Elsevier © 2004.)*

deadweight of the carriage (78 kN per wheel) plus a multiplier of 1.25 to allow for average dynamic loads, giving a total design load of 98 kN. Looking at Figure 19.8, one can see that because the tire is supported on a relatively soft rubber foundation, it will deflect above the loading point, and this will generate a tensile stress at the bore of the tire.

The effect of this deflection is felt all around the circumference of the tire. The stress is a maximum above the loading point, as would be expected, and has a value of 220 MN m$^{-2}$. However, moving around the circumference away from the loading point, the stress falls off rapidly, reaching a minimum of 6 MN m$^{-2}$ after an angular rotation of 45°. Moving around to the top of the wheel, the stress increases again (but to a maximum of only 55 MN m$^{-2}$). The bore of the tire is thus subjected to one complete fatigue cycle with each revolution of the wheel, with stress amplitude 107 MN m$^{-2}$ and mean stress 113 MN m$^{-2}$.

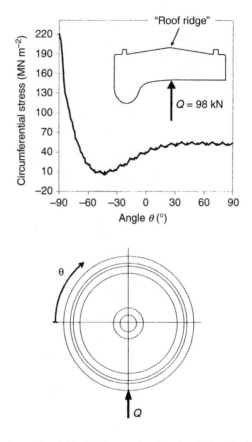

**FIGURE 19.8**

Circumferential tensile stress at "roof-ridge" of tire as a function of angle from the loading point. *(Courtesy of Elsevier © 2005.)*

However, there are two equal stress troughs per cycle, and two unequal stress peaks per cycle.

It should be noted that the rubber blocks did not fill all the space between the tire and the wheel center. Each wheel was fitted with a total of 34 rubber blocks, which were equally spaced around the circumference of the wheel center, with circumferential gaps between them. Had there not been adequate circumferential gaps between the blocks, they would not have been able to expand circumferentially when compressed by the radial load $Q$ (they were not able to expand axially because they were a close fit to the width of the channel in the wheel center). Because Poisson's ratio for rubber is 0.5, it must be allowed to expand sideways in at least one principal direction if it is to be "springy." If sideways expansion were prevented in both principal directions, the block would behave as a very stiff structure indeed, and be useless as a damper (see Example 3.3).

The stress plot in Figure 19.8 was generated using an immensely complex finite-element model. A full three-dimensional model was generated, the elastic behavior of the rubber was modeled with a nonlinear constitutive equation, and the frictional contact between the rubber block and the steel was modeled with a nonlinear frictional law. Such an analysis had never been done at the design stage, which is not surprising given the rapid advances in finite-element modeling and computing power in recent years. Interestingly, the model was so good that the gaps between adjacent rubber blocks gave rise to a periodic "wobble" in the stress plot.

Turning to the fatigue strength of the tire, tests on the steel gave a tensile strength of 828 MN m$^{-2}$, so as a first approximation one would expect a fatigue strength for a long life of approximately 40% of this figure (i.e., 330 MN m$^{-2}$ stress amplitude). However, when designing the actual wheel, the fatigue strength was reduced to 200 MN m$^{-2}$ stress amplitude. This was to allow for the effect of the tensile mean stress and also to have a safety factor. According to the published data, therefore, the tire should not have failed—because the stress amplitude in service (107 MN m$^{-2}$) should only have been half the fatigue strength of the tire with safety factor included. The experts could not agree on the cause for the failure, and as a result, the criminal proceedings were dropped.

The tires of railway wheels wear as a result of contact with the running rails, and as a result they need to be reprofiled regularly. To do this, the wheel-set is removed from the bogie, and the outer circumference of the tire is turned in a lathe to give the correct profile again. Over time, repeated reprofiling reduces the outer diameter of the tire, and the failed tire (diameter 860 mm) had nearly reached the diameter for scrapping. New tires had an outer diameter of 920 mm. After the accident, the lower limit on diameter was increased to 880 mm. Obviously, the larger the diameter, the thicker the tire, and the lower the fatigue stress for a given applied loading.

The disaster raises a number of interesting issues. The first is that a fatigue failure *did* occur, and therefore there *must* be an explanation. The implication is that the investigators must have missed something. This is not to be in any way critical of them. The standard of proof required in a criminal case is to prove beyond all reasonable doubt that the defendants were grossly negligent, which is a much more limited brief than the technical investigation required to ensure the future safety of a fleet of trains. Some technical issues that come to mind are: (a) what was the actual service load spectrum (as distinct from that assumed in the design); (b) was there any fretting between the rubber blocks and the tire; (c) did the rubber blocks "creep" around the circumference of the wheel with service time, resulting in inadequate support? Finally, when an innovative design of wheel brings with it such serious technical difficulties, is it any wonder that the railway industry is so conservative?

# 19.3 CASE STUDY 3: THE SAFETY OF THE STRETHAM ENGINE

The Stretham steam pumping engine (Figure 19.9) was built in 1831 as part of an extensive project to drain the Fens for agricultural use. In its day it was one of the largest beam engines in the Fens, having a maximum horsepower of 105 at

**FIGURE 19.9**

Part of the Stretham pumping engine. In the foreground are the crank and the lower end of the connecting rod. Also visible are the flywheel (with separate spokes and rim segments, all pegged together), the eccentric drive to the valve-gear and, in the background, an early treadle-driven lathe for on-the-spot repairs. – 52 20 03.26 N 0 13 28.0.0 E; *http://www.strethamoldengine.org.uk/; http://en.wikipedia.org/ wiki/Stretham_Old_Engine*

15 rpm (it could lift 30 tons of water per revolution, or 450 tons per minute); it is now the sole surviving steam pump of its type in East Anglia.*

The engine could still be run for demonstration purposes. Suppose that you are called in to assess its safety. We will suppose that a crack 0.02 m deep has been found in the connecting rod—a cast-iron rod, 21 feet long, with a section of 0.04 m². Will the crack grow under the cyclic loads to which the connecting rod is subjected? And what is the likely life of the structure?

## Mechanics

The stress in the crank shaft is calculated approximately from the power and speed as follows. Bear in mind that approximate calculations of this sort may be in error by up to a factor of 2—but this makes no difference to the conclusions reached below. Referring to Figure 19.10:

$$
\begin{aligned}
\text{power} &= 105 \text{ horsepower} = 7.8 \times 10^4 \text{ J s}^{-1} \\
\text{speed} &= 15 \text{ rpm} = 0.25 \text{ rev s}^{-1} \\
\text{stroke} &= 8 \text{ feet} = 2.44 \text{ m}
\end{aligned}
$$

$$\text{Force} \times 2 \times \text{stroke} \times \text{speed} \approx \text{power}$$

$$\therefore \text{ force} \approx \frac{7.8 \times 10^4}{2 \times 2.44 \times 0.25} \approx 6.4 \times 10^4 \text{N}$$

Nominal stress in the connecting rod $= F/A = 6.4 \times 10^4/0.04 = 1.6 \text{ MN m}^{-2}$ approximately.

## Failure by fast fracture

For cast iron, $K_c = 10 \text{ MN m}^{-3/2}$.

First, could the rod fail by fast fracture? The stress intensity is:

$$K = Y\sigma\sqrt{\pi a} = 1.19 \times 1.6\sqrt{\pi \times 0.02} \text{ MN m}^{-3/2} = 0.48 \text{ MN m}^{-3/2}$$

It is so much less than $K_c$ that there is no risk of fast fracture, even at peak load.

## Failure by fatigue

The growth of a fatigue crack is described by

$$\frac{da}{dN} = A(\Delta K)^m \tag{19.1}$$

---

* Until a couple of centuries ago much of the eastern part of England which is now called East Anglia was a vast area of marshes, or fens, which stretched from the North Sea as far inland as Cambridge.

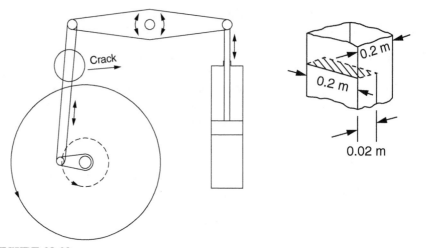

**FIGURE 19.10**
Schematic of the Stretham engine. $Y = 1.19$ for $a = 0.02$ m; $Y = 1.26$ for $a = 0.03$ m.

For cast iron,

$$A = 4.3 \times 10^{-8} \text{m}(\text{MN m}^{-3/2})^{-4}$$
$$m = 4$$

We have that

$$\Delta K = Y\Delta\sigma\sqrt{\pi a}$$

where $\Delta\sigma$ is the range of the tensile stress (Figure 19.11). Although $\Delta\sigma$ is constant (at constant power and speed), $\Delta K$ increases as the crack grows. Substituting in Equation (19.1) gives

$$\frac{da}{dN} = AY^4\Delta\sigma^4\pi^2 a^2$$

and

$$dN = \frac{1}{(AY^4\Delta\sigma^4\pi^2)}\frac{da}{a^2}$$

Integration gives the number of cycles to grow the crack from $a_1$ to $a_2$:

$$N = \frac{1}{(AY^4\Delta\sigma^4\pi^2)}\left\{\frac{1}{a_1} - \frac{1}{a_2}\right\}$$

for a range of $a$ small enough that the crack geometry does not change appreciably. Let us work out how long it would take our crack to grow from 0.02 to 0.03 m. The average value of $Y$ over this interval is 1.23.

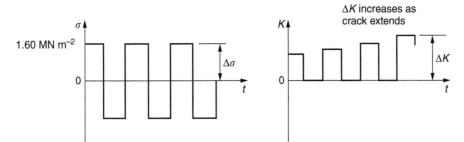

**FIGURE 19.11**
Crack growth by fatigue in the Stretham engine.

$$N = \frac{1}{4.3 \times 10^{-8}1.23^4(1.6)^4\pi^2}\left\{\frac{1}{0.02} - \frac{1}{0.03}\right\}$$

$$= 2.6 \times 10^6 \text{cycles}$$

This is sufficient for the engine to run for 8 h on each of 360 open days for demonstration purposes; that is, to give 8 hours of demonstration each weekend for 7 years. A crack of 0.03 m length is still far too small to go critical, and thus the engine will be perfectly safe after the $2.6 \times 10^6$ cycles. Under demonstration the power delivered will be far less than the full 105 horsepower, and because of the $\Delta\sigma^4$ dependence of $N$, the number of cycles required to make the crack grow to 0.03 m might be as much as 1000 times the one we have calculated.

The estimation of the total lifetime of the structure is more complex—substantial crack growth will make the crack geometry change significantly; this will have to be allowed for in the calculations by treating $Y$ as a variable in the integration.

## EXAMPLES

**19.1** The photographs on the next page show a bicycle crank. The end of the crank has suffered a fatigue failure where the pedal was screwed into the crank (the pedal is missing). In the close-up photograph of the fracture, the bottom of the photograph corresponds to the outside (pedal side) of the crank arm. Answer the following questions:

    **a.** Why do you think the failure was caused by fatigue?

    **b.** Identify the fatigue fracture surface(s). Explain your choice.

    **c.** Identify the final fast fracture surface(s). Explain your choice.

    **d.** Where do you think the fatigue crack(s) initiated?

    **e.** In your opinion, was the level of applied stress at final failure high, moderate, or low?

19.2 The following diagrams show the general arrangement of a swing door and a detailed drawing of the top pivot assembly. The full weight of the door was taken by the bottom pivot. The function of the top pivot was simply to keep the door upright by applying a horizontal force to the top of the door. When the door swung back and forth, it rotated about the top and bottom pivot pins. After a year in service, the top pivot pin failed by fatigue. As a result, the top door housing separated from the top frame housing, and the door fell on a person passing by.

    **a.** Where in the top pivot pin do you think the fatigue fracture occurred?

    **b.** Why do you think the fatigue fracture occurred where it did?

    **c.** What was the origin of the fatigue loading cycle at the failure location?

    **d.** How did the door housing manage to become separated from the frame housing?

[In answering these questions, please note the following information: the pivot pin was a very loose fit in both the door housing and the frame housing; the pivot pin tended not to rotate in the frame housing—instead, the door housing tended to rotate about the pivot pin.]

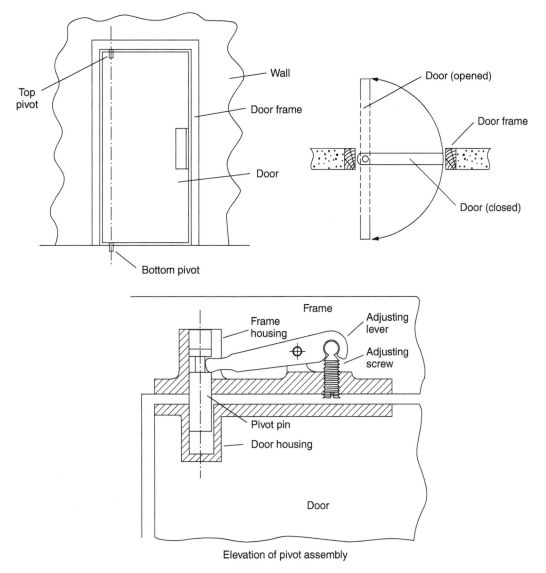

Elevation of pivot assembly

**19.3** With reference to Case Study 3, why do we take into account only the *tensile* part of the stress cycle when calculating $\Delta K$?

**19.4** The diagrams that follow show a process vessel in a chemical plant. The vessel is fabricated from mild steel by welding. The vessel rotates at a speed of 2 revolutions per minute. The weight of the vessel (including contents) is 26 tons, so each trunnion shaft is subjected to an upward reaction force of 13 tons, applied at the center line of the bearing. Because the bearings are self-aligning, the trunnion shafts behave as simple cantilevers.

After 5 years in service (equivalent to $5 \times 10^6$ revolutions), fatigue cracks started to appear, and the vessel had to be taken out of service. Where do you think the fatigue cracks were located?

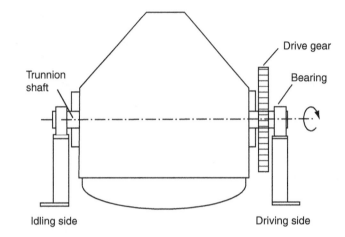

Schematic side elevation of the chemical vessel.

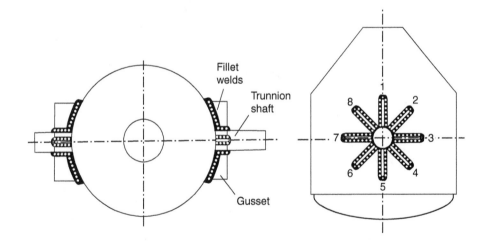

Details of the attachments between the trunnion shafts and the side of the vessel.

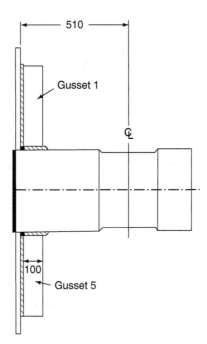

Details of the attachment on the driving side (dimensions in mm 0).

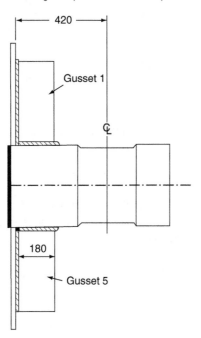

Details of the attachment on the idling side (dimensions in mm).

19.5 The diagram shows a cross-section through a rotating drive shaft from a chemical process vessel. A two-bladed stirrer was bolted to the flange at the lower end of the drive shaft, and the assembly rotated at 60 rpm in service. The stirrer blades were positioned a long distance below the flange, and were only supported by the drive shaft, which was subjected to significant bending loads caused by unequal forces on the stirrer blades.

After about 8 months in service, equivalent to 10 million revolutions, the collar failed from a fatigue crack shown in the diagram. Although there was no information on the fatigue loading produced by the stirrer, it was necessary to return the vessel to service as soon as possible. Dye-penetrant inspection of the 2.6 mm fillet showed no fatigue cracking, even though the surface had obvious circumferential lathe turning marks which would have reduced the fatigue strength. Estimate the radius of fillet which would be required at the location of the fracture to prevent a recurrence of fatigue failure. Data are given in the following table.

| Fillet Radius r (mm) | Notch Sensitivity S | SCF (D/d = 2) | SCF (D/d = 1.11) |
|---|---|---|---|
| 1.0 | 0.72 | | 2.70 |
| 1.5 | 0.77 | 3.42 | 2.47 |
| 2.0 | 0.79 | 2.90 | 2.30 |
| 2.5 | 0.81 | 2.67 | 2.17 |
| 3.0 | 0.82 | 2.50 | 2.06 |
| 3.5 | 0.83 | 2.39 | 1.97 |
| 4.0 | 0.84 | 2.29 | 1.91 |
| 4.5 | 0.85 | 2.20 | 1.83 |

The maximum bending stress in the collar is given approximately by

$$\sigma_{max} = \frac{B}{\phi^3}$$

where $B$ is a constant, and $\phi$ (the outer diameter of the cross section) is either 90 mm or 81 mm. You may assume that the fatigue strength of the steel levels out above 10 million cycles.

The diagram also shows a tubular sleeve mounted on the 81 mm portion of the shaft. The sleeve acts as a spacer between the step and a thrust bearing, and cannot be designed-out without major work. It is considered necessary to have a 2 mm wide contact land between the sleeve and the step, which means that the fillet can only have a radius of 2.5 mm.

Which other methods could you use to increase the resistance of the fillet to fatigue cracking in order to compensate for the required reduction in fillet radius?

**Answer**

4.5 mm

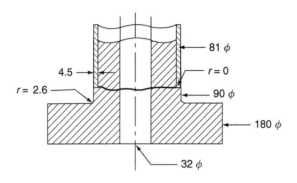

# Creep Deformation and Fracture

## CHAPTER 20

# Creep and Creep Fracture

## 20.1 INTRODUCTION

So far we have concentrated on mechanical properties at room temperature. Many structures—particularly those that are associated with energy conversion (e.g., turbines, reactors, steam, and chemical plant)—operate at much higher temperatures.

At room temperature, most metals and ceramics deform in a way that depends on stress but which, for practical purposes, is independent of time:

$$\varepsilon = f(\sigma) \text{ elastic/plastic solid}$$

As the temperature is raised, loads that give no permanent deformation at room temperature cause materials to *creep*. Creep is slow, continuous deformation with time: the strain, instead of depending only on the stress, now depends on temperature and time as well:

$$\varepsilon = f(\sigma, t, T) \text{ creeping solid}$$

Engineering Materials I: An Introduction to Properties, Applications, and Design, Fourth Edition
© 2012, Michael F. Ashby and David R. H. Jones. Published by Elsevier Ltd. All rights reserved.

**FIGURE 20.1**
A tungsten lamp filament which has sagged undo its own weight owing to creep.

It is common to refer to the former behavior as "low-temperature" behavior, and the latter as "high temperature." But what is a "low" temperature and what is a "high" temperature? Tungsten, used for lamp filaments, has a very high melting point—well over 3000°C. Room temperature, for tungsten, is a very low temperature. If made hot enough, however, tungsten will creep—that is the reason that lamps ultimately burn out. Tungsten lamps run at about 2000°C—this, for tungsten, is a high temperature. If you examine a lamp filament that has failed, you will see that it has sagged under its own weight until the turns of the coil have touched—that is, it has deformed by creep. Figure 20.1 shows a typical example of a sagging filament.

Figure 20.2 and Table 20.1 give melting points for metals and ceramics and softening temperatures for polymers. Most metals and ceramics have high melting points and, because of this, they start to creep only at temperatures well above room temperature—this is why creep is a less familiar phenomenon than elastic or plastic deformation. But the metal *lead*, for instance, has a melting point of 600 K; room temperature, 300 K, is exactly half its absolute melting point. Room temperature for lead is a high temperature, and it creeps—as Figure 20.3 shows (see page 313). And the ceramic *ice* melts at 0°C. Temperate glaciers (those close to 0°C) are at a temperature at which ice creeps rapidly—that is why glaciers move. Even the thickness of the Antarctic ice cap, which controls the levels of the earth's oceans, is determined by the creep spreading of the ice at about –30°C.

The point, then, is that the temperature at which materials start to creep depends on their melting point. As a general rule, it is found that creep starts when

$$T > 0.3 \text{ to } 0.4 \; T_M \text{ for metals}$$
$$T > 0.4 \text{ to } 0.5 \; T_M \text{ for ceramics}$$

where $T_M$ is the melting temperature in degrees kelvin. However, special alloying procedures can raise the temperature at which creep becomes a problem.

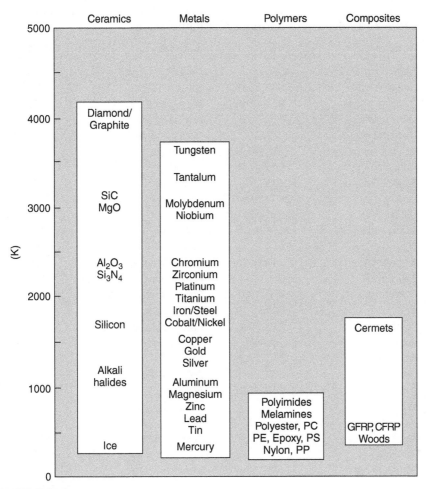

**FIGURE 20.2**
Melting or softening temperature.

Polymers, too, creep—many of them do so at room temperature. As we said in Chapter 5, most common polymers are not crystalline and have no well-defined melting point. For them, the important temperature is the glass temperature, $T_G$, at which the Van der Waals bonds solidify. Above this temperature, the polymer is in a leathery or rubbery state and creeps rapidly under load. Below, it becomes hard (and sometimes brittle) and, for practical purposes, no longer creeps. $T_G$ is near room temperature for most polymers, so creep is a problem.

In design against creep, we select the material and the shape that will carry the design loads, without failure, for the design life at the design temperature. The

**Table 20.1** Melting or Softening[(S)] Temperature

| Material | T(K) | Material | T(K) |
|---|---|---|---|
| Diamond, graphite | 4000 | Gold | 1336 |
| Tungsten alloys | 3500–3683 | Silver | 1234 |
| Tantalum alloys | 2950–3269 | Silica glass | 1100[(S)] |
| Silicon carbide, SiC | 3110 | Aluminum alloys | 750–933 |
| Magnesia, MgO | 3073 | Magnesium alloys | 730–923 |
| Molybdenum alloys | 2750–2890 | Soda glass | 700–900[(S)] |
| Niobium alloys | 2650–2741 | Zinc alloys | 620–692 |
| Beryllia, BeO | 2700 | Polyimides | 580–630[(S)] |
| Iridium | 2682–2684 | Lead alloys | 450–601 |
| Alumina, $Al_2O_3$ | 2323 | Tin alloys | 400–504 |
| Silicon nitride, $Si_3N_4$ | 2173 | Melamines | 400–480[(S)] |
| Chromium | 2148 | Polyesters | 450–480[(S)] |
| Zirconium alloys | 2050–2125 | Polycarbonates | 400[(S)] |
| Platinum | 2042 | Polyethylene, high-density | 300[(S)] |
| Titanium alloys | 1770–1935 | Polyethylene, low-density | 360[(S)] |
| Iron | 1809 | Foamed plastics, rigid | 300–380[(S)] |
| Carbon steels | 1570–1800 | Epoxy, general purpose | 340–380[(S)] |
| Cobalt alloys | 1650–1768 | Polystyrenes | 370–380[(S)] |
| Nickel alloys | 1550–1726 | Nylons | 340–380[(S)] |
| Cermets | 1700 | Polyurethane | 365[(S)] |
| Stainless steels | 1660–1690 | Acrylic | 350[(S)] |
| Silicon | 1683 | GFRP | 340[(S)] |
| Alkali halides | 800–1600 | CFRP | 340[(S)] |
| Beryllium alloys | 1540–1551 | Polypropylene | 330[(S)] |
| Uranium | 1405 | Ice | 273 |
| Copper alloys | 1120–1356 | Mercury | 235 |

meaning of "failure" depends on the application. We distinguish four types of failure, illustrated in Figure 20.4.

1. Displacement-limited applications, in which precise dimensions or small clearances must be maintained (as in the discs and blades of gas turbines).
2. Rupture-limited applications, in which dimensional tolerance is relatively unimportant, but fracture must be avoided (as in pressure-piping).
3. Stress relaxation-limited applications in which an initial tension relaxes with time (as in the pretensioning of bolts).
4. Buckling-limited applications, in which slender columns or panels carry compressive loads (as in structural steelwork exposed to a fire).

To analyze these we need *constitutive equations* which relate the strain-rate $\dot{\varepsilon}$ or time-to-failure $t_f$ to the stress $\sigma$ and temperature $T$.

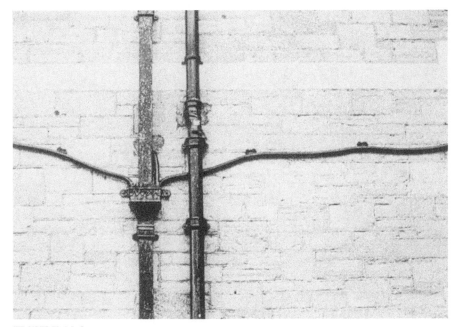

**FIGURE 20.3**
Lead pipes often creep noticeably over the years. (*Source:* M.F. Ashby.)

## 20.2 CREEP TESTING AND CREEP CURVES

Creep tests require careful temperature control. Typically, a specimen is loaded in tension, usually at constant load, inside a furnace maintained at a constant temperature, $T$. The extension is measured as a function of time. Figure 20.5 shows a typical set of results from such a test. Metals, polymers, and ceramics all show creep curves of this shape.

Although the *initial elastic* and the *primary creep* strain cannot be neglected, they occur quickly, and they can be treated in much the way that elastic deflection is allowed for in a structure. But thereafter, the material enters *steady state*, or *secondary* creep, and the strain increases steadily with time. In designing against creep, it is usually this steady accumulation of strain with time that concerns us most.

By plotting the log of the steady creep rate, $\dot{\varepsilon}_{ss}$, against log $\sigma$ at constant $T$, as shown in Figure 20.6, we can establish that

$$\dot{\varepsilon}_{ss} = B\sigma^n \tag{20.1}$$

where $n$, the *creep exponent*, usually lies between 3 and 8. This sort of creep is called "*power-law*" creep. (At low $\sigma$, a different régime is entered where $n \approx 1$; we shall discuss this low-stress deviation from power-law creep in Chapter 22.)

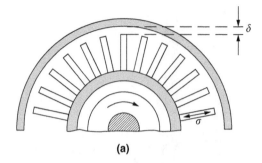

**(a)**

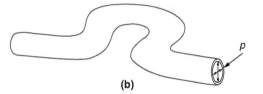

**(b)**

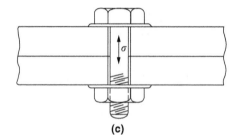

**(c)**

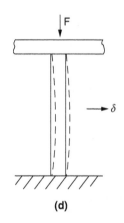

**(d)**

**FIGURE 20.4**

Creep is important in four classes of design: (a) displacement-limited, (b) failure-limited, (c) relaxation-limited, and (d) buckling-limited.

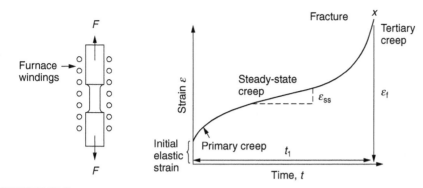

**FIGURE 20.5**

Creep testing and creep curves.

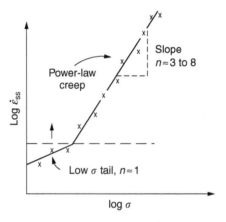

**FIGURE 20.6**

Variation of creep rate with stress.

By plotting the *natural* logarithm (ln) of $\dot{\varepsilon}_{ss}$ against the reciprocal of the *absolute* temperature $(1/T)$ at constant stress, as shown in Figure 20.7, we find that:

$$\dot{\varepsilon}_{ss} = Ce^{-(Q/\bar{R}T)} \tag{20.2}$$

Here $\bar{R}$ is the Universal Gas Constant $(8.31 \text{ J mol}^{-1} \text{ K}^{-1})$ and $Q$ is called the *Activation Energy for Creep*—it has units of J mol$^{-1}$. Note that the creep rate increases exponentially with temperature (Figure 20.7). An increase in temperature of $20°C$ can *double* the creep rate.

Combining these two dependences of $\dot{\varepsilon}_{ss}$ gives

$$\dot{\varepsilon}_{ss} = A\sigma^n e^{-(Q/\bar{R}T)} \tag{20.3}$$

where $A$ is the creep constant. The values of the three constants $A$, $n$, and $Q$ characterize the creep of a material; if you know these, you can calculate the strain rate

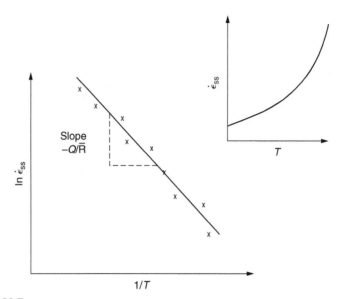

**FIGURE 20.7**
Variation of creep rate with temperature.

at any temperature and stress by using the last equation. They vary from material to material, and have to be found experimentally.

## 20.3 CREEP RELAXATION

At constant displacement, creep causes stresses to relax with time. Bolts in hot turbine casings must be regularly tightened. Plastic paper clips are not, in the long-term, as good as steel ones because, even at room temperature, they slowly lose their grip.

The relaxation time (arbitrarily defined as the time taken for the stress to relax to half its original value) can be calculated from the power-law creep data as follows. Consider a bolt that is tightened onto a rigid component so that the initial stress in its shank is $\sigma_i$. In this geometry (refer to Figure 20.4(c)) the length of the shank must remain constant—that is, the *total* strain in the shank $\varepsilon_{tot}$ must remain constant. But creep strain $\varepsilon_{cr}$ can *replace* elastic strain $\varepsilon_{el}$, causing the stress to relax. At any time $t$

$$\varepsilon_{tot} = \varepsilon_{el} + \varepsilon_{cr} \tag{20.4}$$

But

$$\varepsilon_{el} = \sigma/E \tag{20.5}$$

and (at constant temperature)

$$\dot{\varepsilon}_{cr} = B\sigma^n$$

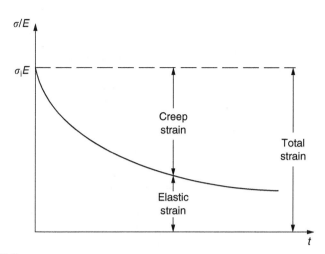

**FIGURE 20.8**
Replacement of elastic strain by creep strain with time at high temperature.

Since $\varepsilon_{tot}$ is constant, we can differentiate Equation (20.4) with respect to time and substitute the other two equations into it to give

$$\frac{1}{E}\frac{d\sigma}{dt} = -B\sigma^n \tag{20.6}$$

Integrating from $\sigma = \sigma_i$ at $t = 0$ to $\sigma = \sigma$ at $t = t$ gives

$$\frac{1}{\sigma^{n-1}} - \frac{1}{\sigma_i^{n-1}} = (n-1)BEt \tag{20.7}$$

Figure 20.8 shows how the initial elastic strain $\sigma_i/E$ is slowly replaced by creep strain, and the stress in the bolt relaxes. If, as an example, it is a casing bolt in a large turbogenerator, it will have to be retightened at intervals to prevent steam leaking from the turbine. The time interval between retightening, $t_r$, can be calculated by evaluating the time it takes for $\sigma$ to fall to (say) one-half of its initial value. Setting $\sigma = \sigma_i/2$ and rearranging gives

$$t_r = \frac{(2^{n-1} - 1)}{(n-1)BE\sigma_i^{n-1}} \tag{20.8}$$

Experimental values for $n$, $A$, and $Q$ for the material of the bolt thus enable us to decide how often the bolt will need retightening. Note that overtightening the bolt does not help because $t_r$ decreases rapidly as $\sigma_i$ increases.

## 20.4 CREEP DAMAGE AND CREEP FRACTURE

During creep, damage accumulates in the form of internal cavities. The damage first appears at the start of the tertiary stage of the creep curve and grows at an increasing rate thereafter. The shape of the tertiary stage of the creep curve

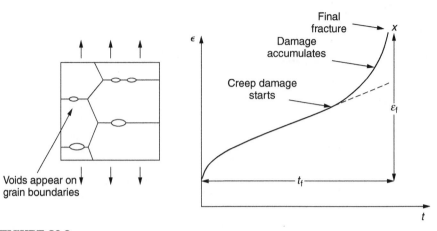

Voids appear on
grain boundaries

**FIGURE 20.9**
Creep damage.

(refer to Figure 20.5) reflects this: as the cavities grow, the section of the sample decreases, and (at constant load) the stress goes up. Since $\dot{\varepsilon} \propto \sigma^n$, the creep rate goes up even faster than the stress does (Figure 20.9).

It is not surprising—since creep causes creep fracture—that the time-to-failure, $t_f$ is described by a constitutive equation which looks very like that for creep itself:

$$t_f = A' \sigma^{-m} e^{+(Q/\bar{R}T)}$$

Here $A'$, $m$, and $Q$ are the creep-failure constants, determined in the same way as those for creep (the exponents have the opposite sign because $t_f$ is a time whereas $\dot{\varepsilon}_{ss}$ is a rate).

In many high-strength alloys this creep damage appears early in life and leads to failure after small creep strains (as little as 1%). In high-temperature design it is important to make sure:

- That the *creep strain* $\varepsilon_{cr}$ during the design life is acceptable
- That the *creep ductility* $\varepsilon_{f,cr}$ (strain to failure) is adequate to cope with the acceptable creep strain
- That the *time-to-failure*, $t_f$, at the design loads and temperatures is longer (by a suitable safety factor) than the design life

Times-to-failure are normally presented as *creep-rupture* diagrams (Figure 20.10). Their application is obvious: if you know the stress and temperature you can read off the life; if you wish to design for a certain life at a certain temperature, you can read off the design stress.

## 20.5 CREEP-RESISTANT MATERIALS

From what we have said so far it should be obvious that the first requirement that we should look for in choosing materials that are resistant to creep is that they should have high melting (or softening) temperatures. If the material can

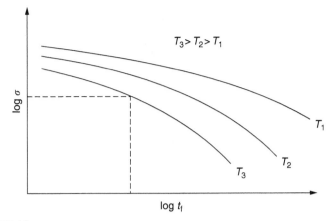

**FIGURE 20.10**
Creep-rupture diagram.

then be used at less than 0.3 of its melting temperature creep will not be a problem. If it has to be used above this temperature, various alloying procedures can be used to increase creep resistance. To understand these, we need to know more about the mechanisms of creep—the subject of the next two chapters.

## WORKED EXAMPLE

The photograph on the next page shows an easy experiment you can rig up at home to demonstrate creep—and even measure the creep exponent, $n$. Take a length of wooden curtain pole or broom handle (mine was 28.5 mm diameter—1⅛"), and wrap electronic solder wire (mine was 1.60 mm diameter—1/16") around the pole to make a coil that has no gaps between the turns. I wound 36 turns of wire on to the pole. Then hook the end of the coil over the end of the pole as shown, and put the axis of the pole vertical (hold the lower end in a vice or clamp). Do the experiment indoors in the warm, and within a few minutes the coil will start to open out. After about 15 minutes it will look like the one in the photograph. Obviously the solder (an alloy of lead and tin) has experienced considerable creep deformation. This is exactly what you would expect—electronic solder melts at 183°C (456 K), so a room temperature of 20°C (293 K) is 0.64 $T_M$—well into the régime where creep will occur.

You can see from the photograph that the coils have opened out much more at the top (where the wire supports the maximum load) than they have toward the bottom (where the wire supports a much smaller load). This tells you that the creep rate is increasing with stress, as it should do. But how fast does it increase? The value of the creep exponent $n$ can easily be found as follows.

Starting from the top of the coil, measure the positions of the crests of the wire (e.g., 1, 18, 34, 49, 63 mm). List the differences between these measurements which is the pitch of each turn (e.g., 17, 16, 15, 14 mm) and correlate these pitches with the number of turns of wire they support (e.g., 36, 35, 34, 33). The opening-out of each turn is found by subtracting the wire diameter (1.6 mm) from the pitch, giving 15.5, 14.5, 13.5, 12.5 mm, and so on (rounded to the nearest 0.5 mm).

Then plot log (opening out) versus log (number of turns supported), e.g., log 15.5 versus log 36, log 14.5 versus log 35, etc. The full graph is shown in the following diagram. Over most of the range, there is an excellent fit to a straight line with a slope of 2.1, so $n = 2.1$. Only toward the bottom of the coil do the data points go off, but this is because it is very difficult to make accurate measurements of pitch from the photograph when the turns have opened out so little.

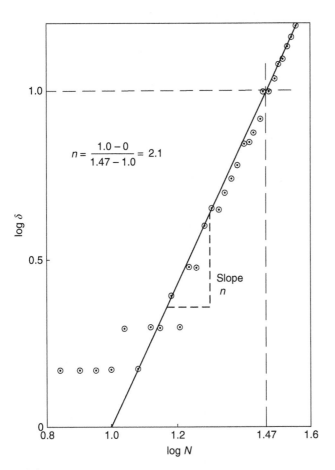

$$n = \frac{1.0 - 0}{1.47 - 1.0} = 2.1$$

Slope
$n$

log $\delta$

log $N$

## EXAMPLES

**20.1** A cylindrical tube in a chemical plant is subjected to an excess internal pressure of 6 MN m$^{-2}$, which leads to a circumferential stress in the tube wall. The tube wall is required to withstand this stress at a temperature of 510°C for 9 years. A designer has specified tubes of 40 mm bore and 2 mm wall thickness made from a stainless alloy of iron with 15% by weight of chromium. The manufacturer's specification for this alloy gives the following information.

| Temperature (°C) | 618 | 640 | 660 | 683 | 707 |
|---|---|---|---|---|---|
| Steady-state creep rate $\dot{\varepsilon}$ (s$^{-1}$), for an applied tensile stress $\sigma$ of 200 MN m$^{-2}$ | $1.0 \times 10^{-7}$ | $1.7 \times 10^{-7}$ | $4.3 \times 10^{-7}$ | $7.7 \times 10^{-7}$ | $2.0 \times 10^{-6}$ |

Over the present ranges of stress and temperature the alloy can be considered to creep according to the equation

$$\dot{\varepsilon} = A\sigma^5 e^{-Q/\bar{R}T}$$

where $A$ and $Q$ are constants, $\bar{R}$ is the universal gas constant, and $T$ is the absolute temperature. Given that failure is imminent at a creep strain of 0.01 for the present alloy, comment on the safety of the design.

**Answer**

Strain over 9 years = 0.00057; design safe

20.2 An alloy tie bar in a chemical plant has been designed to withstand a stress, $\sigma$, of $25 \text{ MN m}^{-2}$ at 620°C. Creep tests carried out on specimens of the alloy under these conditions indicated a steady-state creep rate, $\dot{\varepsilon}$, of $3.1 \times 10^{-12} \text{ s}^{-1}$. In service it was found that, for 30% of the running time, the stress and temperature increased to $30 \text{ MN m}^{-2}$ and 650°C. Calculate the average creep rate under service conditions. It may be assumed that the alloy creeps according to the equation

$$\dot{\varepsilon} = A\sigma^5 e^{-Q/\bar{R}T}$$

where $A$ and $Q$ are constants, $\bar{R}$ is the universal gas constant, and $T$ is the absolute temperature. $Q$ has a value of $160 \text{ kJ mol}^{-1}$.

**Answer**

$6{:}82 \times 10^{-12} \text{s}^{-1}$

20.3 The window glass in old buildings often has an uneven surface, with features that look like flow marks. The common explanation is that the glass has suffered creep deformation over the years under its own weight. Explain why this scenario is complete rubbish (why do you think the glass does appear to have "flow marks"?).

20.4 Why do bolted joints in pressure vessels operating at high temperature have to be retorqued at regular intervals?

20.5 Why are creep–rupture diagrams useful when specifying materials for high-temperature service?

20.6 Structural steelwork in high-rise buildings is generally protected against fire by a thick coating of refractory material even though a fire would not melt the steel. Explain.

20.7 Aluminum alloy ($T_M = 920$ K) pistons in large diesel engines are subjected to working temperatures of around 150°C (423 K). When the engine is started up, thermal gradients in the pistons produce thermal strains (coefficient of thermal expansion × temperature difference) of the order of 0.3%. Young's modulus for the alloy at 150°C is $74 \text{ GN m}^{-2}$, so assuming linear elastic material behavior the thermal stress is calculated to be $220 \text{ MN m}^{-2}$ ($74{,}000 \times 0.003$). The tensile ductility (failure strain) of the alloy is 0.7% at 150°C, so the pistons should not fail (and in fact they don't). However, the tensile strength of the alloy is $170 \text{ MN m}^{-2}$ at 150°C, so the linear elastic stress analysis indicates that the pistons *should* fail! Explain the reason for this apparent discrepancy.

## CHAPTER 21

# Kinetic Theory of Diffusion

## CONTENTS

## 21.1 INTRODUCTION

We saw in the last chapter that the rate of steady-state creep, $\dot{\varepsilon}_{ss}$, varies with temperature as

$$\dot{\varepsilon}_{ss} = Ce^{-(Q/\bar{R}T)}. \tag{21.1}$$

Here $Q$ is the activation energy for creep (J mol$^{-1}$ or, more usually, kJ mol$^{-1}$), $\bar{R}$ is the universal gas constant (8.31 J mol$^{-1}$ K$^{-1}$) and $T$ is the absolute temperature (K). This is an example of *Arrhenius's law*. Like all good laws, it has great generality. Thus it applies not only to the rate of creep, but also to the rate of oxidation (Chapter 24), the rate of corrosion (Chapter 26), the rate of diffusion (this chapter), even to the rate at which bacteria multiply and milk turns sour. It states that the *rate of the process* (creep, corrosion, diffusion, etc.) *increases exponentially with temperature* (or, equivalently, that the time for a given amount of creep, or of oxidation, *decreases* exponentially with temperature) in the way shown in Figure 21.1. If the rate of a process that follows Arrhenius's law is plotted on a log$_e$ scale against $1/T$, a straight line with a slope of $-Q/\bar{R}$ is obtained (Figure 21.2). The value of $Q$ characterizes the process—a feature we have already used in Chapter 20.

Engineering Materials I: An Introduction to Properties, Applications, and Design, Fourth Edition

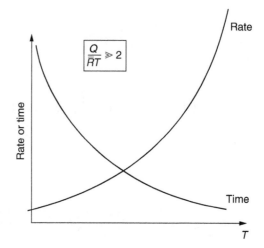

**FIGURE 21.1**
Consequences of Arrhenius's law.

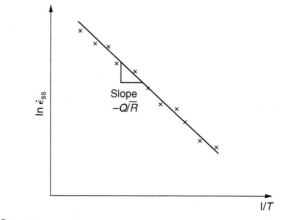

**FIGURE 21.2**
Creep rates follow Arrhenius's law.

In this chapter, we discuss the origin of Arrhenius's law and its application to *diffusion*. In the next, we examine how it is that the rate of diffusion determines that of creep.

## 21.2 DIFFUSION AND FICK'S LAW

First, what do we mean by diffusion? A good way of demonstrating it is to take a glass Petri dish, and run a very shallow layer of water into it. Wait until the water is totally still, and then drop a very small crystal of *potassium permanganate* into it.

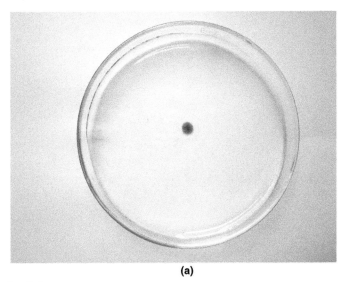

(a)

**FIGURE 21.3(a)**

Elapsed time ≈ 0 minutes. Ring radius = 2.6 mm. Temperature = 18°C.

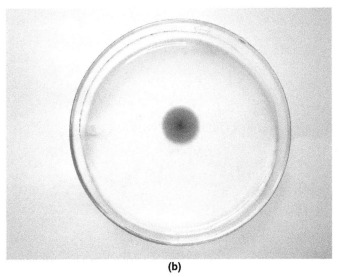

(b)

**FIGURE 21.3(b)**

Elapsed time = 3 minutes. Ring radius = 8.3 mm.

The edge of the crystal will immediately begin to dissolve in the water, producing a dark pink ring of solution centered on the crystal—see Figure 21.3(a). The ring then begins to spread radially outward into the layer of water—initially

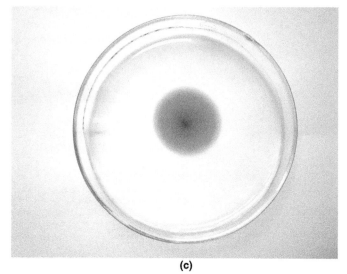

**(c)**

**FIGURE 21.3(c)**
Elapsed time = 9 minutes. Ring radius = 14.9 mm.

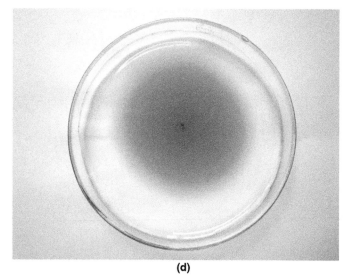

**(d)**

**FIGURE 21.3(d)**
Elapsed time = 57 minutes. Ring radius = 30.3 mm.

very quickly, but progressively more slowly as time passes. Figures 21.3(a) to 21.3(d) are time-lapse photographs showing the ring spreading. This is caused by the movement of potassium permanganate ions by random exchanges with

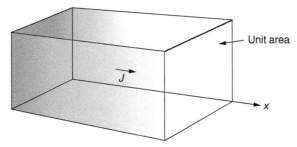

**FIGURE 21.4**
Diffusion down a concentration gradient.

the water molecules. The ions move from regions where they are concentrated to regions where they are less concentrated—the ions diffuse down the *concentration gradient*. This behavior is described by *Fick's first law of diffusion*:

$$J = -D\frac{dc}{dx} \tag{21.2}$$

$J$ is the number of ions diffusing down the concentration gradient per second per unit area; it is called the *flux* of ions (Figure 21.4). $c$ is the concentration of ions in the water—the number of ions (in this case permanganate ions) per unit volume of solution. $x$ is the distance measured along the flux lines, and $D$ is the *diffusion coefficient* for permanganate ions in solution—it has units of $m^2\ s^{-1}$.

This diffusive behavior is not just limited to ions in water—it occurs in all liquids, and more remarkably, in all solids as well. As an example, in the alloy *brass*—a mixture of zinc in copper—zinc atoms diffuse through the solid copper in just the way that ions diffuse through water. Because the materials of engineering are mostly solids, we shall now confine ourselves to talking about diffusion in the solid state.

Physically, diffusion occurs because atoms, even in a solid, are able to move—to jump from one atomic site to another. Figure 21.5 shows a solid in which there is a concentration gradient of black atoms: there are more to the left of the broken line than there are to the right. If atoms jump across the broken line at random, then there will be a *net flux* of black atoms to the right (simply because there are more on the left to jump), and, of course, a net flux of white atoms to the left. Fick's law describes this. It is derived in the following way.

The atoms in a solid vibrate, or oscillate, about their mean positions, with a frequency $\upsilon$ (typically about $10^{13}\ s^{-1}$). The crystal lattice defines these mean positions. At a temperature $T$, the average energy (kinetic plus potential) of a vibrating atom is $3kT$ where $k$ is Boltzmann's constant ($1.38 \times 10^{-23}$ J atom$^{-1}$ K$^{-1}$). But this is only the average energy. As atoms (or molecules) vibrate, they collide, and energy is continually transferred from one to another.

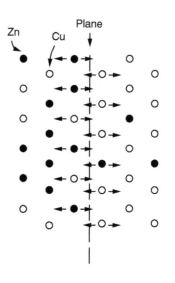

**FIGURE 21.5**

Atom jumps across a plane.

Although the *average* energy is $3kT$, at any instant, there is a certain probability that an atom has more or less than this. A very small fraction of the atoms have, at a given instant, much more—enough, in fact, to jump to a neighboring atom site. It can be shown from statistical mechanical theory that the probability, $p$, that an atom will have, at any instant, an energy $\geq q$ is

$$p = e^{-q/kT} \tag{21.3}$$

Why is this relevant to the diffusion of zinc in copper? Imagine two adjacent lattice planes in the brass with two slightly different zinc concentrations, as shown in exaggerated form in Figure 21.6. Let us denote these two planes as A and B. Now for a zinc atom to diffuse from A to B, down the concentration gradient, it has to "squeeze" between the copper atoms (a simplified statement—but we shall elaborate on it in a moment). This is another way of saying: the zinc atom has to overcome an *energy barrier* of height $q$, as shown in Figure 21.6. Now, the number of zinc atoms in layer A is $n_A$, so the number of zinc atoms that have enough energy to climb over the barrier from A to B at any instant is

$$n_A p = n_A e^{-q/kT} \tag{21.4}$$

For these atoms *actually* to climb over the barrier from A to B, they must of course be moving in the correct direction. The number of times each zinc atom oscillates toward B is $\approx v/6 \text{ s}^{-1}$ (there are six possible directions in which the zinc atoms can move in three dimensions, only one of which is from A to B). Thus the *number* of atoms that *actually* jump from A to B per second is

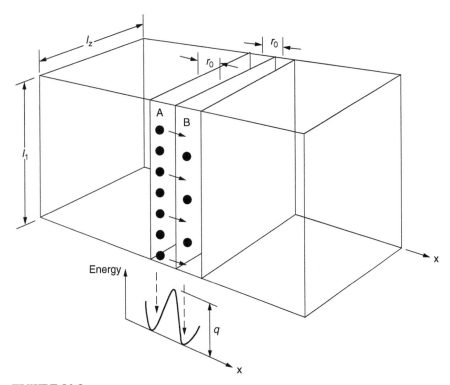

**FIGURE 21.6**
Diffusion requires atoms to cross the energy barrier $q$.

$$\frac{v}{6}n_A e^{-q/kT} \tag{21.5}$$

But, meanwhile, some zinc atoms jump back. If the number of zinc atoms in layer B is $n_B$, the number of zinc atoms that can climb over the barrier from B to A per second is

$$\frac{v}{6}n_B e^{-q/kT} \tag{21.6}$$

Thus the net number of zinc atoms climbing over the barrier per second is

$$\frac{v}{6}(n_A - n_B)e^{-q/kT} \tag{21.7}$$

The area of the surface across which they jump is $l_1\, l_2$, so the net *flux* of atoms, using the definition given earlier, is:

$$J = v\frac{(n_A - n_B)}{6l_1 l_2}e^{-q/kT} \tag{21.8}$$

Concentrations, $c$, are related to numbers, $n$, by

$$c_A = \frac{n_A}{l_1 l_2 r_0}, c_B = \frac{n_B}{l_1 l_2 r_0} \tag{21.9}$$

where $c_A$ and $c_B$ are the zinc concentrations at A and B and $r_0$ is the atom size. Substituting for $n_A$ and $n_B$ in Equation (21.8) gives

$$J = \frac{v}{6} r_0 (c_A - c_B) e^{-q/kT} \tag{21.10}$$

But, $(c_A - c_B)/r_0$ is the concentration gradient, $dc/dx$. The quantity $q$ is inconveniently small and it is better to use the larger quantities $Q = N_A q$ and $\bar{R} = N_A k$, where $N_A$, is Avogadro's number. The quantity $v r_0^2/6$ is usually written as $D_0$. Making these changes gives:

$$J = -D_0 e^{-Q/\bar{R}T} \left(\frac{dc}{dx}\right) \tag{21.11}$$

and this is just Fick's law (Equation 21.2) with

$$D = D_0 e^{-Q/\bar{R}T} \tag{21.12}$$

This method of writing $D$ emphasizes its exponential dependence on temperature, and gives a conveniently sized activation energy (expressed per mole of diffusing atoms rather than per atom). Thinking again of creep, the thing about Equation (21.12) is that the exponential dependence of $D$ on temperature has exactly the same form as the dependence of $\dot{\varepsilon}_{ss}$ on temperature that we seek to explain.

## 21.3 DATA FOR DIFFUSION COEFFICIENTS

Diffusion coefficients are best measured in the following way. A thin layer of a *radioactive isotope* of the diffusing atoms or molecules is plated onto the bulk material (e.g., radioactive zinc onto copper). The temperature is raised to the diffusion temperature for a measured time, during which the isotope diffuses into the bulk. The sample is cooled and sectioned, and the concentration of isotope measured as a function of depth by measuring the radiation it emits. $D_0$ and $Q$ are calculated from the diffusion profile.

Materials handbooks list data for $D_0$ and $Q$ for various atoms diffusing in metals and ceramics—Table 21.1 gives some useful values. Diffusion occurs in polymers, composites, and glasses, too, but the data are less reliable. The last column of the table shows the *normalized activation energy*, $Q/\bar{R}T_M$, where $T_M$ is the melting point (in degrees kelvin). It has been found that, for a given *class of material* (e.g., f.c.c. metals, or refractory oxides), the diffusion parameter $D_0$ for mass transport— the one that is important in creep—is roughly constant; and the activation energy is proportional to the melting temperature $T_M$ (K)

## Table 21.1 Data for Bulk Diffusion

| Material | $D_0(m^2\ s^{-1})$ | $Q(kJ\ mol^{-1})$ | $Q/\bar{R}\ T_M$ |
|---|---|---|---|
| BCC metals | | | |
| Tungsten | $5.0 \times 10^{-4}$ | 585 | 19.1 |
| Molybdenum | $5.0 \times 10^{-5}$ | 405 | 14.9 |
| Tantalum | $1.2 \times 10^{-5}$ | 413 | 16.9 |
| Alpha-iron | $2.0 \times 10^{-4}$ | 251 | 16.6 |
| CPH metals | | | |
| Zinc | $1.3 \times 10^{-5}$ | 91.7 | 15.9 |
| Magnesium | $1.0 \times 10^{-4}$ | 135 | 17.5 |
| Titanium | $8.6 \times 10^{-10}$ | 150 | 9.3 |
| FCC metals | | | |
| Copper | $2.0 \times 10^{-5}$ | 197 | 17.5 |
| Aluminum | $1.7 \times 10^{-4}$ | 142 | 18.3 |
| Lead | $1.4 \times 10^{-4}$ | 109 | 21.8 |
| Gamma-iron | $1.8 \times 10^{-5}$ | 270 | 17.9 |
| Oxides | | | |
| MgO | $1.4 \times 10^{-6}$ | 460 | 17.7 |
| $Al_2O_3$ | $3.0 \times 10^{-2}$ | 556 | 28.0 |
| FeO | $1.0 \times 10^{-2}$ | 326 | 23.9 |
| Interstitial diffusion in iron | | | |
| C in $\alpha$ Fe | $2.0 \times 10^{-6}$ | 84 | 5.6 |
| C in $\gamma$ Fe | $2.3 \times 10^{-5}$ | 147 | 9.7 |
| N in $\alpha$ Fe | $3.0 \times 10^{-7}$ | 76 | 5.1 |
| N in $\gamma$ Fe | $9.1 \times 10^{-5}$ | 168 | 11.6 |
| H in $\alpha$ Fe | $1.0 \times 10^{-7}$ | 14 | 1.0 |

## Table 21.2 Average Values of $D_0$ and $Q/\bar{R}T_M$ for Material Classes

| Material Class | $D_0(m^2\ s^{-1})$ | $Q/\bar{R}\ T_M$ |
|---|---|---|
| BCC metals (W, Mo, Fe below 911°C, etc.) | $1.6 \times 10^{-4}$ | 17.8 |
| CPH metals (Zn, Mg, Ti, etc.) | $5 \times 10^{-5}$ | 17.3 |
| FCC metals (Cu, Al, Ni, Fe above 911°C, etc.) | $5 \times 10^{-5}$ | 18.4 |
| Alkali halides (NaCl, LiF, etc.) | $2.5 \times 10^{-3}$ | 22.5 |
| Oxides (MgO, FeO, $Al_2O_3$, etc.) | $3.8 \times 10^{-4}$ | 23.4 |

so $Q/\bar{R}T_M$, too, is a constant (which is why creep is related to the melting point). This means that many diffusion problems can be solved approximately using the data given in Table 21.2, which shows the average values of $D_0$ and $Q/\bar{R}\ T_M$, for material classes.

## 21.4 MECHANISMS OF DIFFUSION

In our discussion so far we have sidestepped the question of how the atoms in a solid move around when they diffuse. There are several ways in which this can happen. For simplicity, we shall talk only about crystalline solids, although diffusion occurs in amorphous solids as well, and in similar ways.

### Bulk diffusion: Interstitial and vacancy diffusion

Diffusion in the bulk of a crystal can occur by two mechanisms. The first is *interstitial* diffusion. Atoms in all crystals have spaces, or *interstices*, between them, and *small* atoms dissolved in the crystal can diffuse by squeezing between atoms, jumping—when they have enough energy—from one interstice to another (Figure 21.7). Carbon, a small atom, diffuses through steel in this way; in fact C, O, N, B, and H diffuse interstitially in most crystals. These small atoms diffuse very quickly. This is reflected in their very small values of $Q/\bar{R}T_M$, seen in the last column of Table 21.1.

The second mechanism is that of *vacancy* diffusion. When zinc diffuses in brass, for example, the zinc atom (comparable in size to the copper atom) cannot fit into the interstices—the zinc atom has to wait until a *vacancy*, or missing atom, appears next to it before it can move. This is the mechanism by which most diffusion in crystals takes place (Figures 21.8 and 10.4).

### Fast diffusion paths: Grain boundary and dislocation core diffusion

Diffusion in the bulk crystals may sometimes be *short circuited* by diffusion down grain boundaries or dislocation cores. The boundary acts as a planar channel, about two atoms wide, with a local diffusion rate which can be as

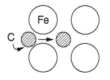

**FIGURE 21.7**
Interstitial diffusion.

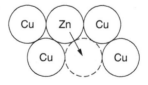

**FIGURE 21.8**
Vacancy diffusion.

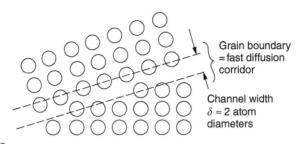

Grain boundary
= fast diffusion
corridor

Channel width
$\delta \approx 2$ atom
diameters

**FIGURE 21.9**
Grain-boundary diffusion.

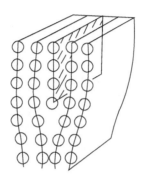

Dislocation core
= fast diffusion
tube, area $(2b)^2$

**FIGURE 21.10**
Dislocation-core diffusion.

much as $10^6$ times greater than in the bulk (Figures 21.9 and 10.4). The dislocation core, too, can act as a high conductivity "wire" of cross-section about
$(2b)^2$, where $b$ is the atom size (Figure 21.10). Of course, their contribution
to the total diffusive flux depends also on how many grain boundaries or
dislocations there are: when grains are small or dislocations numerous, their
contribution becomes important.

## A useful approximation
Because the diffusion coefficient $D$ has units of $m^2\ s^{-1}$, $\sqrt{D \times \text{times(in seconds)}}$
has units of meters. The length scale of $\sqrt{Dt}$ is in fact a good approximation to the
distance that atoms diffuse over a period of time $t$. The proof of this result involves nonsteady diffusion (Fick's second law of diffusion) and we will not go
through it here. However, the approximation

$$x = \sqrt{Dt}$$

is a useful result for estimating diffusion distances in a wide range of situations.

## EXAMPLES

**21.1** It is found that a force $F$ will inject a given weight of a thermosetting polymer into an intricate mold in 30 s at 177°C and in 81.5 s at 157°C. If the viscosity of the polymer follows an *Arrhenius* law, with a rate of process proportional to $e^{-Q/\bar{R}T}$, calculate how long the process will take at 227°C.

**Answer**

3.5

**21.2** Explain what is meant by *diffusion* in materials. Account for the variation of diffusion rates with (a) temperature, (b) concentration gradient and (c) grain size.

**21.3** Use your knowledge of diffusion to account for the following observations:

**a.** Carbon diffuses fairly rapidly through iron at 100°C, whereas chromium does not.

**b.** Diffusion is more rapid in polycrystalline silver with a small grain size than in coarse-grained silver.

**21.4** Give an approximate expression for the time $t$ required for significant diffusion to take place over a distance $x$ in terms of $x$, and the diffusion coefficient, $D$. A component is made from an alloy of copper with 18% by weight of zinc. The concentration of zinc is found to vary significantly over distances of 10 μm. Estimate the time required for a substantial leveling out of the zinc concentration at 750°C. The diffusion coefficient for zinc in the alloy is given by

$$D = D_0 e^{-Q/\bar{R}T}$$

where $\bar{R}$ is the universal gas constant and $T$ is the absolute temperature. The constants $D_0$ and $Q$ have the values 9.5 mm$^2$ s$^{-1}$ and 159 kJ mol$^{-1}$, respectively.

**Answers**

$t = x^2/D$; 23 min.

## CHAPTER 22

# Mechanisms of Creep, and Creep-Resistant Materials

## 22.1 INTRODUCTION

In Chapter 20 we showed that when a material is loaded at high temperature it creeps, that is, it deforms continuously and permanently at a stress that is less than the stress that would cause permanent deformation at room temperature. In order to understand how we can make engineering materials more resistant to creep deformation and creep fracture, we must first look at how creep and creep-fracture take place on an atomic level, that is, we must identify and understand the *mechanisms* by which they take place.

There are two mechanisms of creep: *dislocation creep* (which gives power-law behavior) and *diffusional creep* (which gives linear-viscous creep). The rate of both is usually limited by diffusion, so both follow Arrhenius's law. Creep fracture, too, depends on diffusion. Diffusion becomes appreciable at about $0.3T_M$—that is why materials start to creep above this temperature.

Engineering Materials I: An Introduction to Properties, Applications, and Design, Fourth Edition

**337**

## 22.2 CREEP MECHANISMS: METALS AND CERAMICS

### Dislocation creep (giving power-law creep)

As we saw in Chapter 10, the stress required to make a crystalline material deform plastically is that needed to make the dislocations in it move. Their movement is resisted by (a) the intrinsic lattice resistance, and (b) the obstructing effect of obstacles (e.g., dissolved solute atoms, precipitates formed with undissolved solute atoms, or other dislocations). Diffusion of atoms can "unlock" dislocations from obstacles in their path, and the movement of these unlocked dislocations under the applied stress is what leads to dislocation creep.

How does this unlocking occur? Figure 22.1 shows a dislocation that cannot glide because a precipitate blocks its path. The glide force $\tau b$ per unit length is balanced by the reaction $f_0$ from the precipitate. But unless the dislocation hits the precipitate at its mid-plane (an unlikely event) there is a component of force left over. It is the component $\tau b \tan \theta$, which tries to push the dislocation *out of its slip plane*.

The dislocation cannot *glide* upward by the shearing of atom planes—the atomic geometry is wrong—but the dislocation *can* move upward if atoms at the bottom of the half-plane are able to diffuse away (Figure 22.2). We have come across Fick's law in which diffusion is driven by differences in *concentration*. A *mechanical* force can do exactly the same thing, and this is what leads to the diffusion of atoms away from the "loaded" dislocation, eating away its extra half-plane of atoms until it can clear the precipitate. The process is called "climb," and since it requires diffusion, it can occur only when the temperature is above $0.3T_M$ or so. At the lower end of the creep régime ($0.3–0.5T_M$) core diffusion tends to be the dominant mechanism; at the higher end ($0.5T_M–0.99T_M$) it is bulk diffusion (Figure 22.2).

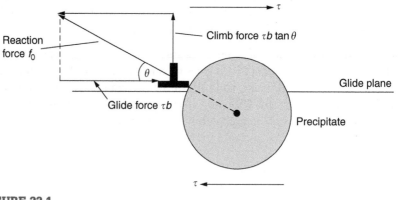

**FIGURE 22.1**

The climb force on a dislocation.

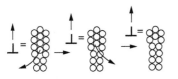 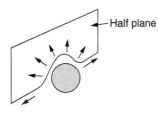

Atoms diffuse away from the bottom of the
half plane. At high $T/T_M$ this takes place
mainly by bulk diffusion through the crystal.

Core diffusion of atoms is
important at lower $T/T_M$.

**FIGURE 22.2**

How diffusion leads to climb.

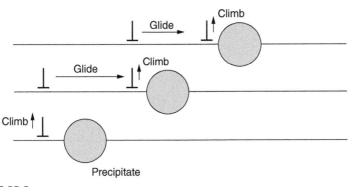

Precipitate

**FIGURE 22.3**

How the climb-glide sequence leads to creep.

Climb unlocks dislocations from the precipitates that pin them and further slip
(or "glide") can then take place (Figure 22.3). Similar behavior takes place for
pinning by solute, and by other dislocations. After a little glide, of course, the
unlocked dislocations bump into the next obstacles, and the whole cycle re-
peats itself. This explains the *progressive, continuous* nature of creep. The role
of diffusion, with diffusion coefficient

$$D = D_0 e^{-Q/\bar{R}T}$$

explains the dependence of creep rate on *temperature*, with

$$\dot{\varepsilon}_{ss} = A\sigma^n e^{-Q/\bar{R}T} \tag{22.1}$$

The dependence of creep rate on applied *stress* $\sigma$ is due to the climb force: the
higher $\sigma$, the higher the climb force $\tau b \tan \theta$, the more dislocations become
unlocked per second, the more dislocations glide per second, and the higher
is the strain rate.

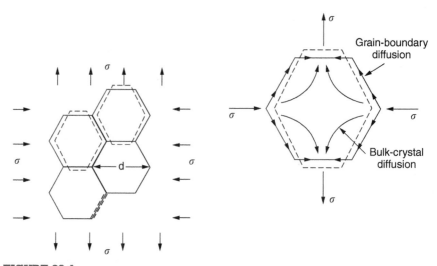

**FIGURE 22.4**
How creep takes place by diffusion.

## Diffusion creep (giving linear-viscous creep)

As the stress is reduced, the rate of power-law creep (Equation (22.1)) falls quickly (remember $n$ is between 3 and 8). But creep does not stop; instead, an alternative mechanism takes over. As Figure 22.4 shows, a polycrystal can extend in response to the applied stress, $\sigma$, by grain elongation; here, $\sigma$ acts again as a mechanical driving force, but this time atoms *diffuse* from one set of the grain faces to the other, and dislocations are not involved. At high $T/T_M$, this diffusion takes place through the crystal itself, that is, by bulk diffusion.

The rate of creep is then obviously proportional to the *diffusion coefficient D* (refer to data in Table 21.1) and to the stress $\sigma$ (because $\sigma$ drives diffusion in the same way that $dc/dx$ does in Fick's law). The creep rate varies as $1/d^2$ where $d$ is the grain size (because when $d$ gets larger, atoms have to diffuse further). Assembling these facts leads to the constitutive equation

$$\dot{\varepsilon}_{ss} = C\frac{D\sigma}{d^2} = \frac{C'\sigma e^{-Q/\bar{R}T}}{d^2} \tag{22.2}$$

where $C$ and $C' = CD_0$ are constants. At lower $T/T_M$, when bulk diffusion is slow, grain-boundary diffusion takes over, but the creep rate is still proportional to $\sigma$. In order that holes do not open up between the grains, grain-boundary *sliding* is required as an accessory to this process.

## Deformation mechanism diagrams

This competition between mechanisms is conveniently summarized on deformation mechanism diagrams (e.g., Figure 22.5). They show the range of stress

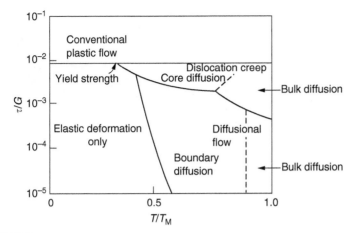

**FIGURE 22.5**

Deformation mechanisms at different stresses and temperatures. $\tau$ is the equivalent shear stress, and $G$ is the shear modulus.

and temperature in which we expect to find each sort of creep (they also show where plastic yielding occurs, and where deformation is simply elastic). Diagrams like these are available for metals and ceramics.

Sometimes creep *is* desirable. Extrusion, hot rolling, hot pressing, and forging are carried out at temperatures at which power-law creep is the dominant mechanism of deformation. Then raising the temperature reduces the pressures required for the operation. The change in forming pressure for a given change in temperature can be calculated from Equation (22.1).

## Creep fracture

Diffusion gives creep. It also gives creep fracture. If you stretch anything for long enough, it will break. You might think that a creeping material would—like toffee—stretch a long way before breaking in two but, for crystalline materials, this is very rare. Indeed, creep fracture (in tension) can happen at unexpectedly small strains, often only 2 to 5%, by the mechanism shown in Figure 22.6. Voids appear on grain boundaries that lie normal to the tensile stress. These are the boundaries to which atoms diffuse to give diffusional creep, coming from the boundaries that lie parallel to the stress. But if the tensile boundaries have voids on them, they act as sources of atoms too, and in doing so, they grow. The voids cannot support load, so the *stress* rises on the remaining intact bits of boundary, the voids grow more and more quickly, until finally they link and fracture takes place.

The lifetime of a component—its time-to-failure, $t_f$—is related to the rate at which it creeps. The equation on the next page applies as a general rule.

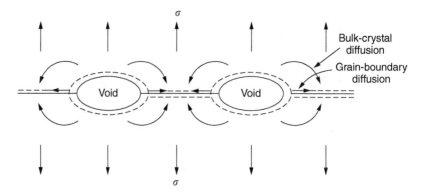

**FIGURE 22.6**
The growth of voids on grain boundaries by diffusion.

$$\dot{\varepsilon}_{ss} t_f = C$$

where $C$ is a constant, roughly 0.1. So, knowing the creep rate, the life can be estimated.

## Designing metals and ceramics to resist power-law creep

If you are asked to select, or even to design, a material that will resist power-law creep, the criteria are:

- Choose a material with a high-melting point, since diffusion (and thus creep-rates) scale as $T/T_M$.
- Maximize obstructions to dislocation motion by alloying to give a solid solution and precipitates—as much of both as possible; the precipitates must, of course, be stable at the service temperature.
- Choose, if this is practical, a solid with a large lattice resistance: this means covalent bonding (as in many oxides, and in silicates, silicon carbide, silicon nitride, and related compounds).

Current creep-resistant materials are successful because they satisfy these criteria.

## Designing metals and ceramics to resist diffusional flow

Diffusional flow is important when grains are small (as they often are in ceramics) and when the component is subject to high temperatures at low loads. To select a material that resists it, you should:

- Choose a material with a high melting temperature.
- Arrange for it to have a large grain size so that diffusion distances are long and grain boundaries do not help diffusion much—single crystals are best of all.

- Arrange for precipitates at grain boundaries to impede grain-boundary sliding.

Metallic alloys are usually designed to resist power-law creep: diffusional flow is only rarely considered. One major exception is the range of directionally solidified ("DS") alloys described in the Case Study of Chapter 23: here special techniques are used to obtain very large grains.

Ceramics, on the other hand, often deform predominantly by diffusional flow (because their grains are small, and the high lattice resistance already suppresses power-law creep). Special heat treatments to increase the grain size can make them more creep-resistant.

## 22.3 CREEP MECHANISMS: POLYMERS

Creep of polymers is a major design problem. The glass temperature $T_G$ for a polymer is a criterion of creep resistance, in much the way that $T_M$ is for a metal or a ceramic. For most polymers, $T_G$ is close to room temperature. Well below $T_G$, the polymer is a glass (often containing crystalline regions—Chapter 5) and is a brittle, elastic solid; rubber, cooled in liquid nitrogen, is an example. Above $T_G$ the Van der Waals bonds within the polymer melt, and it becomes a rubber (if the polymer chains are cross-linked) or a viscous liquid (if they are not). Thermoplastics, which can be molded when hot, are a simple example: well below $T_G$ they are elastic; well above, they are viscous liquids, and flow like treacle.

Viscous flow is a sort of creep. Like diffusion creep, its rate increases linearly with stress and exponentially with temperature, with

$$\dot{\varepsilon}_{ss} = C\sigma e^{-Q/\bar{R}T} \tag{22.3}$$

where $Q$ is the activation energy for viscous flow.

The exponential term appears for the same reason as it does in diffusion; it describes the rate at which molecules can slide past each other, permitting flow. The molecules have a lumpy shape (see Figure 5.9) and the lumps key the molecules together. The activation energy, $Q$, is the energy it takes to push one lump of a molecule past that of a neighboring molecule. If we compare the last equation with that defining the *viscosity* (for the tensile deformation of a viscous material)

$$\eta = \frac{\sigma}{3\dot{\varepsilon}} \tag{22.4}$$

we see that the viscosity is

$$\eta = \frac{1}{3C}e^{+Q/\bar{R}T} \tag{22.5}$$

(The factor 3 appears because the viscosity is defined for shear deformation—as is the shear modulus $G$. For tensile deformation we want the viscous equivalent of Young's modulus $E$. The answer is $3\eta$, for much the same reason that $E \approx (8/3)G \approx 3G$—see Chapter 3.) Data giving $C$ and $Q$ for polymers are available from suppliers. Then Equation (22.3) allows injection molding or pressing temperatures and loads to be calculated.

The temperature range in which most polymers are used is that near $T_G$ when they are neither simple elastic solids nor viscous liquids; they are *visco-elastic* solids. If we represent the elastic behavior by a spring and the viscous behavior by a dashpot, then visco-elasticity (at its simplest) is described by a coupled spring and dashpot (Figure 22.7). Applying a load causes creep, but at an ever-decreasing rate because the spring takes up the tension. Releasing the load allows slow reverse creep, caused by the extended spring.

## Designing polymers to resist creep

The glass temperature of a polymer increases with the degree of cross-linking; heavily cross-linked polymers (e.g., epoxies) are therefore more creep-resistant at room temperature than those that are less cross-linked (e.g., polyethylene). The viscosity of polymers above $T_G$ increases with molecular weight, so the rate of creep there is reduced by having a high molecular weight. Finally, crystalline or partly crystalline polymers (e.g., high-density polyethylene) are more creep-resistant than those that are entirely glassy (e.g., low-density polyethylene).

The creep rate of polymers is reduced by filling them with glass or silica powders, roughly in proportion to the amount of filler added (PTFE on saucepans and polypropylene used for automobile components are both strengthened in this way). Much better creep resistance is obtained with composites containing continuous fibers (GFRP and CFRP) because much of the load is now carried by the fibers which, being very strong, do not creep at all.

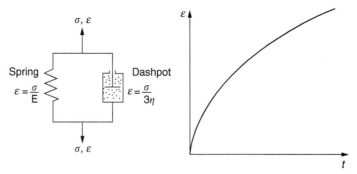

**FIGURE 22.7**
A simple model to describe creep in polymers.

## 22.4 SELECTING MATERIALS TO RESIST CREEP

Classes of industrial applications tend to be associated with certain characteristic temperature ranges. There is the cryogenic range, between –273°C and –20°C, associated with the use of liquid gases such as hydrogen, oxygen, nitrogen, or LPG. There is the régime at and near room temperature (–20 to +150°C) associated with conventional mechanical and civil engineering: household appliances, sporting goods, aircraft structures, and bridges are examples. Above this is the range 150 to 400°C, associated with automobile engines and with food and industrial processing. Higher still are the régimes of steam turbines and superheaters (typically 400–650°C) and of gas turbines and chemical reactors (650–1000°C). Special applications (lamp filaments, rocket nozzles) require materials that withstand even higher temperatures, extending as high as 2800°C.

Materials have evolved to fill the needs of each of these temperature ranges (see Table 22.1 on the next page). Certain polymers, and composites based on them, can be used in applications up to 250°C, and now compete with magnesium and aluminum alloys and with the much heavier cast irons and steels, traditionally used in those ranges. Temperatures above 400°C require special creep resistant alloys: ferritic steels, titanium alloys (lighter, but more expensive), and certain stainless steels.

Stainless steels and ferrous super-alloys really come into their own in the temperature range above this, where they are widely used in steam turbines and heat exchangers. Gas turbines require, in general, nickel-based or cobalt-based super-alloys. Above 1000°C, the refractory metals and ceramics become the only candidates.

## WORKED EXAMPLE

Figure 22.5 introduced the concept of the deformation mechanism diagram, which has fields showing the ranges of stress and temperature (normalized by shear modulus and melting temperature) in which the four distinct *mechanisms* of creep dominate. These creep maps can also be used to display data for creep *rates* as a function of stress and temperature, by adding *strain rate contours* to them.

The figure that follows on page 347 shows an example for pure nickel with a grain size of 1 mm. Because we are now dealing with a specific material, the melting temperature is known, so a *real* temperature scale can be added—see the $T$ (°C) scale at the top of the diagram. With a specific material, the shear modulus is also known, so *real* shear stresses can be added too. These are represented by the dashed lines labeled with stress values. As the temperature increases, these lines curve upward across the diagram—this is because the shear moduli of metals decrease as the temperature increases.

**Table 22.1 Temperature Ranges and Associated Materials**

| Temperature Range | Principal Materials* | Applications |
|---|---|---|
| –273 to –20°C | Austenitic (stainless) steels<br>Aluminum alloys Copper | Liquid $H_2$, $O_2$, $N_2$, LPG equipment |
| –20 to 150°C | Most polymers (max temp: 60 to 150°C)<br>Magnesium alloys (up to 150°C)<br>Aluminum alloys (up to 150°C)<br>Monels and steels | Civil construction<br>Household appliances<br>Automotive<br>Aerospace |
| 150 to 400°C | PEEK, PEK, PI, PPD, PTFE, and PES<br>(up to 250°C)<br>Fiber-reinforced polymers<br>Copper alloys (up to 400°C)<br>Nickel, monels and nickel-silvers | Food processing<br>Automotive (engine) |
| 400 to 575°C | Low-alloy ferritic steels<br>Titanium alloys (up to 450°C)<br>Inconels and nimonics | Heat exchangers<br>Steam turbines<br>Gas turbine compressors |
| 575 to 650°C | Iron-based super-alloys<br>Ferritic stainless steels<br>Austenitic stainless steels<br>Inconels and nimonics | Steam turbines<br>Superheaters<br>Heat exchangers |
| 650 to 1000°C | Austenitic stainless steels<br>Nichromes, nimonics<br>Nickel-based super-alloys<br>Cobalt-based super-alloys | Gas turbines<br>Chemical and petrochemical reactors<br>Furnace components<br>Nuclear construction |
| Above 1000°C | Refractory metals: Mo, W, Ta<br>Alloys of Nb, Mo, W, Ta<br>Ceramics: Oxides<br>$Al_2O_3$, MgO, etc.<br>Nitrides, carbides: $Si_3N_4$, SiC | Special furnaces<br>Lamp filaments<br>Spacecraft heat shields<br>Rocket nozzles |

*Copper alloys include brasses (Cu–Zn alloys), bronzes (Cu–Sn alloys), cupronickels (Cu–Ni alloys), and nickel-silvers (Cu–Sn–Ni–Pb alloys).

Titanium alloys generally mean those based on Ti–V–Al alloys.

Nickel alloys include monels (Ni–Cu alloys), nichromes (Ni–Cr alloys), nimonics, and nickel-based super-alloys (Ni–Fe–Cr–Al–Co–Mo alloys).

Stainless steels include ferritic stainless (Fe–Cr–Ni alloys with <6% Ni) and austenitic stainless (Fe–Cr–Ni alloys with >6.5% Ni). Low-alloy ferritic steels contain up to 4% of Cr, Mo, and V.

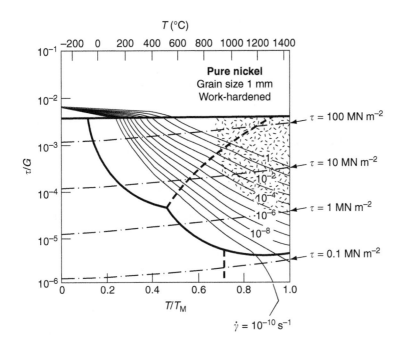

The contours for creep rates can now be added to the diagram—they show the equivalent shear strain rate, in units of $s^{-1}$ (strain per second). These range from the very low—$10^{-10}$ $s^{-1}$—to the very high—$10^0$ (or 1) $s^{-1}$.

These maps can be very useful for seeing how changes to the composition, microstructure, and processing affect the creep mechanism and the creep rate. The four maps for nickel in the next figures have one thing that is different: the grain size. It is 1, 0.1, 0.01 and 0.001 mm in the four maps. As the grain size is reduced, creep by diffusion along the grain boundaries becomes more important. This field expands, and the fields for power-law creep and bulk-diffusion creep contract. Eventually (grain size 0.001 mm) the field for bulk-diffusion creep vanishes altogether.

As long as the *operating point* of the component (stress–temperature) remains within the power-law creep field, the creep rate stays the same. However, as this field contracts and the operating point is taken over by the diffusion-creep field, the creep rates increase dramatically. For example, at 10 MN $m^{-2}$ and 800°C (grain size 0.1 mm) the strain rate is $10^{-6}$ $s^{-1}$ (still inside the power-law creep field). At 10 MN $m^{-2}$ and 800°C (grain size 0.01 mm) the strain rate is $10^{-5}$ $s^{-1}$—ten times greater, because the operating point is now inside the diffusion creep field. The creep rate inside the diffusion creep field is dominated by the grain size (see Equation 22.2), so the small grain size of 0.01 mm makes diffusion creep very fast.

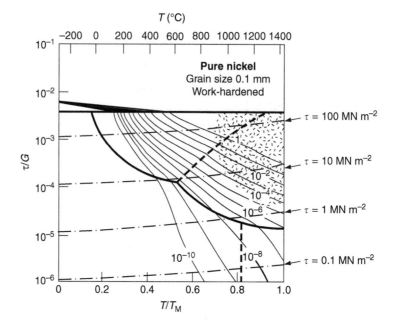

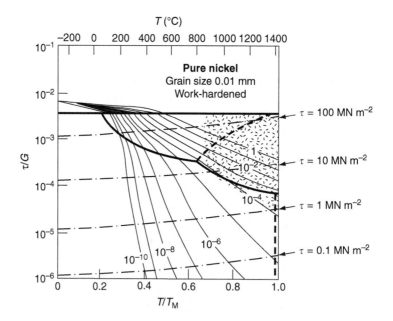

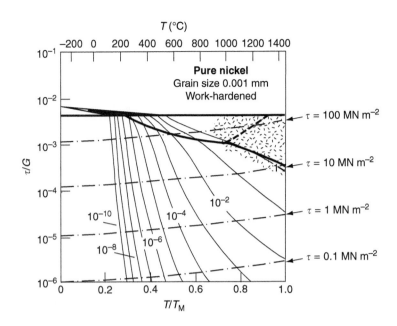

# EXAMPLES

**22.1** What is climb? Why does it require diffusion?

**22.2** How do the processes of climb and glide produce creep deformation?

**22.3** How does the rate of dislocation (power-law) creep depend on stress and temperature?

**22.4** What is diffusion creep? How does the rate of diffusion creep depend on stress, temperature, and grain size?

**22.5** Why does core diffusion take over from bulk diffusion at lower temperatures in power-law creep?

**22.6** Why does grain-boundary diffusion take over from bulk diffusion at lower temperatures in diffusion creep?

**22.7** How would you design metals and ceramics to resist creep?

**22.8** How do polymers creep? How would you design polymers to resist creep?

## CHAPTER 23

# The Turbine Blade—A Case Study in Creep-Limited Design

## CONTENTS

## 23.1 INTRODUCTION

In the last chapter we saw how a basic knowledge of the mechanisms of creep was an important aid to the development of materials with good creep properties. An impressive example is in the development of materials for the high-pressure stage of a modern aircraft gas turbine (see Figure 1.3). Here we examine the properties such materials must have, the way in which the present generation of materials has evolved, and the likely direction of their future development.

The *ideal* thermodynamic efficiency of a heat engine is given by

$$\frac{T_1 - T_2}{T_1} = 1 - \frac{T_2}{T_1} \tag{23.1}$$

where $T_1$ and $T_2$ are the absolute temperatures of the heat source and heat sink respectively. Obviously the greater $T_1$, the greater the maximum efficiency that can be derived from the engine. In practice the efficiency is a good deal less than

Engineering Materials I: An Introduction to Properties, Applications, and Design, Fourth Edition

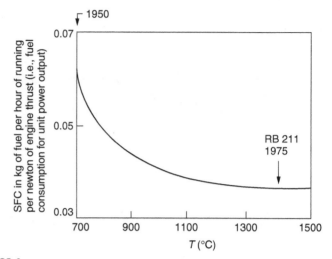

**FIGURE 23.1**

Turbofan efficiency at different inlet temperatures.

ideal, but an increase in combustion temperature in a turbofan engine will, nevertheless, generate an increase in engine efficiency.

Figure 23.1 shows the variation in efficiency of a turbofan engine plotted as a function of the turbine inlet temperature. In 1950 a typical aeroengine operated at 700°C. The incentive then to increase the inlet temperature was strong, because of the steepness of the fuel-consumption curve at that temperature. By 1975 a typical engine (the RB211, for instance) operated at 1350°C, with a 50% saving in fuel per unit power output over the 1950 engines. But is it worth raising the temperature further?

The shallowness of the consumption curve at 1400°C suggests that it might not be profitable; but there is a second factor: power-to-weight ratio. Figure 23.2 shows a typical plot of the power output of a particular engine against turbine inlet temperature. This increases *linearly* with the temperature. If the turbine could both run at a higher temperature and be made of a lighter material there would be a double gain, with important financial benefits of increased payload.

## 23.2 PROPERTIES REQUIRED OF A TURBINE BLADE

Let us first examine the development of turbine-blade materials to meet the challenge of increasing engine temperatures. Although so far we have been stressing the need for excellent creep properties, a turbine-blade alloy must satisfy other criteria too. They are listed in Table 23.1.

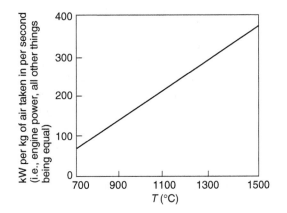

**FIGURE 23.2**

Turbofan power at different inlet temperatures.

| Table 23.1 Alloy Requirements |
| --- |
| **Criteria** |
| Resistance to creep |
| Resistance to high-temperature oxidation |
| Toughness |
| Thermal fatigue resistance |
| Thermal stability |
| Low density |

The first—creep—is our interest here. The second—resistance to oxidation—is the subject of Chapter 24. Toughness and fatigue resistance (Chapters 13 and 17) are obviously important: blades must be tough enough to withstand the impact of birds and hailstones; and changes in the power level of the engine produce mechanical and thermal stresses which—if the blade material is wrongly chosen—cause *thermal fatigue*. The alloy composition and structure must remain *stable* at high temperature—precipitate particles can dissolve away if the alloy is overheated and the creep properties will then degrade significantly. Finally, the *density* must be as low as possible—not so much because of blade weight but because of the need for stronger and hence heavier turbine discs to take the radial load.

These requirements severely limit our choice of creep-resistant materials. For example, ceramics, with their high softening temperatures and low densities, are ruled out for aeroengines because they are far too brittle (they are under evaluation for use in land-based turbines, where the risks and consequences of sudden failure are less severe—see the following). Cermets offer no great advantage because their metallic matrices soften at much too low a temperature. The materials that best fill present needs are the *nickel-based super-alloys*.

## 23.3 NICKEL-BASED SUPER-ALLOYS

The alloy used for turbine blades in the high-pressure stage of an aircraft turbo-fan engine is a classic example of a material designed to be resistant to dislocation (power-law) creep at high stresses and temperatures. At take-off, the blade is subjected to stresses approaching 250 MN m$^{-2}$, and the design specification requires that this stress shall be supported for 30 h at 850°C without more than a 0.1% irreversible creep strain. In order to meet these stringent requirements, an alloy based on nickel has evolved with the rather mind-boggling specification given in Table 23.2.

No one tries to remember exact details of this or similar alloys. But the point of all these complicated additions of foreign atoms to the nickel is straightforward. It is: (a) to have as many atoms in solid solution as possible (the cobalt, the tungsten, and the chromium); (b) to form stable, hard precipitates of compounds such as $Ni_3Al$, $Ni_3Ti$, MoC, TaC to obstruct the dislocations; and (c) to form a protective surface oxide film of $Cr_2O_3$ to protect the blade itself from attack by oxygen (we shall discuss this in Chapter 25). Figure 23.3(a, b) shows a piece of a nickel-based super-alloy cut open to reveal its complicated structure.

These super-alloys are remarkable materials. They resist creep so well that they can be used at 850°C—and since they melt at 1280°C, this is 0.72 of their (absolute) melting point. They are so hard that they cannot be machined easily by normal methods, and must be precision-cast to their final shape. This is done by *investment casting*: a precise wax model of the blade is embedded in an alumina paste which is then fired; the wax burns out leaving an accurate mold from which one blade can be made by pouring liquid super-alloy into it (Figure 23.4).

Cast in this way, the grain size of such a blade is small (Figure 23.4). The strengthening caused by alloying successfully suppresses power-law creep,

**Table 23.2** Composition of Typical Creep-Resistant Blade

| Metals | Wt.% | Metals | Wt.% |
|---|---|---|---|
| Ni | 59 | Mo | 0.25 |
| Co | 10 | C | 0.15 |
| W | 10 | Si | 0.1 |
| Cr | 9 | Mn | 0.1 |
| Al | 5.5 | Cu | 0.05 |
| Ta | 2.5 | Zr | 0.05 |
| Ti | 1.5 | B | 0.015 |
| Hf | 1.5 | S | <0.008 |
| Fe | 0.25 | Pb | <0.0005 |

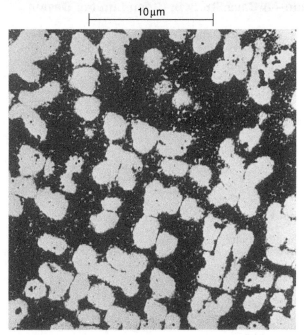

**FIGURE 23.3(a)**

A piece of a nickel-based super-alloy cut open to show the structure: there are two sizes of precipitates in the alloy—the large white precipitates, and the much smaller black precipitates in between.

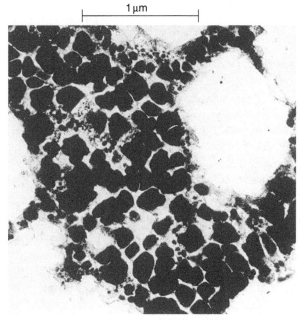

**FIGURE 23.3(b)**

As displayed in Figure 23.3(a), but showing a much more magnified view of the structure, in which the small precipitates are more clearly identifiable.

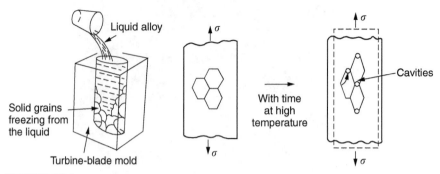

**FIGURE 23.4**

Investment casting of turbine blades. This produces a fine-grained material which may undergo a fair amount of diffusion creep, and which may fail rather soon by cavity formation.

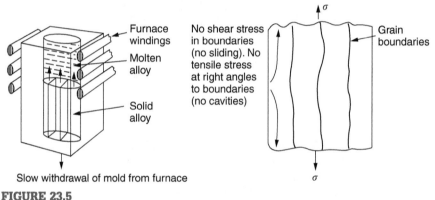

**FIGURE 23.5**

Directional solidification of turbine blades.

but at $0.72T_M$, diffusional flow then becomes a problem. The way out is to increase the grain size, or even make single-crystal blades with no grain boundaries at all. In addition, *creep damage* (Chapter 22) accumulates at grain boundaries; we can obviously delay failure by eliminating grain boundaries, or aligning them parallel to the applied stress.

To do this, we *directionally solidify* the alloys (see Figure 23.5) to give long grains with grain boundaries parallel to the applied stress (or single-crystal blades with no grain boundaries). The diffusional distances required for diffusional creep are then very large (greatly cutting down the rate of diffusional creep); in addition, there is no driving force for grain-boundary sliding or for cavitation at grain boundaries. Directionally solidified (DS) and single crystal (SX) alloys are standard in all high-performance engines. The improved creep properties of these alloys allows the engine to run at a flame temperature approximately 200°C higher than before.

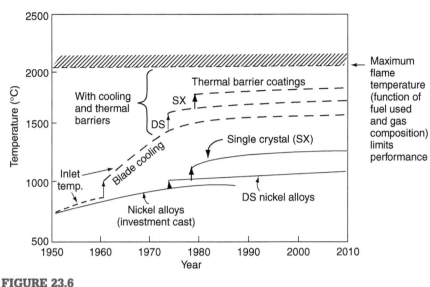

**FIGURE 23.6**
Temperature evolution and materials trends in turbine blades.

How was this type of alloy discovered in the first place? Well, the fundamental principles of creep-resistant materials design that we talked about help us to select the more promising alloy recipes and discard the less promising ones fairly easily. Thereafter, the approach is an empirical one. Alloys having different recipes are made up in the laboratory and tested for creep, oxidation, toughness, thermal fatigue, and stability. The choice eventually narrows down to a few and these are subjected to more stringent testing, coupled with judicious tinkering with the alloy recipe. Small improvements are continually made in alloy composition and in the manufacture of finished blades, which evolve by a sort of creepy Darwinism, the fittest (in the sense of Table 23.1) surviving.

Figure 23.6 shows how this evolutionary process has resulted in a continual improvement of creep properties of nickel alloys over time. The figure also shows how improvements in alloy manufacture—directional solidification and growing single crystals—have helped to increase the operating temperature. Nevertheless, it is clear from the graph that improvements in nickel alloys are now nearing the point of diminishing returns.

## 23.4 ENGINEERING DEVELOPMENTS—BLADE COOLING

Figure 23.6 shows that up until 1960 turbine inlet temperatures were virtually the same as the metal temperatures. After 1960 there was a sharp divergence, with inlet temperatures substantially above the temperatures of the blade metal itself—indeed, the gas temperature is greater than the *melting point* of the blades. Impossible? Not at all. It is done by air-cooling the blades, and in a cunning way. In the earliest form of cooled blade, cooling air from the compressor stage of the engine was fed through ports passing along the full length of the blade, and was ejected into the gas stream at the blade end (see Figure 23.7). This *internal cooling* of the blade enabled the inlet temperature to be increased immediately by 100°C with no change in alloy composition.

A later improvement was *film cooling*, in which the air was ejected over the surface of the blade, through little holes connected to the central channel, giving a cool boundary layer between the blade and the hot gases. More recently, blades have been coated with thin layers of ceramics—such as zirconium oxide (zirconia). These protect the blades from oxidation and hot corrosion, but because they are also poor conductors of heat, they add a thin layer of thermal insulation between the hot gases and the metal of the blade—they are "thermal barrier coatings." A program of continuous improvement in the efficiency of heat transfer by refinements of this type has made it possible for modern turbofans to operate at temperatures that are no longer dominated by the properties of the material.

But blade cooling has reached a limit: ducting still more cold air through the blades will begin to *reduce* the thermal efficiency by taking too much heat away from the combustion chamber. What do we do now?

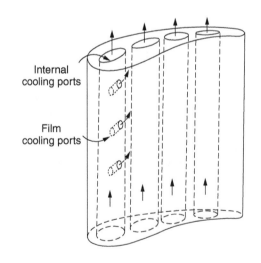

Internal cooling ports

Film cooling ports

**FIGURE 23.7**
Air-cooled blades.

# 23.5 FUTURE DEVELOPMENTS: HIGH-TEMPERATURE CERAMICS

The ceramics best suited for structural use at high temperatures ($>1000°C$) are listed in Table 23.3 and compared with nickel-based super-alloys. The comparison shows that all the ceramics have attractively low densities, high moduli, and high melting points (and thus excellent creep strength at $1000°C$); and many are completely resistant to oxidation—they are oxides already. But several have poor thermal conductivity (leading to high thermal stresses) and all have very low toughness.

Alumina ($Al_2O_3$) was one of the first pure oxides to be produced in complex shapes, but its combination of high expansion coefficient, poor conductivity, and low toughness gives it bad thermal-shock resistance.

Glass ceramics are made by forming a complex silicate glass and then causing it to crystallize partly. They are widely used for ovenware and for heat exchangers for small engines. Their low thermal expansion gives them much better thermal shock resistance than most other ceramics, but the upper working temperature of $900°C$ (when the glass phase softens) limits their use.

The covalently-bonded silicon carbide, silicon nitride, and sialons (alloys of $Si_3N_4$ and $Al_2O_3$) seem to be the best bet for high-temperature structural use. Their creep resistance is outstanding up to $1300°C$, and their low expansion and high conductivity (better than nickel alloys!) makes them resist thermal shock well in spite of their typically low toughness. They can be formed by hot-pressing fine powders, by vapor deposition, or by nitriding silicon which is already pressed to shape: in this way, precise shapes (e.g., turbine blades) can be formed without the need for machining (they are much too hard to machine). Their brittleness, however, creates major design problems. It might be overcome by creating ceramics with a fibrous structure. In an effort to do this, ceramic-matrix components are under development: they combine strong, exceptionally perfect, fibers (e.g., silicon carbide or alumina, grown by special techniques) in a matrix of ceramic (silicon carbide or alumina again).

# 23.6 COST EFFECTIVENESS

Any major materials development program, such as that on DS/SX super-alloys, can only be undertaken if a successful outcome would be cost effective. As Figure 23.8 shows, the costs of development can be colossal. Even before a new material is out of the laboratory, $200 million can have been spent, and failure in an engine test can be expensive. Because the performance of a new alloy cannot finally be verified until it has been extensively flight-tested, at each

**Table 23.3 Ceramics for High-Temperature Structures**

| Material | Density (Mg m$^{-3}$) | Melting or Decomposition (D) Temperature (K) | Modulus (GN m$^{-2}$) | Expansion Coefficient × 10$^{+6}$ (K$^{-1}$) | Thermal Conductivity at 1000 K (Wm$^{-1}$ K$^{-1}$) | Fracture Toughness K$_c$(MN m$^{-3/2}$) |
|---|---|---|---|---|---|---|
| Alumina, Al$_2$O$_3$ | 4.0 | 2320 | 360 | 6.9 | 7 | ≈5 |
| Glass-ceramics (pyrocerams) | 2.7 | >1700 | ≈120 | ≈3 | ≈3 | ≈3 |
| Hot-pressed silicon nitride, Si$_3$N$_4$ | 3.1 | 2173 (D) | 310 | 3.1 | 16 | ≈5 |
| Hot-pressed silicon carbide, SiC | 3.2 | 3000 (D) | ≈420 | 4.3 | 60 | ≈3.5 |
| Nickel alloys (nimonics) | 8.0 | 1600 | 200 | 12.5 | 12 | ≈100 |

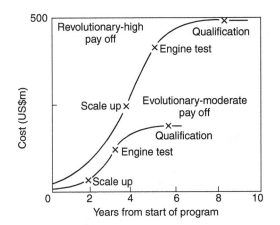

**FIGURE 23.8**
Development costs of new turbine-blade materials.

stage of development risk decisions have to be taken whether to press ahead, or cut losses and abandon the program.

One must consider, too, the cost of the materials themselves. Some of the metals used in conventional nickel alloys—such as hafnium—are hideously expensive and extremely scarce; and the use of greater and greater quantities of exotic materials in an attempt to improve the creep properties will drive the cost up. A single-crystal (SC) turbine blade currently costs US$10,000, so a set of blades for a high-pressure turbine disc typically runs to US$1 million! But expensive though it is, the cost of the turbine blades is still only a fraction of the cost of an engine, or of the fuel it will consume in its lifetime. Blade costs *are* important, but if new alloys offer improved life or inlet temperature, there is a strong incentive to pursue them.

## WORKED EXAMPLE

The diagrams that follow show deformation-mechanism maps for pure nickel and a nickel-based superalloy (MAR–M200), both having a grain size of 0.1 mm. The effect of the alloying additions in the superalloy (solid solution and precipitate strengthening) is dramatic—the yield stress is increased by more than 10 times, and the field of power-law creep is reduced to a small zone at the top right corner of the diagram. At 1000°C, the bottom of the power-law creep field is situated at a shear stress of 1 MN m$^{-2}$ for nickel, but 100 MN m$^{-2}$ for the superalloy—a factor of 100 times higher in stress. The result is that the creep of the superalloy in a gas turbine application is governed by diffusion creep. However, with a grain size as small as 0.1 mm, the creep rates are far too high. At a shear stress of 10 MN m$^{-2}$ and 1000°C, the shear strain rate is $10^{-7}$ s$^{-1}$—in one year of operation, this would give a creep strain of 100%!

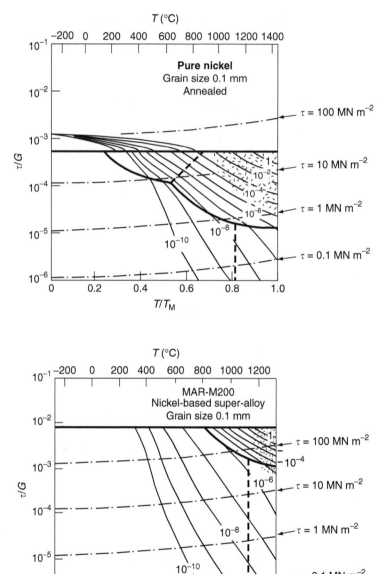

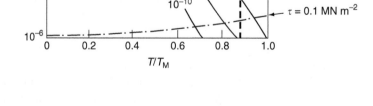

Since the creep rate scales as $1/d^2$ (Equation 22.2), increasing the grain size to 1 mm would decrease the strain rate by 100 times. This would give a creep strain of 1%—better, but still not good enough. This is why it is so important for hot-end applications to have a large grain size—or no grain size at all (SX). A 5 mm grain size (effectively through the thickness of the blade) would decrease the strain rate by 2500 times—to 0.04% per year, which is OK. Using DS or SX blades would suppress diffusion creep even more, and might allow the blade to be run at a higher stress.

## EXAMPLES

**23.1** Why is there an incentive to maximize the turbine inlet temperature in aircraft engines?

**23.2** Why are DS and single-crystal turbine blades better than conventional investment-cast blades?

**23.3** What are the advantages and disadvantages of high-performance ceramics as turbine-blade materials?

**23.4** How can blade cooling be used to increase engine performance without changing the blade material?

**23.5** How can coating blades with a thin layer of oxide ceramic increase engine performance?

PART

# Oxidation and Corrosion

## CHAPTER 24

# Oxidation of Materials

## CONTENTS

## 24.1 INTRODUCTION

In the last chapter we said that one of the requirements of a high-temperature material—in a turbine blade, or a super-heater tube, for example—was that it should resist attack by gases at high temperatures and, in particular, that it should resist oxidation. Turbine blades *do* oxidize in service, and react with $H_2S$, $SO_2$, and other combustion products. Excessive attack of this sort is obviously undesirable in such a highly stressed component. Which materials best resist oxidation, and how can the resistance to gas attack be improved?

The earth's atmosphere is oxidizing. We can get some idea of oxidation-resistance by using the earth as a laboratory, and looking for materials that survive well in its atmosphere. All around us we see ceramics: the earth's crust (Chapter 2) is almost entirely made of oxides, silicates, aluminates, and other compounds of oxygen; and being oxides already, they are completely stable. Alkali halides, too, are stable: NaCl, KCl, NaBr—all are widely found in nature. By contrast, metals are not stable: only gold is found in "native" form under

Engineering Materials I: An Introduction to Properties, Applications, and Design, Fourth Edition

normal circumstances (it is completely resistant to oxidation at all temperatures); all the others in our data sheets will oxidize in contact with air.

Polymers are not stable either: most will burn if ignited, meaning that they oxidize readily. Coal and oil (the raw materials for polymers), it is true, are found in nature, but that is only because geological accidents have sealed them off from all contact with air. A few polymers, among them PTFE (a polymer based on $-CF_2-$), are so stable that they survive long periods at high temperatures, but they are the exceptions. And polymer-based composites, of course, are just the same: wood is not noted for its high-temperature oxidation resistance.

## 24.2 THE ENERGY OF OXIDATION

This tendency of many materials to react with oxygen can be quantified by laboratory tests that measure the energy needed for the reaction

$$\text{Material} + \text{Oxygen} + \text{Energy} \rightarrow \text{Oxide of material}$$

If this energy is *positive*, the material is stable; if *negative*, it will oxidize. The bar-chart of Figure 24.1 shows the energies of oxide formation for our four categories of materials; numerical values are given in Table 24.1.

## 24.3 RATES OF OXIDATION

When designing with oxidation-prone materials, it is obviously vital to know how *fast* the oxidation process is going to be. Intuitively one might expect that, the larger the energy released in the oxidation process, the faster the *rate* of oxidation. For example, one might expect aluminum to oxidize 2.5 times faster than iron from the energy data in Figure 24.1. In fact, aluminum oxidizes much more slowly than iron. Why should this happen?

If you heat a piece of bright iron in a gas flame, the oxygen in the air reacts with the iron at the surface of the metal where the oxygen and iron atoms can contact, creating a thin layer of iron oxide on the surface, and making the iron turn black. The layer grows in thickness, quickly at first, and then more slowly because iron atoms now have to diffuse through the film before they make contact and react with oxygen. If you plunge the piece of hot iron into a dish of water the shock of the quenching breaks off the iron oxide layer, and you can see the pieces of layer in the dish. The iron surface now appears bright again, showing that the shock of the quenching has completely stripped the metal of the oxide layer which formed during the heating; if it were reheated, it would oxidize at the old rate.

The important thing about the oxide film is that it acts as a *barrier* which keeps the oxygen and iron atoms apart and cuts down the rate at which these atoms

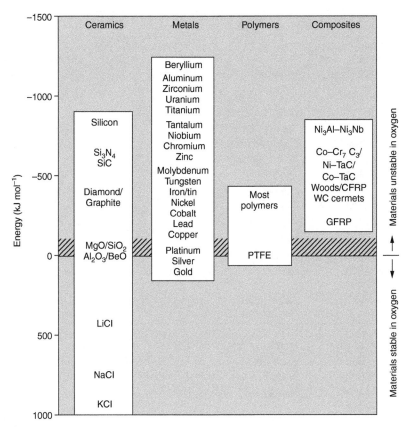

**FIGURE 24.1**

Energies of formation of oxides at 273 K in kJ mol$^{-1}$ of oxygen $O_2$.

react to form more iron oxide. Aluminum, and most other materials, form oxide barrier layers in just the same sort of way—but the oxide layer on aluminum is a *much* more effective barrier than the oxide film on iron is.

How do we measure rates of oxidation in practice? Because oxidation proceeds by the addition of oxygen atoms to the surface of the material, the *weight* of the material usually goes up in proportion to the amount of material that has become oxidized. This weight increase, $\Delta m$, can be monitored continuously with time $t$ in the way illustrated in Figure 24.2. Two types of behavior are usually observed at high temperature. The first is *linear oxidation*, with

$$\Delta m = k_L t \tag{24.1}$$

where $k_L$ is a *kinetic constant*. Naturally, $k_L$ is usually positive. (In a few materials, however, the oxide evaporates away as soon as it has formed; the material then *loses* weight and $k_L$ is then negative.)

**Table 24.1** Energies of Formation of Oxides at 273 K

| Material | Oxide | Energy (kJ mol$^{-1}$ of Oxygen, $O_2$) | Material | Oxide | Energy (kJ mol$^{-1}$ of Oxygen, $O_2$) |
|---|---|---|---|---|---|
| Beryllium | BeO | −1182 | Woods, most polymers, CFRP | | ≈ −400 |
| Magnesium | MgO | −1162 | | | |
| Aluminum | $Al_2O_3$ | −1045 | Diamond, graphite | $CO_2$ | −389 |
| Zirconium | $ZrO_2$ | −1028 | Tungsten carbide cermet (mainly WC) | $WO_3+$ $CO_2$ | −349 |
| Uranium | $U_3O_8$ | ≈ −1000 | | | |
| Titanium | TiO | −848 | Lead | $Pb_3O_4$ | −309 |
| Silicon | $SiO_2$ | −836 | Copper | CuO | −254 |
| Tantalum | $Ta_2O_5$ | −764 | GFRP | | ≈ −200 |
| Niobium | $Nb_2O_5$ | −757 | Platinum | $PtO_2$ | ≈ −160 |
| Chromium | $Cr_2O_3$ | −701 | Silver | $Ag_2O$ | −5 |
| Zinc | ZnO | −636 | PTFE | | ≈ zero |
| Silicon nitride, $Si_3N_4$ | $3SiO_2+2N_2$ | ≈ −629 | Gold | $Au_2O_3$ | +80 |
| Silicon carbide, SiC | $SiO_2+CO_2$ | ≈ −580 | Alkali halides | | ≈ +400 to ≈ +1400 |
| Molybdenum | $MoO_2$ | −534 | | | |
| Tungsten | $WO_3$ | −510 | Magnesia, MgO | | |
| Iron | $Fe_3O_4$ | −508 | Silica, $SiO_2$ | Higher oxides | Large and positive |
| Tin | SnO | −500 | Alumina, $Al_2O_3$ | | |
| Nickel | NiO | −439 | Beryllia, BeO | | |
| Cobalt | CoO | −422 | | | |

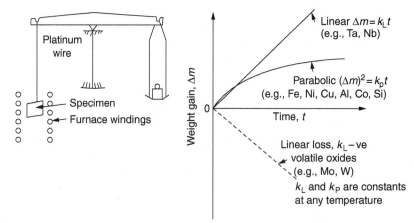

**FIGURE 24.2**
Measurement of oxidation rates.

The second type of oxidation behavior is *parabolic oxidation*, with

$$(\Delta m)^2 = k_p t \tag{24.2}$$

where $k_p$ is another kinetic constant, this time always positive.

Oxidation rates follow Arrhenius's law (Chapter 21), that is, the kinetic constants $k_L$ and $k_p$ increase exponentially with temperature:

$$k_L = A_L e^{-Q_L/\bar{R}T} \text{ and } k_p = A_p e^{-Q_P/\bar{R}T} \tag{24.3}$$

$A_L$ and $A_p$, $Q_L$, and $Q_p$ are constants. As the temperature is increased, the rate of oxidation increases exponentially (Figure 24.3).

Finally, oxidation rates obviously increase with increasing partial pressure of oxygen, although rarely in a simple way. The partial pressure of oxygen in a gas turbine atmosphere, for example, may well be very different from that in air, and it is important to conduct oxidation tests on high-temperature components under the right conditions.

## 24.4 DATA

It is important from the design standpoint to know how much material is replaced by oxide. The mechanical properties of the oxide are usually grossly inferior to the properties of the material (e.g., oxides are comparatively brittle), and even if the layer is firmly attached to the material—which is certainly not always the case—the effective section of the component is reduced. The reduction in the section of a component can obviously be calculated from data for $\Delta m$.

Table 24.2 gives the times for a range of materials required to oxidize them to a depth of 0.1 mm from the surface when exposed to air at $0.7 T_M$ (a typical figure for the operating temperature of a turbine blade or similar component): these times vary by many orders of magnitude, and clearly show that there is no correlation between oxidation rate and energy needed for the reaction (see Al, W

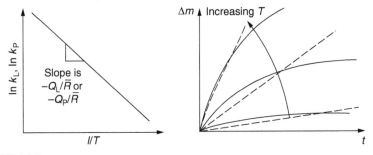

**FIGURE 24.3**
Oxidation rates increase with temperature according to Arrhenius's law.

**Table 24.2** Time in Hours for Material to Be Oxidized to a Depth of 0.1 mm at 0.7 $T_M$ in Air

| Material | Time | Melting Point (K) | Material | Time | Melting Point (K) |
|---|---|---|---|---|---|
| Au | Infinite? | 1336 | Ni | 600 | 1726 |
| Ag | Very long | 1234 | Cu | 25 | 1356 |
| Al | Very long | 933 | Fe | 24 | 1809 |
| $Si_3N_4$ | Very long | 2173 | Co | 7 | 1765 |
| SiC | Very long | 3110 | Ti | <6 | 1943 |
| Sn | Very long | 505 | WC cermet | <5 | 1700 |
| Si | $2 \times 10^6$ | 1683 | Ba | «0.5 | 983 |
| Be | $10^6$ | 1557 | Zr | 0.2 | 2125 |
| Pt | $1.8 \times 10^5$ | 2042 | Ta | Very short | 3250 |
| Mg | $>10^5$ | 923 | Nb | Very short | 2740 |
| Zn | $>10^4$ | 692 | U | Very short | 1405 |
| Cr | 1600 | 2148 | Mo | Very short | 2880 |
| Na | >1000 | 371 | W | Very short | 3680 |
| K | >1000 | 337 | | | |

Note: Data subject to considerable variability due to varying degrees of material purity, prior surface treatment, and presence of atmospheric impurities such as sulfur.

for extremes: Al, *very* slow—energy $= -1045$ kJ mol$^{-1}$ of $O_2$; W, very fast—energy $= -510$ kJ mol$^{-1}$ of $O_2$).

## 24.5 MICROMECHANISMS

Figure 24.4 illustrates the mechanism of parabolic oxidation. The reaction

$$M + O \rightarrow MO$$

(M is the oxidizing material and O is oxygen) goes in two steps. First M forms an ion, releasing electrons, *e*:

$$M \rightarrow M^{++} + 2e$$

These electrons are then absorbed by oxygen to give an oxygen ion:

$$O + 2e \rightarrow O^{--}$$

*Either* the $M^{++}$ and the two *e*s diffuse outward through the film to meet the $O^{--}$ at the outer surface, *or* the oxygen diffuses inward (with two electron holes) to meet the $M^{++}$ at the inner surface. The concentration gradient of oxygen is simply the concentration in the gas, *c*, divided by the film thickness, *x*; and the rate of growth of the film $dx/dt$ is obviously proportional to the flux of atoms diffusing through the film. So, from Fick's law (Equation 21.2):

$$\frac{dx}{dt} \, \alpha \, D \, \frac{c}{x}$$

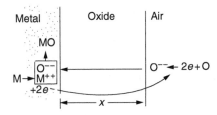

**Case 1.** $M^{++}$ diffuses very slowly in oxide. Oxide grows at metal–oxide interface.
*Examples:* Ti, Zr, U

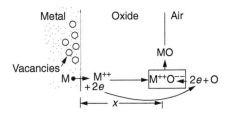

**Case 2.** $O^{--}$ diffuses very slowly in oxide. Oxide grows at oxide–air interface.
Vacancies form between metal and oxide.
*Examples:* Cu, Fe, Cr, Co

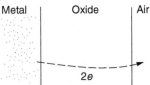

**Case 3.** Electrons move very slowly. Oxide can grow (slowly) at metal–oxide interface
or oxide–air interface depending on whether $M^{++}$ diffuses faster than $O^{--}$ or not.
*Example:* Al

**FIGURE 24.4**
How oxide layers grow to give parabolic oxidation behavior.

Integrating gives

$$x^2 = k_p t \qquad (24.4)$$

where

$$k_p \, \alpha \, cD_0 e^{-Q/\bar{R}T} \qquad (24.5)$$

This growth law has exactly the form of Equation (24.2) and the kinetic constant is analogous to* that of Equation (24.3). This shows us why some films are more protective than others: protective films are those with low diffusion

---

\* It does not have the same value, however, because Equation (24.5) refers to thickness gain and not *mass* gain; the two can be easily related if quantities such as the density of the oxide are known.

coefficients—and thus high melting points. That is one reason why $Al_2O_3$ protects aluminum, $Cr_2O_3$ protects chromium, and $SiO_2$ protects silicon so well, whereas $Cu_2O$ and even FeO (which have lower melting points) are less protective. But there is an additional reason: electrons must also pass through the film and these films are insulators (the electrical resistivity of $Al_2O_3$ is $10^9$ times greater than that of FeO).

Although our simple oxide film model explains most of the experimental observations we have mentioned, it does not explain the linear laws. How, for example, can a material *lose* weight linearly when it oxidizes as is sometimes observed (see Figure 24.2)? Well, some oxides (e.g., $MoO_3$, $WO_3$) are very volatile. During oxidation of Mo and W at high temperature, the oxides evaporate as soon as they are formed, and offer no barrier at all to oxidation. Oxidation, therefore, proceeds at a rate that is independent of time, and the material loses weight because the oxide is *lost*. This behavior explains the catastrophically rapid section loss of Mo and W shown in Table 24.2.

The explanation of a linear weight *gain* is more complex. Basically, as the oxide film thickens, it develops cracks, or partly lifts away from the material, so the barrier between material and oxide does not become any more effective as oxidation proceeds. Figure 24.5 shows how this can happen. If the volume of the oxide is much less than that of the material from which it is formed, it will crack to relieve the strain (oxide films are usually brittle).

If the volume of the oxide is much greater, on the other hand, the oxide will tend to release the strain energy by breaking the adhesion between material and oxide, and springing away. For protection, then, we need an oxide skin that is neither too small and splits open (like the bark on a fir tree) nor one that is too big and wrinkles up (like the skin of a rhinoceros), but one which is just right. Then, and only then, do we get protective parabolic growth.

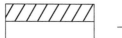 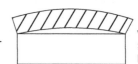

Volume oxide
$\leqslant$ volume material

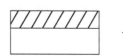

Volume oxide
$\geqslant$ volume material
Examples: Ta, Nb

**FIGURE 24.5**
Breakdown of oxide films, leading to linear oxidation behavior.

# EXAMPLES

**24.1** The oxidation of a particular metal in air is limited by the outward diffusion of metallic ions through an unbroken surface film of one species of oxide. Assume that the concentration of metallic ions in the film immediately next to the metal is $c_1$, and that the concentration of ions in the film immediately next to the air is $c_2$, where $c_1$ and $c_2$ are constants. Use Fick's law to show that the oxidation of the metal should satisfy parabolic kinetics, with weight gain $\Delta m$ given by

$$(\Delta m)^2 = k_p t$$

**24.2** The oxidation of another metal is limited by the outward flow of electrons through a uniform, unbroken oxide film. Assume that the electrical potential in the film immediately next to the metal is $V_1$, and the potential immediately next to the free surface is $V_2$, where $V_1$ and $V_2$ are constants. Use Ohm's law to show that parabolic kinetics should apply in this case also.

**24.3** The kinetics of oxidation of mild steel at high temperature are parabolic, with

$$k_p (\mathrm{kg^2\,m^{-4}\,s^{-1}}) = 37\exp\left\{ -\frac{138\mathrm{kJ\ mol^{-1}}}{\bar{R}T} \right\}$$

Find the depth of metal lost from the surface of a mild steel tie bar in a furnace at 500°C after 1 year. You may assume that the oxide scale is predominantly FeO. The atomic weight and density of iron are 55.9 kg kmol$^{-1}$ and 7.87 Mg m$^{-3}$; the atomic weight of oxygen is 16 kg kmol$^{-1}$. What would be the loss at 600°C?

**Answers**
0.33 mm at 500°C; 1.13 mm at 600°C

**24.4** Of the many metals found in the earth's crust, why is gold the only one that is found "native," that is, as pieces of pure metal?

**24.5** The energy released when silver reacts with oxygen at room temperature is small—only 5 kJ mol$^{-1}$ of $O_2$. At 230°C, the energy is zero, and above 230°C the energy is positive (i.e., silver is stable and does not oxidize). Use this information to explain why small electrical contacts are usually plated with a thin layer of silver. Why is gold not used instead?

**24.6** Why does the rate at which a metal oxidizes not necessarily correlate with the energy released in the oxidation reaction?

**24.7** Explain why some metals show either a linear weight gain or a linear weight loss with time during oxidation.

## CHAPTER 25

# Case Studies in Dry Oxidation

## CONTENTS

## 25.1 INTRODUCTION

In this chapter we look first at an important class of alloys designed to resist corrosion: the stainless steels. We then examine a more complicated problem: that of protecting the most advanced gas turbine blades from gas attack. The basic principle applicable to both cases is to coat the steel or the blade with a stable ceramic: usually $Cr_2O_3$ or $Al_2O_3$. But the ways this is done differ widely.

## 25.2 CASE STUDY 1: MAKING STAINLESS ALLOYS

Mild steel is an excellent structural material—cheap, easily formed, and strong. But at low temperatures it rusts, and at high temperatures, it oxidizes. There is a large demand, for high-temperature applications ranging from chemical reactors to superheater tubes, for oxidation-resistant steel. In response to this demand, a range of stainless irons and steels has been developed. When mild steel is exposed to hot air, it oxidizes quickly to form FeO (or higher oxides). But if one of the elements near the top of Table 24.1 with a large energy of oxidation is dissolved in the steel, then this element oxidizes preferentially (because it is more stable than FeO), forming a layer of its oxide on the surface.

And if this oxide is a protective one, like $Cr_2O_3$, $Al_2O_3$, $SiO_2$, or BeO, it stifles further growth, and protects the steel.

A considerable quantity of this foreign element is needed to give adequate protection. The best is chromium, 18% of which gives a very protective oxide film: it cuts down the rate of attack at 900°C, for instance, by more than 100 times.

Other elements, when dissolved in steel, cut down the rate of oxidation, too. $Al_2O_3$ and $SiO_2$ both form in preference to FeO (refer to Table 24.1) and form protective films (refer to Table 24.2). Thus 5% Al dissolved in steel decreases the oxidation rate by 30 times, and 5% Si by 20 times. The same principle can be used to impart oxidation resistance to other metals. We shall discuss nickel and cobalt in the next Case Study—they can be alloyed in this way. So, too, can copper; although it will not dissolve enough chromium to give a good $Cr_2O_3$ film, it *will* dissolve enough aluminum, giving a range of stainless alloys called "aluminum bronzes." Even silver can be prevented from tarnishing (reaction with sulfur) by alloying it with aluminum or silicon, giving protective $Al_2O_3$ or $SiO_2$ surface films.

Ceramics themselves are sometimes protected in this way. Silicon carbide, SiC, and silicon nitride, $Si_3N_4$, both have large negative energies of oxidation (meaning that they oxidize easily). But when they do, the silicon in them turns to $SiO_2$ which quickly forms a protective skin and prevents further attack.

Protection by alloying has one great advantage over protection by surface coating (like chromium plating or gold plating): it repairs itself when damaged. If the protective film is scored or abraded, fresh metal is exposed, and the chromium (or aluminum or silicon) it contains immediately oxidizes, healing the break in the film.

## 25.3 CASE STUDY 2: PROTECTING TURBINE BLADES

As we saw in Chapter 23, the materials at present used for turbine blades consist chiefly of nickel, with various foreign elements added to get the creep properties right. With the advent of DS and SX blades, such alloys will normally operate around 950°C, which is close to $0.7T_M$ for Ni (1208 K, 935°C). If we look at Table 24.2 we can see that at this temperature, nickel loses 0.1 mm of metal from its surface by oxidation in 600 h. The thickness of the metal between the outside of the blade and the integral cooling ports is about 1 mm, so in 600 h a blade would lose about 10% of its cross-section in service.

This represents a serious loss in mechanical integrity and makes no allowance for statistical variations in oxidation rate—which can be quite large—or for preferential oxidation (at grain boundaries, for example) leading to pitting. Because of the large cost of replacing a set of blades, they are expected to last for more than 5000 h. Nickel oxidizes with parabolic kinetics (Equation (24.4)) so after a time $t_2$, the loss in section $x_2$ is given by substituting the data into:

$$\frac{x_2}{x_1} = \left(\frac{t_2}{t_1}\right)^{1/2}$$

giving

$$x_2 = 0.1 \left(\frac{5000}{600}\right)^{1/2} = 0.29 \text{ mm}$$

Obviously this sort of loss is not OK, but how do we stop it?

As we saw in Chapter 23, the alloys used for turbine blades contain large amounts of chromium, dissolved in solid solution in the nickel matrix. If we look at Table 24.1, which gives the energies released when oxides are formed from materials, we see that the formation of $Cr_2O_3$ releases much more energy ($701 \text{ kJ mol}^{-1}$) than $NiO$ ($439 \text{ kJ mol}^{-1}$). This means that $Cr_2O_3$ will form in *preference* to $NiO$ on the surface of the alloy. Obviously, the more Cr there is in the alloy, the greater is the preference for $Cr_2O_3$. At the 20% level, enough $Cr_2O_3$ forms on the surface of the turbine blade to make the material act a bit as though it were chromium.

Suppose for a moment that our material *is* chromium. Table 24.2 shows that Cr would lose 0.1 mm in 1600 h at $0.7T_M$. Of course, we have forgotten about one thing. $0.7T_M$ for Cr is 1504 K (1231°C), whereas for Ni, it is 1208 K (935°C). We should probably consider how $Cr_2O_3$ would act as a barrier to oxidation at 1208 K rather than at 1504 K (Figure 25.1). The oxidation of Cr follows parabolic kinetics with an activation energy of $330 \text{ kJ mol}^{-1}$. Then the ratio of the times required to remove 0.1 mm (from Equation 24.3) is

$$\frac{t_2}{t_1} = \frac{\exp - (Q/\bar{R}T_1)}{\exp - (Q/\bar{R}T_2)} = 0.65 \times 10^3$$

Thus the time at 1208 K is

$$t_2 = 0.65 \times 10^3 \times 1600 \text{ h}$$
$$= 1.04 \times 10^6 \text{ h}$$

Of course, there is only at most 20% Cr in the alloy, so the alloy behaves only *partly* as if it were protected by $Cr_2O_3$. Experimentally, we find that 20% Cr increases the time for a given metal loss by only about 10 times; that is, the time taken to lose 0.1 mm at blade working temperature becomes $600 \times 10 \text{ h} = 6000 \text{ h}$ rather than $10^6$ hours.

Why this large difference? Whenever you consider an *alloy* rather than a pure material, the oxide layer—whatever its nature ($NiO$, $Cr_2O_3$, etc.)—has foreign

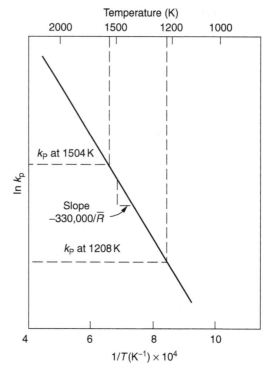

**FIGURE 25.1**
The way in which $k_p$ varies with temperature.

elements contained in it, too. Some of these will greatly increase the diffusion coefficients or electrical conductivity of the layer, and make the rate of oxidation much more than it would be in the absence of foreign element contamination.

One therefore has to be very careful in transferring data on film protectiveness from a pure material to an alloyed one, but the approach does give us an *idea* of what to expect. As in all oxidation work, however, experimental data on actual alloys are *essential* for design.

This 0.1 mm loss in 6000 h from a 20% Cr alloy at 935°C, though better than pure nickel, is still not good enough. What is worse, we saw in Chapter 23 that, to improve the creep properties, the quantity of Cr has been reduced to 10%, and the resulting oxide film is even less protective. The obvious way out of this problem is to *coat* the blades with a protective layer (Figure 25.2). This is usually done by spraying molten droplets of aluminum on to the blade surface to form a layer, some microns thick. The blade is then heated in a furnace to allow the Al

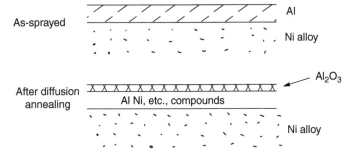

**FIGURE 25.2**
Protection of turbine blades by sprayed-on aluminum.

to diffuse into the surface of the Ni. During this process, some of the Al forms compounds such as AlNi, which are themselves good barriers to oxidation, while the rest becomes oxidized up to give $Al_2O_3$—which, as we can see from our oxidation-rate data—should be a very good barrier. An incidental benefit of the AlNi layer is its poor thermal conductivity—this helps insulate the metal of the cooled blade from the hot gases, and allows a slight extra increase in blade working temperature.

Other coatings, though more difficult to apply, are even more attractive. It is possible to diffusion-bond a layer of Ni–Cr–Al alloy to the blade surface to give a ductile coating which forms a very protective film of oxide. And zirconium oxide (zirconia) is now used as a combined thermal barrier and oxidation protection coating on many hot end components.

## Influence of coatings on mechanical properties

So far, we have been considering the advantage of an oxide layer in reducing the rate of oxidation. Oxide films, however, have some *disadvantages*.

Because oxides are usually quite brittle at the temperatures encountered on a turbine blade surface, they can crack, especially when the temperature of the blade changes and differential thermal contraction and expansion stresses are set up between alloy and oxide. These can act as initiation sites for thermal fatigue cracks, and because oxide layers in nickel alloys are well bonded to the underlying alloy, the crack can spread into the alloy itself (Figure 25.3). The properties of the oxide film are therefore important in affecting the fatigue properties of the whole component.

## Protecting future blade materials

You may wonder why we did not mention the "refractory" metals Nb, Ta, Mo, and W in the chapter on turbine-blade materials. These metals have very high melting temperatures and should therefore have very good creep properties.

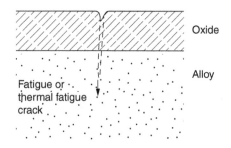

**FIGURE 25.3**
Fatigue cracks can spread from coatings into the material itself.

$$\left.\begin{array}{ll} \text{Nb} & \text{2740K} \\ \text{Ta} & \text{3250K} \\ \text{Mo} & \text{2880K} \\ \text{W} & \text{3680K} \end{array}\right\} T_M$$

But they all oxidize very rapidly indeed (refer to Table 24.2), and are useless without coatings. The problem with coated refractory metals is that if a break occurs in the coating (e.g., by thermal fatigue, erosion by dust particles, etc.), catastrophic oxidation of the underlying metal will take place, leading to rapid failure. The "unsafeness" of this situation is a major problem that has to be solved before we can use these otherwise excellent materials.

The ceramics SiC and $Si_3N_4$ do not share this problem. They oxidize readily (refer to Table 24.1); but in doing so, a surface film of $SiO_2$ forms which gives adequate protection up to 1300°C. And because the film forms by oxidation of the material itself, it is self-healing.

## 25.4 A NOTE ON JOINING OPERATIONS

One might imagine that it is always a good thing to have a protective oxide film on a material. Not always—if you wish to join materials by brazing or soldering, the protective oxide film can be a problem. It is this that makes stainless steel hard to braze and almost impossible to solder; even spot-welding and diffusion bonding become difficult. Protective films create poor electrical contacts—that is why aluminum is not more widely used as a conductor. And production of components by powder metallurgy, which involves compaction and sintering (diffusion bonding) of powdered material to the desired shape, is made difficult by protective surface films.

# EXAMPLES

**25.1** Diffusion of aluminum into the surface of a nickel super-alloy turbine blade reduced the rate of high-temperature oxidation. Explain.

**25.2** Explain why many high-temperature components, e.g., furnace flues and heat exchanger shells, are made from stainless steel rather than mild steel.

**25.3** Give examples of the use of alloying elements to reduce the rate at which metals oxidize.

**25.4** Why are the refractory metals (Nb, Ta, Mo, W) of very limited use at high temperature? What special steps are taken to ensure that tungsten electric light filaments can operate for long times at white heat?

**25.5** Why is oxidation a problem when soldering or brazing metal parts? Why are the connection tabs on small electrical components usually supplied pre-tinned with electrical solder?

**25.6** Why is nichrome wire (Ni + 30Fe + 20 Cr), rather than mild steel, used for heating elements in electric fires?

**25.7** Explain, giving examples, why ceramics are good at resisting high-temperature oxidation.

## CHAPTER 26

# Wet Corrosion of Materials

## CONTENTS

## 26.1 INTRODUCTION

In the last two chapters we showed that most materials that are unstable in oxygen tend to oxidize. We were principally concerned with loss of material at high temperatures, in dry environments, and found that, under these conditions, oxidation was usually controlled by the diffusion of ions or the conduction of electrons through oxide films that formed on the material surface. Because of the thermally activated nature of the diffusion and reaction processes we saw that the rate of oxidation was much greater at high temperature than at low, although even at room temperature, very thin films of oxide *do* form on all unstable metals. This minute amount of oxidation is important: it protects, preventing further attack; it causes tarnishing; it makes joining difficult; and (as we shall see in Chapters 28 and 29) it helps keep sliding surfaces apart, and so influences the coefficient of friction. But the *loss* of material by oxidation at room temperature under these *dry* conditions is very slight.

Engineering Materials I: An Introduction to Properties, Applications, and Design, Fourth Edition

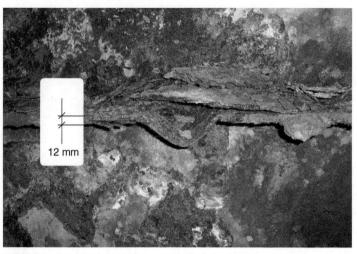

12 mm

**FIGURE 26.1**
Framing inside the seawater ballast tanks of a large cargo ship. The material is mild steel. A horizontal stiffener (12 mm thick when new) has rusted in places to almost nothing, and also buckled. Large slabs must have fallen on top of the stiffener from corroded steelwork above. The steel was originally coated, but the coating has broken down. Ship repair yard, Antwerp, Belgium. – 51 15 06.70 N 4 21 22.00 E

Under *wet* conditions, the situation is different. When mild steel is exposed to oxygen and water at room temperature, it rusts and the loss of metal quickly becomes appreciable. Unless precautions are taken, the life of many things, from cars to bridges, from ships to aircraft, is limited by wet corrosion. The annual bill worldwide for either replacing corroded components, or preventing corrosion (e.g., by painting Scotland's Forth Bridge), is around US$2.2 trillion—more than 3% of Britain's GDP. Figure 26.1 shows a graphic example.

## 26.2 WET CORROSION

Why the dramatic effect of water on the rate of loss of material? As an example we shall look at *iron*, immersed in *aerated water* (Figure 26.2). Iron atoms pass into solution in the water as $Fe^{++}$, leaving behind two electrons each (the *anodic* reaction). These are conducted through the metal to a place where the "oxygen reduction" reaction can take place to consume the electrons (the *cathodic* reaction).

This reaction generates $OH^-$ ions which then combine with the $Fe^{++}$ ions to form a *hydrated iron oxide* $Fe(OH)_2$ (really $FeO \cdot H_2O$); but instead of forming on the surface where it might give some protection, it often forms as a precipitate in the water itself. The reaction can be summarized by the following, just as in the case of dry oxidation:

$$\text{Material} + \text{Oxygen} \rightarrow \text{(Hydrated) Material Oxide}$$

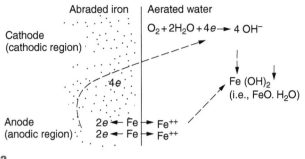

**FIGURE 26.2**
Wet corrosion.

The formation and solution of $Fe^{++}$ is analogous to the formation and diffusion of $M^{++}$ in an oxide film under dry oxidation; and the formation of $OH^-$ is closely similar to the reduction of oxygen on the surface of an oxide film. However, the much faster attack found in wet corrosion is due to the following:

- The $Fe(OH)_2$ either deposits *away* from the corroding material; or, if it deposits on the surface, it does so as a loose deposit, giving little or no protection.
- Consequently $M^{++}$ and $OH^-$ usually diffuse in the *liquid* state, and therefore do so very rapidly.
- In *conducting* materials, the electrons can move very easily as well.

The result is that the oxidation of iron in aerated water (rusting) goes on at a rate that is millions of times faster than that in dry air.

## 26.3 VOLTAGE DIFFERENCES AS THE DRIVING FORCE FOR WET OXIDATION

In dry oxidation we quantified the tendency for a material to oxidize in terms of the energy needed, in $kJ\ mol^{-1}$ of $O_2$, to manufacture the oxide from the material and oxygen. Because wet oxidation involves electron flow in conductors, which is easier to measure, the tendency of a metal to oxidize in solution is described by using a *voltage* scale rather than an *energy* one.

Suppose we could separate the cathodic and the anodic regions of a piece of iron, as shown in Figure 26.3. Then at the cathode, oxygen is reduced to $OH^-$, absorbing electrons, and the metal therefore becomes positively charged. The reaction continues until the potential rises to $+0.8$ V. Then the coulombic attraction between the positive charged metal and the negative charged $OH^-$ ion becomes so large that the $OH^-$ is pulled back to the surface, and reconverted to $H_2O$ and $O_2$; in other words, the reaction stops.

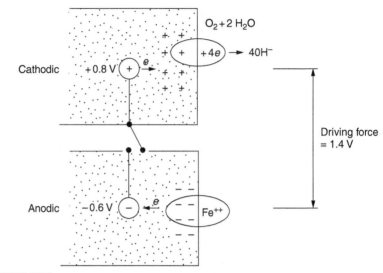

**FIGURE 26.3**
The voltages that drive wet corrosion.

At the anode, $Fe^{++}$ forms, leaving electrons behind in the metal which acquires a negative charge. When its potential falls to −0.6 V, that reaction, too, stops (for the same reason as before). If the anode and cathode are now *connected*, electrons flow from the one to the other, the potentials *converge*, and *both reactions start up again*. The difference in voltage of 1.4 V is the driving potential for the oxidation reaction. The bigger it is, the bigger the tendency to corrode.

## 26.4 POURBAIX (ELECTROCHEMICAL EQUILIBRIUM) DIAGRAMS

Figure 26.4 shows a typical Pourbaix diagram (or electrochemical equilibrium diagram)—in this case for copper. The diagrams are maps that show the conditions under which a metal:

- Cannot corrode—because there is no voltage driving force, or a negative one
- May corrode—because there is a voltage driving force, and a stable oxide film does not form on the surface
- May not corrode—although there is a voltage driving force, a stable oxide film forms on the surface (this may or may not be an effective barrier to corrosion)

The axes of the diagram are the electrochemical potential of the metal (in volts) and the pH, which indicates how acidic or alkaline the water is. The three

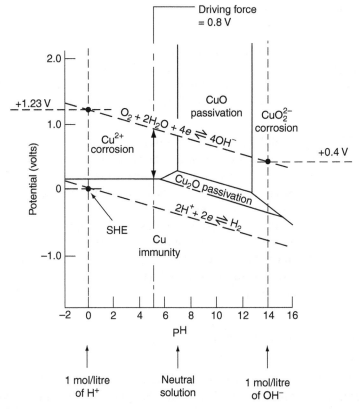

**FIGURE 26.4**

Pourbaix diagram for copper at 25°C. The vertical axis is the electrochemical potential relative to the standard hydrogen electrode (SHE). The horizontal axis is the pH of the water-based solution.

conditions are then represented on the diagram by fields of *immunity*, *corrosion*, and *passivation*.

The diagram also shows the line for the oxygen reduction reaction. Note that this slopes, from a potential of $+1.23$ V when the water is very acid (pH $= 0$) to $+0.4$ V when the water is very alkaline (pH $= 14$). This is because the oxygen reduction reaction produces $OH^-$ ions, so increasing the concentration of these ions in the solution (which is what we do when we go from acid to alkaline conditions) shifts the equilibrium position of the reaction.

In addition, Figure 26.4 shows that when copper is immersed in water with a pH less than about 6 to 7, it may corrode if the potential of the copper is greater than about 0.2 to 0.3 V. So if there is air in the water, the copper may corrode, because the oxygen reduction line is positioned at a much higher potential.

For example, if the pH is 5, the voltage driving force for corrosion is 0.8 V, and because there is no stable oxide film, corrosion may occur.

If the pH of the water is between 7 and 12 (from neutral to strongly alkaline), the copper will grow a stable film of copper oxide, which may protect the metal against further corrosion—even though there will still be a driving force of about 0.8 V between the potentials of the copper and the oxygen reduction reaction.

It should be emphasized that these diagrams cannot give any indication of the *actual* rates of corrosion in any situation—they just show where corrosion cannot occur, and where a surface film should form which might prevent or slow down corrosion. Experimental data and field experience in the real environment are essential for corrosion design—even more so because corrosion processes in general, and the barrier properties of surface films in particular, can be very strongly affected by the presence of other ions. These speed up corrosion processes and can attack surface films, which is why seawater—high in chloride ion, $Cl^-$—is such bad news for many metals (see the heavily corroded seawater ballast tank in Figure 26.1).

## 26.5 SOME EXAMPLES

### Copper

Provided the water is neutral or alkaline (and low in other ions) the rate of corrosion is usually extremely low. This would be expected from the Pourbaix diagram if the oxide film acted as an effective barrier. This is why copper is much used for water pipes. However, if the water is mildly acidic, corrosion "pinholes" can form in the tube wall (the Pourbaix diagram shows that there is no oxide film below a pH of about 6). Copper is much used for architectural purposes, particularly in the United States and Australia.

Figure 26.5 shows copper cladding on the outside of a modern home. The thin sheet copper has not corroded significantly, but a protective film has formed on exposure to the elements, which is why the color is darker than freshly polished copper. Older architectural copper usually develops a bright green color, because the carbon dioxide in the air forms a copper carbonate film. This is just as well, because carbon dioxide also makes the rain water acidic! Copper alloys like brass and bronze are also resistant to corrosion—the fire hydrant shown in Figure 26.6 is as shiny as the day it was made.

### Steel

The Pourbaix diagram for iron is shown in Figure 26.7. In order to have a stable surface film, the pH of the water needs to be above 9, so in most applications (neutral or slightly acidic water) steel will rust unless it is protected. One way of

**FIGURE 26.5**
Copper cladding on a home in Muston Street, Mosman, NSW, Australia. – 33 49 32.50 S 151 14 41.00 E

**FIGURE 26.6**
Brass fire hydrant at 1 W 75th Street, New York City (corner of Central Park West and W 75th).
– 40 46 41.44 N 73 58 28.94 W

making steel resistant to corrosion is to alloy it with foreign elements which will provide a stable protective film on the surface. Stainless steel is the best known example—adding 18% chromium produces an invisible film of $Cr_2O_3$ on the surface, which completely prevents corrosion in most situations (we saw in Chapter 25 that stainless steel is also very resistant to dry oxidation). Another way is to keep any moisture and/or oxygen away from the surface of the steel

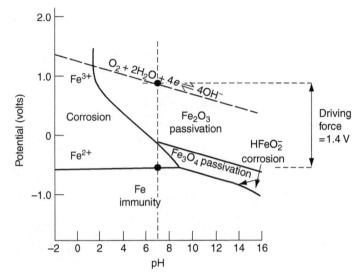

**FIGURE 26.7**
Pourbaix diagram for iron at 25°C.

**FIGURE 26.8**
The Goodwill Bridge over the Brisbane River in Queensland, Australia. – 27 28 49.20 S 153 01 38.04 E

(e.g., using paint or epoxy coatings). The Goodwill Bridge (over the river in Brisbane, Australia) has an ingenious way of doing this (Figures 26.8 and 26.9). The main span is supported by two steel arches made from steel plates welded into a box section. The problem with this is that the box sections are too small to be accessed for painting the insides after fabrication is completed—and if they were painted before fabrication the heat of the welding would burn off

**FIGURE 26.9**
The nitrogen charging value on one of the box section arches of the Goodwill Bridge.

some of the paint anyway. The solution adopted is to make the box sections pressure tight, and then fill them with dry nitrogen gas under a positive gauge pressure—the result is no moisture and no oxygen either, so the insides cannot corrode!

## Aluminum

The Pourbaix diagram for aluminum is shown in Figure 26.10. Between pH 4 and 8.5, a thin and very stable film of hydrated aluminum oxide forms, protecting the metal. The corrosion rate of aluminum in pure water is extremely low, even though the driving force for corrosion is very large—about 2.8 volts! However, over time there is a tendency for attack to occur at weak points in the oxide film, so for most outdoor uses—such as window frames, roofs, and cladding— aluminum is first *anodized*—given a surface treatment that artificially thickens the oxide film to make it even more protective. In the anodizing process, the metal is put into a bath of water containing various additives to promote compact film growth (e.g., boric acid). It is then made positive electrically, which attracts the oxygen atoms in the polar molecules (see Chapter 4).

The attached oxygen atoms react with the metal to give more oxide film as shown in Figure 26.11. The film can even be colored for decorative purposes, by adding coloring agents to the bath toward the end of the process. However, in water with lots of other ions, even this anodized film may break down. This is why aluminum corrodes in seawater—the chloride ions are very aggressive to the oxide film.

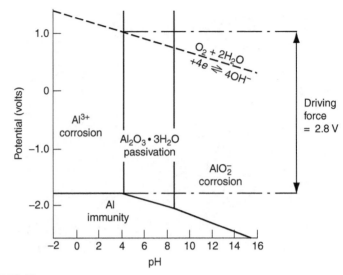

**FIGURE 26.10**
Pourbaix diagram for aluminum at 25°C.

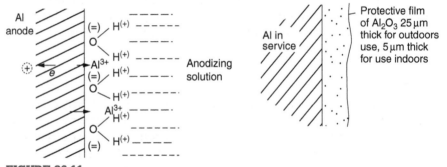

**FIGURE 26.11**
Protecting aluminum by anodizing it.

## 26.6 A NOTE ON STANDARD ELECTRODE POTENTIALS

Many books list standard electrode potentials for metals (see Table 26.1). These show the voltage when the metal is in equilibrium with a solution of its ions having a concentration of 1 mol per liter. These are essentially the same voltages as those shown in the Pourbaix diagrams for the horizontal lines at the top of the immunity field when there is no oxide film. The driving force for corrosion can be estimated by subtracting the potential for the metal from that for the oxygen reduction reaction. The potentials for gold, platinum, and palladium

**Table 26.1** Standard Electrode Potentials

| Reaction | Volts | Reaction | Volts |
|---|---|---|---|
| $Au = Au^{3+} + 3e$ | +1.50 | $Ni = Ni^{2+} + 2e$ | −0.25 |
| | | $Co = Co^{2+} + 2e$ | −0.28 |
| $Pt = Pt^{2+} + 2e$ | +1.2 | $Cd = Cd^{2+} + 2e$ | −0.40 |
| $Pd = Pd^2 + 2e$ | +0.99 | $Fe = Fe^{2+} + 2e$ | −0.44 |
| $O_2 + 2H_2O + 4e$ $= 4OH^-$ | +0.82 | | |
| $Ag = Ag^+ + e$ | +0.80 | $Cr = Cr^{3+} + 3e$ | −0.74 |
| | | $Zn = Zn^{2+} + 2e$ | −0.76 |
| | | $Ti = Ti^{2+} + 2e$ | −1.63 |
| $Cu = Cu^{2+} + 2e$ | +0.34 | $Al = Al^{3+} + 3e$ | −1.66 |
| | | $Mg = Mg^{2+} + 2e$ | −2.36 |
| $2H^+ + 2e = H_2$ | 0 | $Na = Na^+ + e$ | −2.71 |
| $Pb = Pb^{2+} + 2e$ | −0.13 | $K = K^+ + e$ | −2.93 |
| $Sn = Sn^{2+} + 2e$ | −0.14 | | |

Note: *Volts at 25°C relative to the standard hydrogen electrode. All solutions have a concentration of dissolved ions of 1 mol per liter. All gases have partial pressures of 1 atmosphere. Note that the potential of the oxygen reduction reaction is given for a solution pH of 7 (neutral).*

are greater than the potential for the oxygen reduction reaction, so these metals cannot corrode—there is no voltage driving force. The other metals have potentials that are less than the potential for the oxygen reduction reaction, so in theory they can corrode. Whether they will in practice depends to a large extent on the barrier properties of any surface films. Because of this, electrode potentials are a poor indicator of actual corrosion rates.

## 26.7 LOCALIZED ATTACK

Corrosion often attacks metals *selectively*, and this can lead to component failure much more rapidly and insidiously than one would infer from average corrosion rates. Three important kinds of selective attack are *pitting corrosion*, *intergranular attack*, and *stress corrosion cracking* (see Figure 26.12).

In pitting corrosion, preferential attack starts at breaks or weaknesses in the oxide film (sometimes where there are precipitates of compounds in alloys). Once pitting has started, it is virtually impossible to stop it. Figure 26.13 shows a good example: an underground water pipe has corroded from the *outside*, and although the uniform rate of corrosion is relatively small, there are many areas of pitting corrosion—in the worst, the bottom of the corrosion pit has actually gone through the wall of the pipe.

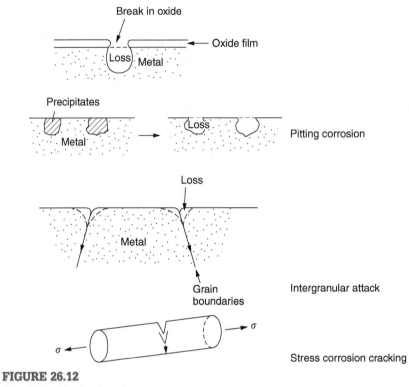

FIGURE 26.12

Some kinds of localized attack.

FIGURE 26.13

Pitting corrosion in an underground water pipe.

Intergranular attack can occur when the grain boundaries have a lower corrosion resistance than the grains themselves. In extreme cases, the material can literally fall apart at the grain boundaries.

In stress corrosion cracking, a critical combination of stress, material, and corrosive environment can lead to cracks forming and growing under static stress—at stress intensities much less than $K_c$. Examples are brass in ammonia, mild steel in caustic soda, and austenitic stainless steels in hot chloride solutions.

## EXAMPLES

**26.1** Measurements of the rate of crack growth in brass exposed to ammonium sulfate solution and subjected to a constant tensile stress gave the following data.

| Nominal Stress $\sigma$ (MN m$^{-2}$) | Crack Depth $a$ (mm) | Crack Growth Rate $da/dt$ (mm year$^{-1}$) |
|---|---|---|
| 4 | 0.25 | 0.3 |
| 4 | 0.50 | 0.6 |
| 8 | 0.25 | 1.2 |

Show that these data are consistent with a relationship of form

$$\frac{da}{dt} = AK^n$$

where $K = Y\sigma\sqrt{\pi a}$ is the stress intensity factor. Find the values of the integer $n$ and the constant $A$. Assume $Y = 1$.

**Answer**

$$n = 2, A = 0.0239 \text{ mm year}^{-1}(\text{MPa}\sqrt{\text{mm}})^{-2}$$

**26.2** The diagram at the top of the next page shows a mild steel pipe which carried water around a sealed central heating system. The pipes were laid in the concrete floor slabs of the building at construction, and thermally insulated with blocks of expanded polystyrene foam. After 15 years, the pipes started to leak. When the floor was dug up, it was found that the pipe had rusted through from the *outside*. It was known that rainwater had been finding its way into the building along joints between the concrete slabs, and consequently the foam in the pipe ducts would have been wet. Explain why the corrosion occurred, and also why the pipe did not rust from the inside.

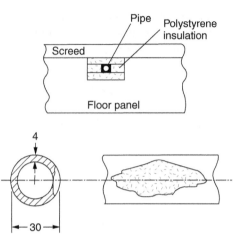

**26.3** This photograph shows the inside of a centrifugal pump used for pumping industrial effluent. The material is cast iron. Identify the type of corrosion attack.

**26.4** Why is resistance to stress corrosion cracking sometimes more important than resistance to uniform corrosion?

**26.5** Because polymer molecules are covalently bonded (so do not ionize in water), plastics are essentially immune to wet corrosion (although they can suffer environmental degradation from UV light, ozone, and some chemicals). Give examples from your own experience where polymers are used in applications where corrosion resistance is one of the prime factors in the choice of material.

**26.6** This photograph shows the surface of an expansion bellows made from austenitic stainless steel, which carried wet steam at 184°C. The stress in the steel was approximately 50% of the yield stress. The system had been treated with zinc chloride corrosion inhibitor. Identify the type of corrosion attack.

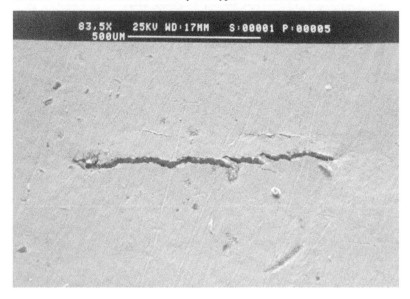

**26.7** Explain the following observations.
  **a.** Gold, platinum, and palladium do not corrode in water-based solutions.
  **b.** There is a poor correlation between the standard electrode potential of a metal and its corrosion resistance.
  **c.** Aluminum is only resistant to corrosion when the pH is between 4 and 8.5.
  **d.** Iron is only resistant to corrosion when the pH is between 9 and 14.

**26.8** Answer the following questions.
  **a.** On a Pourbaix diagram, what do the terms (i) immunity, (ii) corrosion, (iii) passivation mean?
  **b.** Why is stainless steel very resistant to corrosion?
  **c.** Why is aluminum anodized for outdoors applications?
  **d.** Why does excluding oxygen prevent corrosion?

## RATES OF UNIFORM METAL LOSS

- Mils per year (mpy): 1 mpy $= 0.0254$ mm year$^{-1} = 25.4$ μm year$^{-1}$.
- Penetration rate in μm year$^{-1} = 87,600 \, (w/\rho At)$, where $w$ is the weight of metal lost in mg, $\rho$ is the density of the metal in Mg m$^{-3}$ or g cm$^{-3}$, $A$ is the corroding area in cm$^2$ and $t$ is the time of exposure in hours.
- Penetration rate in μm year$^{-1} = 3.27 \, (ai/z\rho)$, where $a$ is the atomic weight of the corroding metal, $i$ is the current density at the corroding surface in

$\mu A\ cm^{-2}$, $z$ is the number of electrons lost by each metal atom in the corrosion reaction, and $\rho$ is the density of the metal in $Mg\ m^{-3}$ or $g\ cm^{-3}$.

| Metal | Atomic Weight | Density (Mg m$^{-3}$) |
|---|---|---|
| Ag | 107.9 | 10.5 |
| Al | 26.98 | 2.70 |
| Au | 197.0 | 19.3 |
| Cd | 112.4 | 8.6 |
| Co | 58.93 | 8.9 |
| Cr | 52.00 | 7.1 |
| Cu | 63.54 | 8.96 |
| Fe | 55.85 | 7.87 |
| K | 39.10 | 0.86 |
| Mg | 24.31 | 1.74 |
| Na | 22.99 | 0.97 |
| Ni | 58.71 | 8.9 |
| Pb | 207.2 | 11.7 |
| Pd | 106.4 | 12.0 |
| Pt | 195.9 | 21.5 |
| Sn | 118.7 | 7.3 |
| Ti | 47.90 | 4.5 |
| Zn | 65.37 | 7.14 |

## CHAPTER 27

# Case Studies in Wet Corrosion

## CONTENTS

## 27.1 CASE STUDY 1: PROTECTING SHIPS' HULLS FROM CORROSION

Figure 27.1shows a large cargo ship in the final stages of being scrapped. Because the hull is sitting on the beach, it dries out at low tide, exposing details that would not normally be visible. Figures 27.2 and 27.3 show a *magnesium sacrificial anode* bolted to the steel hull just above the keel—similar anodes are regularly spaced along the length and breadth of the vessel below the waterline. But what are these for?

Figure 27.4 shows how the sacrificial anode works. Table 26.1 shows that the standard electrode potentials for iron and magnesium are –0.44 V and –2.36 V. The SEP for magnesium is therefore 1.9 V less than that for iron. If these two metals are wired together *in a conducting medium* (e.g., seawater), the magnesium becomes the anode and corrodes. The iron becomes the cathode, where the oxygen reduction reaction takes place, and does not corrode.

Engineering Materials I: An Introduction to Properties, Applications, and Design, Fourth Edition

**FIGURE 27.1**
A large cargo ship in the final stages of being scrapped.

**FIGURE 27.2**
Magnesium sacrificial anode attached to the steel hull below the waterline.

Although this difference in SEP suggests that the magnesium *should* protect the steel, we still need to check that no stable protective film will form on the surface of the magnesium, because this would prevent it from corroding. Figure 27.5 (see page 402) shows the Pourbaix diagram for magnesium (with

**FIGURE 27.3**
Magnesium sacrificial anode showing heavy corrosion of magnesium.

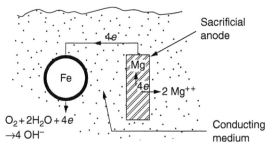

**FIGURE 27.4**
Sacrificial protection. Typical materials used are Mg, Al, and Zn.

part of the Pourbaix diagram for iron superposed on it). The pH of seawater is about 8, so magnesium does not form a protective oxide film in seawater.

Figure 27.5 also shows why the steel is protected. The potential of the steel is pulled down by the corrosion of the magnesium so that it ends up well within the region of Fe immunity on the Pourbaix diagram for iron. Because this turns the steel into the cathode, protecting it using sacrificial anodes is also called *cathodic protection* (referred to in the trade as "CP").

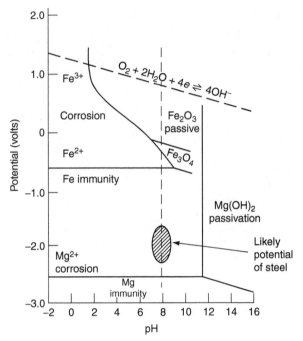

**FIGURE 27.5**
Superposition of Pourbaix diagrams for magnesium and iron.

Figure 27.3 shows the anode being eaten away by corrosion. In order to min-imize the loss of anode metal, it is important to protect the steel by *painting* it. However, paint does not provide complete protection—it gets scratched and abraded in service (for example, when removing marine organisms like the bar-nacles shown in Figure 27.2), and it is also attacked by long term exposure to seawater. The role of CP is to protect the steel from corrosion where the paint breaks down.

Zinc could also be used to protect the steel. Table 26.1 shows that the standard electrode potentials for iron and zinc are −0.44 V and −0.76 V. The difference of 0.3 V is still enough to make the iron the cathode, because zinc does not form a stable oxide film below a pH of 8.5 (see Figure 27.6). Aluminum (SEP = −1.66) is less straightforward. As Figure 26.10 shows, aluminum forms a stable pro-tective film in the pH range 4 to 8.5, so in theory it should not corrode in seawater (pH = 8). However the high concentration of chloride ions in seawater attacks the film, allowing the anode to corrode. Alloying the aluminum with 5% zinc also helps it to corrode. Titanium (SEP = −1.63) does not work, because the surface film stops it corroding.

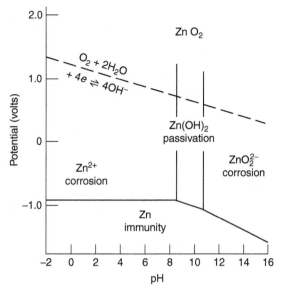

**FIGURE 27.6**
Pourbaix diagram for zinc.

## 27.2 CASE STUDY 2: RUSTING OF A STAINLESS STEEL WATER FILTER

Figure 27.7 and 27.8 show part of a water filter. The filter consisted of a perforated tube with a diameter of 200 mm. The end of the tube was welded to a short screwed section and this was used to couple the filter to a length of ordinary nonperforated pipe. The filter was intended for use in a water-supply network on an irrigation project. The components were made from an austenitic stainless steel of type AISI 304. The perforated tube was made by assembling a tubular cage of steel rods and welding the end of each rod to the screwed coupling. A helix of steel wire was then wound around the outside of the cage to complete the perforated wall. The wire was fixed to the support rods by electrical-resistance spot welding. For some reason the connection between the end of the helix and the coupling was not satisfactory. To correct this an extra weld had been made on the outside of the coupling. Finally the weld and the adjoining helix were leveled by grinding.

The filters were transported to their destination by sea. When they were unloaded it was noticed that some of the repair welds had corroded. A closer inspection revealed that corrosion had occurred not just on the weld bead itself but also on the parts of the helix that had been ground flat. The surface was not pitted, but was covered with a thin uniform deposit of red rust.

**FIGURE 27.7**
The stainless steel water filter.

**FIGURE 27.8**
The stainless steel has rusted where it has been ground and machined.

The standard range of austenitic stainless steels is prone to a number of corrosion problems—the rather vague description "stainless" cannot necessarily be taken to mean "immune to corrosion." The iron in stainless steel wants to react with the environment and the metal depends for its corrosion resistance on the very thin protective film of chromium oxide. If the passive film breaks down for any reason then stainless steel can corrode very rapidly indeed. When the filters were unloaded from their containers they were found to be running with condensation. The film of condensed water would have been saturated with air and would probably have contained a significant concentration of chloride ions picked up from the salty atmosphere. The solution should then have been an ideal medium for corrosion, with an ample supply of oxygen and a reasonable electrical conductivity. Chloride ions are also very effective at breaking down the protective films that form on most metals. In view of this it is not surprising that the weak areas were identified in the surface film on the filters.

When a fresh, dry surface of stainless steel is exposed to the oxygen in the atmosphere the passive oxide film rapidly forms of its own accord. For critical applications (e.g., water pipes in the nuclear industry) the film can even be thickened artificially by treating the surface with an oxidizing agent such as nitric acid. However, problems can arise when stainless steel is welded. Because the surface of the weld bead is exposed to the atmosphere at high temperature it oxidizes. A layer of black oxide scale forms on the surface of the weld bead. Unfortunately this high-temperature oxide protects the metal much less well than the normal passive film. The problem can be solved by removing the oxide with a pickling solution of nitric and hydrofluoric acids. This produces a fresh clean surface which is rinsed and allowed to passivate naturally in air.

What seems less obvious is why the outside of the filter rusted even though the oxide scale had been ground off. The answer is that the rough, cold-worked surface produced by grinding is more liable to corrode than a smooth stress-free surface.

Indeed, stainless steel components for critical applications are often "cleaned" by *electropolishing*. This dissolves away the cold-worked layer, producing a surface that is smooth, clean, and stress-free and which forms an optimum base for the passive film. Electropolished stainless steel is much in demand in the medical, pharmaceutical, and food-handling industries where freedom from contamination is essential.

AISI 304 is the most common and basic stainless steel—it contains to 18 to 20% Cr (and 8–10.5% Ni to make it f.c.c.). Its resistance to corrosion (especially pitting) can be improved a lot by adding molybdenum, which helps to stabilize the passive film. The most common stainless steel containing Mo is AISI 316, which has 2 to 3% Mo, as well as 16 to 18% Cr (and 10–14% Ni). More recently the super austenitics have been developed, such as 254 SMO and AL-6XN. These contain even more Mo (6%), Cr (20%), and Ni (20%), plus about 0.2% nitrogen. The N is absorbed as an interstitial solid solution—it is 16 times as effective as Cr in resisting pitting (and, incidentally, raises the yield strength by 25% because it pins dislocations). These alloys are also highly resistant to stress corrosion cracking. Finally, the new super duplex steels such as SAF 2507 have more Cr (25%), and less Mo (4%) and Ni (7%) than the super austenitics. The low Ni content gives them a mixed (or duplex) f.c.c./b.c.c. structure, which has a yield strength twice that of 304 or 316. Because Ni is expensive, they are cheaper than the super austenitics yet are just as resistant to pitting and stress corrosion cracking. Such are the complications of developing alloys to resist corrosion!

## 27.3 CASE STUDY 3: CORROSION IN REINFORCED CONCRETE

Figure 27.9 shows a reinforced concrete pile at the Inkerman Street Wharf in Sydney Harbour. The steel reinforcement has rusted and caused cracking and splitting of the surrounding concrete (the dark stains are rust deposits). Why has this happened? When steel rusts, the volume of rust produced is greater

**FIGURE 27.9**

Reinforced concrete pile at Inkerman Street Wharf, Mosman, NSW, Australia. − 33 48 59. 08 S 151 14 16.28 E

than the volume of steel that is lost. This is partly because the density of iron oxide is less than that of steel ($5.26$ Mg m$^{-3}$ for $Fe_2O_3$ compared to $7.8$ Mg m$^{-3}$ for steel). But because the rust deposits are hydrated ($FeO \cdot H_2O$) and full of voids, their density is usually much less again—as a rough guide, if the surface of the steel corrodes back by 1 mm, it will produce a layer of rust scale 5 to 10 mm thick. So if the steel reinforcement bars inside the concrete rust, they "expand," and crack open the surrounding concrete.

This type of attack is most likely to occur when there is not enough thickness of concrete over the steelwork ("cover") to protect it from the environment, or when there is a lot of chloride ion around. This can come from an incorrect cement mix (too much calcium chloride). Or it can come from exposure to salt–immersion in seawater, exposure to a marine atmosphere, or de-icing salt on highways. The pile at the Inkerman Street Wharf cracked because of a combination of insufficient cover and a very salty environment. Figures 27.10 and 27.11 show more examples, from a structure on South Head, Sydney—a place where there is a lot of salt-water spray in the air. Chunks of concrete have come away from the structure, exposing the steel reinforcement bars (a phenomenon known as *spalling*).

The concrete in these examples is beyond repair, but there are many cases where repairs are technically possible and cost-effective—for example, highway bridges that have suffered superficial spalling of concrete where the underlying reinforcement is not too badly corroded. The spalled areas are repaired using cement mortar, and a *cathodic protection* system is installed to prevent further corrosion of the steelwork. Holes are drilled into the surface of the concrete

**FIGURE 27.10**
Spalled reinforced concrete at South Head, Sydney. – 33 50 00.48 S 151 16 51.62 E

**FIGURE 27.11**
More spalled concrete at South Head.

at regular intervals, and plugs of zinc are cemented into them. The zinc anodes are electrically connected to the steelwork by copper wires.

A good example is given at *http://www.vector-corrosion.com/2009/09/ten-year-results-of-galvanic-sacrificial-anodes-in-steel-reinforced-concrete/*, which describes cement repairs made to a highway bridge in the United Kingdom in 1999. When some zinc anodes were removed for examination in 2009, it was found that only 35% of the zinc had been used up in protecting the steelwork, indicating that the repairs should have a life of at least 25 years before the anodes need renewing.

## 27.4 A NOTE ON SMALL ANODES AND LARGE CATHODES

Figure 27.12 shows a classic corrosion design error—fixing a sheet copper roof in place using steel screws. The SEPs for copper and iron are +0.34 and −0.44 V. Putting these two metals into electrical contact in a corrosive environment makes the iron the anode and the copper the cathode. The steel will rust of

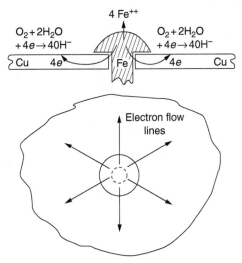

**FIGURE 27.12**

Large cathodes or small anodes can lead to very rapid corrosion.

course—but *much* faster than it would do on its own. The corrosion current through both anode and cathode are the same, but because the anode has a small surface area, and the cathode a large surface area, the *current density* $(A\ m^{-2})$ at the anode is much higher than at the cathode. This means that the anode will corrode rapidly—it is very easy for the electrons generated by the corrosion of the steel screw to get away to the large copper cathode. Even when a metal is intentionally made the anode—as in cathodic protection—the protected (cathodic) metal should be shielded from the environment (e.g., by painting, or covering in cement) in order to reduce the corrosion current and maximize the life of the anode.

## WORKED EXAMPLE

The following photograph shows a steel gate at the Cruising Yacht Club of Australia in Sydney. The surface of the steel looks strange—as if it were covered in large grains of metal. This is indeed the case—the steel gate was dipped in a bath of molten zinc after it was made, and the thin layer of molten zinc that was left behind on the surface when the gate was removed from the bath soon cooled down and solidified, producing very large (but very thin) grains of zinc.

This is the basis of *galvanizing*—where steel is coated with a thin uniform layer of zinc in order to protect it from corrosion. The diagram shows how this works. The zinc acts as a physical barrier between the steel and the corrosive environment. But if it is scratched or damaged, any exposed areas of steel will be protected by CP. The large area of zinc anode and small area of steel cathode mean

Steel gate at the Cruising Yacht Club of Australia, Rushcutters Bay. – 33 52 27.25 S 151 14 02.37 E

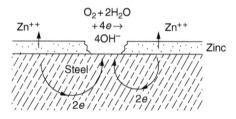

that any exposed steel is protected very effectively. However, the zinc does corrode on its own (typically 0.1 mm in 20 years), and because the zinc is only about 0.15 mm thick, after about 30 years most of the zinc will disappear, and the steel will corrode at the normal rate.

## EXAMPLES

**27.1** Mild-steel radiators in a sealed (oxygen free) central-heating system were found to have undergone little corrosion after several years service. Explain.

**27.2** To prevent the corrosion of a mild-steel structure immersed in seawater, a newly qualified engineer suggested the attachment of titanium plates in the expectation of powerful cathodic action. He later found that the structure had corroded badly. Explain.

**27.3** The corrosion of an underground steel pipeline was greatly reduced when the pipeline was connected to a buried bar of magnesium alloy. Explain.

**27.4** Under aggressive corrosion conditions it is estimated that the maximum corrosion current density in a galvanized steel sheet will be $6 \times 10^{-3}$ A m$^{-2}$. Estimate the thickness of the galvanized layer needed to give a rust-free life of at least 5 years. The density of zinc is 7.13 Mg m$^{-3}$, and its atomic weight is 65.4. Assume that the zinc corrodes to give $Zn^{2+}$ ions.

**Answer**

0.045 mm

**27.5** A sheet of steel of thickness 0.50 mm is tinplated on both sides and subjected to a corrosive environment. During service, the tinplate becomes scratched, so that steel is exposed over 0.5% of the area of the sheet. Under these conditions it is estimated that the current consumed at the tinned surface by the oxygen-reduction reaction is $2 \times 10^{-3}$ A m$^{-2}$. Will the sheet rust through within 5 years in the scratched condition? The density of steel is 7.87 Mg m$^{-3}$. Assume that the steel corrodes to give $Fe^{2+}$ ions. The atomic weight of iron is 55.9.

**Answer**

Yes

**27.6** Steel nails used to hold copper roofing sheet in position failed rapidly by wet corrosion. Explain.

**27.7** A marine harbor wall was cased with sheet piling made from steel. Five months after the piles had been driven a flaking corrosion product typically 0.8 mm thick had formed on the surface of the piles where they went through the tidal zone. Eight separate flakes were analyzed for iron with the following results.

| Sample No. | Area of Flake (mm$^2$) | Mass of Iron (mg) |
|---|---|---|
| 1 | 174 | 207.9 |
| 2 | 68 | 43.0 |
| 3 | 68 | 43.8 |
| 4 | 56 | 42.6 |
| 5 | 73 | 39.4 |
| 6 | 149 | 146.4 |
| 7 | 78 | 52.2 |
| 8 | 107 | 92.4 |

The piling was originally covered with a dense layer of iron oxide mill scale from the hot-rolling operation. The scale was estimated to have been 0.10 mm. thick. It is assumed that the scale was incorporated in the flakes of corrosion product. Estimate the thickness of steel lost by wet corrosion in the marine environment assuming that the mill scale layer is composed of $Fe_2O_3$, with a density of 5.26 g cm$^{-3}$. The atomic weight of oxygen is 16.00, the atomic weight of iron is 55.85, and the density of steel is 7.8 g cm$^{-3}$.

**Answer**

0.05 mm

27.8  A pipe 300 mm in diameter was made by forming a long strip of austenitic stainless steel into a helix and butt welding the helical joint. The pipe was used to carry hot nitric acid across a river estuary. The temperature of the acid was 80°C and the pipe was therefore lagged with fiber insulation which was clad with a waterproof skin of aluminum. After several years in service the pipeline supports suffered significant corrosion. In order to repair them the cladding and lagging were removed. Leakage of acid was discovered at a number of locations. Numerous cracks were found in the wall of the pipe. Deposits on the external surface of the pipe were rich in chloride ions. It was also found that rainwater had been seeping under the cladding. Give a likely explanation for the failure. Note that welding introduces large tensile residual stresses. Why was the acid not the cause of the cracking?

# Friction, Abrasion, and Wear

# Friction and Wear

## CONTENTS

## 28.1 INTRODUCTION

We now come to the final properties that we shall be looking at in this book: the frictional properties of materials in contact, and the wear that results when such contacts slide. This is of considerable importance in mechanical design. Frictional forces are undesirable in bearings because of the power they waste; and wear is bad because it leads to poor working tolerances, and ultimately to failure.

On the other hand, when selecting materials for clutch and brake linings—or even for the soles of shoes—we aim to maximize friction but still to minimize wear, for obvious reasons. But wear is not always bad: in operations such as grinding and polishing, we try to achieve maximum wear with the minimum of energy expended in friction; and without wear you could not write with chalk on a blackboard, or with a pencil on paper. In this chapter and the next we examine the origins of friction and wear and then explore case studies that illustrate the influence of friction and wear on component design.

Engineering Materials I: An Introduction to Properties, Applications, and Design, Fourth Edition

## 28.2 FRICTION BETWEEN MATERIALS

As you know, when two materials are placed in contact, any attempt to cause one of the materials to slide over the other is resisted by a *friction force* (Figure 28.1). The force that will just cause sliding to start, $F_s$, is related to the force $P$ acting normal to the contact surface by

$$F_s = \mu_s P \qquad (28.1)$$

Where $\mu_s$ is the *coefficient of static friction*. Once sliding starts, the limiting frictional force decreases slightly and we can write

$$F_k = \mu_k P \qquad (28.2)$$

where $\mu_k$ ($<\mu_s$) is the *coefficient of kinetic friction* (Figure 28.1). The work done in sliding against kinetic friction appears as *heat*.

These results at first sight run counter to our intuition—how is it that the friction between two surfaces can depend only on the *force P* pressing them together and not on their area? In order to understand this behavior, we must first look at the geometry of a typical surface.

If the surface of a fine-turned bar of metal is examined by making an oblique slice through it (a "taper section" which magnifies the height of any asperities), or if its profile is measured with a profilometer, it is found that the surface looks like Figure 28.2. The figure shows a large number of projections or *asperities*—it looks rather like a cross-section through Switzerland. If the metal is abraded

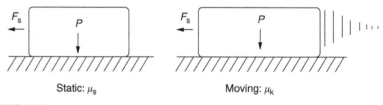

**FIGURE 28.1**
Static and kinetic coefficients of friction.

**FIGURE 28.2**
What a finely machined metal surface looks like at high magnification (the heights of the asperities are plotted on a much more exaggerated scale than the distances between asperities).

with the finest abrasive paper, the scale of the asperities decreases but they are still there—just smaller. Even if the surface is polished for a long time using the finest type of metal polish, micro-asperities still survive.

So it follows that, if two surfaces are placed in contact, no matter how carefully they have been machined and polished, they will contact only at the occasional points where one set of asperities meets the other. It is rather like turning Austria upside down and putting it on top of Switzerland. The load pressing the surfaces together is supported solely by the contacting asperities. The *real* area of contact, $a$, is very small and because of this the stress $P/a$ (load/area) on each asperity is very large.

Initially, at very low loads, the asperities deform elastically where they touch. However, for realistic loads, the high stress causes extensive *plastic* deformation at the tips of asperities. If each asperity yields, forming a junction with its partner, the total load transmitted across the surface (Figure 28.3) is

$$P \approx a\sigma_y \qquad (28.3)$$

where $\sigma_y$ is the compressive yield stress. In other words, the real area of contact is given by

$$a \approx \frac{P}{\sigma_y} \qquad (28.4)$$

Obviously, if we double $P$ we double the real area of contact, $a$.

Let us now look at how this contact geometry influences friction. If you attempt to slide one of the surfaces over the other, a shear stress $F_s/a$ appears at the asperities. The shear stress is greatest where the cross-sectional area of asperities is least, that is, at or very near the contact plane. Now, the intense plastic deformation in the regions of contact presses the asperity tips together so well that there is atom-to-atom contact across the junction. The junction, therefore, can withstand a shear stress as large as $k$ approximately, where $k$ is the shear-yield strength of the material (Chapter 11).

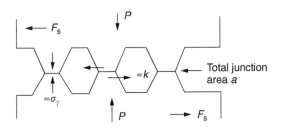

**FIGURE 28.3**
The real contact area between surfaces is much less than it appears to be, because the surfaces touch only where asperities meet.

The asperities will give way, allowing sliding, when

$$\frac{F_s}{a} \geq k$$

or, since $k \approx \sigma_y/2$, when

$$F_s \approx ak \approx a\sigma_y/2 \tag{28.5}$$

Combining this with Equation (28.3), we have

$$F_s \approx \frac{P}{2} \tag{28.6}$$

This is just the empirical Equation (28.1) we started with, with $\mu_s \approx 1/2$, but this time it is not empirical—we derived it from a model of the sliding process. The value $\mu_s \approx 1/2$ is close to the value of coefficients of static friction between unlubricated metal, ceramic, and glass surfaces—a considerable success.

How do we explain the lower value of $\mu_k$? Well, once the surfaces are sliding, there is less *time* available for atom-to-atom bonding at the asperity junctions than when the surfaces are in static contact, and the contact area over which shearing needs to take place is correspondingly reduced. As soon as sliding stops, creep allows the contacts to grow a little, diffusion allows the bond there to become stronger, and $\mu$ rises again to $\mu_s$.

## 28.3 DATA FOR COEFFICIENTS OF FRICTION

If *metal* surfaces are thoroughly cleaned in vacuum it is almost impossible to slide them over each other. Any shearing force causes further plasticity at the junctions, which quickly grow, leading to complete seizure ($\mu > 5$). This is a problem in outer space, and in atmospheres (e.g., $H_2$) that remove any surface films from the metal. A little oxygen or $H_2O$ greatly reduces $\mu$ by creating an oxide film that prevents these large metallic junctions forming.

We said in Chapter 24 that all metals except gold have a layer, no matter how thin, of metal oxide on their surfaces. Experimentally, it is found that for some metals the junction between the oxide films formed at asperity tips is weaker in shear than the metal on which it grew (Figure 28.4). In this case, sliding of the surfaces will take place in the thin oxide layer, at a stress less than in the metal itself, and lead to a corresponding reduction in $\mu$ to a value between 0.5 and 1.5.

When soft metals slide over each other (e.g., lead on lead, Figure 28.5) the junctions are weak but their area is large so $\mu$ is large. When hard metals slide (e.g., steel on steel) the junctions are small, but they are strong, and again friction is large (Figure 28.5). Many bearings are made of a thin film of a soft metal between two hard ones, giving weak junctions of small area. *White metal* bearings, for example, consist of soft alloys of lead or tin supported in a matrix of stronger

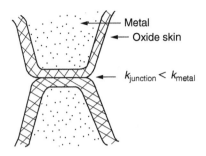

**FIGURE 28.4**
Oxide-coated junctions can often slide more easily than ones that are clean.

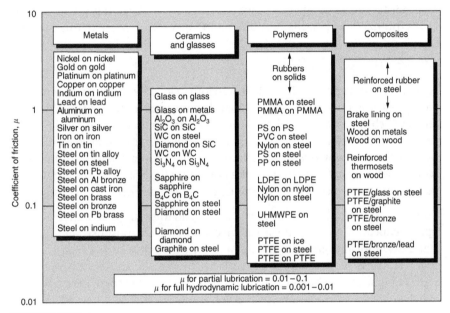

**FIGURE 28.5**
Bar chart showing the coefficient of static friction for various material combinations.

phases; *bearing bronzes* consist of soft lead particles (which smear out to form the lubricating film) supported by a bronze matrix; and polymer-impregnated *porous bearings* are made by partly sintering copper with a polymer (usually PTFE) forced into its pores. Bearings such as these are not designed to run dry—but if lubrication does break down, the soft component gives a coefficient of friction of 0.1 to 0.2 which may be low enough to prevent catastrophic overheating and seizure.

When ceramics slide on ceramics (Figure 28.5), friction is lower. Most ceramics are very hard—good for resisting wear—and, because they are stable in air and water (metals, except gold, are not genuinely stable, even if they appear so)—they have less tendency to bond, and shear more easily.

When metals slide on bulk polymers, friction is still caused by adhesive junctions, transferring a film of polymer to the metal. And any plastic flow tends to orient the polymer chains parallel to the sliding surface, and in this orientation they shear easily, so $\mu$ is low—0.05 to 0.5 (Figure 28.5). Polymers make attractive low-friction bearings, although they have some drawbacks: polymer molecules peel easily off the sliding surface, so wear is heavy; and because creep allows junction growth when the slider is stationary, the coefficient of static friction, $\mu_s$, is sometimes much larger than that for sliding friction, $\mu_k$.

Composites can be designed to have high friction (brake linings) or low friction (PTFE/bronze/lead bearings), as shown in Figure 28.5. More of this presently.

## 28.4 LUBRICATION

As we said in the introduction, friction absorbs a lot of work in machinery and as well as wasting power, this work is mainly converted to heat at the sliding surfaces, which can damage and even melt the bearing. In order to minimize frictional forces we need to make it as easy as possible for surfaces to slide over one another. The obvious way to try to do this is to contaminate the asperity tips with something that: (a) can stand the pressure at the bearing surface and so prevent atom-to-atom contact between asperities; (b) can itself shear easily.

Polymers and soft metal, as we have said, can do this; but we would like a much larger reduction in $\mu$ than these can give, and then we must use *lubricants*. The standard lubricants are oils, greases, and fatty materials such as soap and animal fats. These "contaminate" the surfaces, preventing adhesion, and the thin layer of oil or grease shears easily, obviously lowering the coefficient of friction. What is *not* so obvious is why the very fluid oil is not squeezed out from between the asperities by the enormous pressures generated there.

One reason is that nowadays oils have added to them small amounts ($\approx 1\%$) of active organic molecules. One end of each molecule reacts with the metal oxide surface and sticks to it, while the other ends attract one another to form an oriented "forest" of molecules (Figure 28.6), rather like mold on cheese. These forests can resist very large forces normal to the surface (and hence separate the asperity tips very effectively) while the two layers of molecules can shear over each other quite easily. This type of lubrication is termed *partial* or *boundary lubrication*, and is capable of reducing $\mu$ by a factor of 10 (Figure 28.5). *Hydrodynamic lubrication* is even more effective: we shall discuss it in the next chapter.

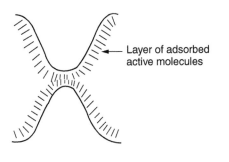

Layer of adsorbed
active molecules

**FIGURE 28.6**
Boundary lubrication.

Even the best boundary lubricants cease to work above about 200°C. Soft metal bearings such as those described earlier can cope with *local* hot-spots: the soft metal melts and provides a local lubricating film. But when the entire bearing is designed to run hot, special lubricants are needed. The best are a suspension of PTFE in special oils (good to 320°C); graphite (good to 600°C); and molybdenum disulfide (good to 800°C).

## 28.5 WEAR OF MATERIALS

Even when solid surfaces are protected by oxide films and boundary lubricants, some solid-to-solid contact occurs at regions where the oxide film breaks down under mechanical loading, and adsorption of active boundary lubricants is poor. This intimate contact will generally lead to *wear*. Wear is normally divided into two main types: *adhesive* and *abrasive wear*.

### Adhesive wear

Figure 28.7 shows that, if the adhesion between A and B atoms is good enough, wear fragments will be removed from the softer material A. If materials A and B are the same, wear takes place from *both* surfaces—the wear bits fall off and are lost or get trapped between the surfaces and cause further trouble (see below). The size of the bits depends on how far away from the junction the shearing takes place: if work-hardening extends well into the asperity, the tendency will be to produce large pieces. To minimize the *rate of wear* we obviously need to minimize the size of each piece removed.

The obvious way to do this is to minimize the area of contact $a$. Since $a \approx P/\sigma_y$ reducing the loading on the surfaces will reduce the wear, as would seem intuitively obvious. Try it with chalk on a chalkboard (see Figure 28.8): the higher the pressure, the stronger the line (a wear track). The second way to reduce $a$ is to increase $\sigma_y$ (i.e., the *hardness*). This is why hard pencils write with a lighter line than soft pencils.

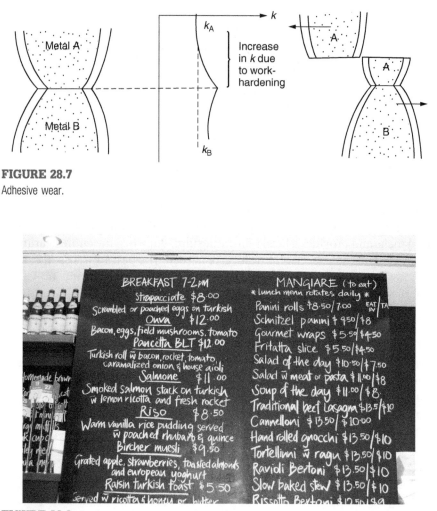

**FIGURE 28.7**
Adhesive wear.

**FIGURE 28.8**
Adhesive wear. Menu at Bertoni's café, Balmoral, NSW, Australia. – 33 49 36.40S 151 15 05.20 E

## Abrasive wear

Wear fragments produced by adhesive wear often become detached from their asperities during further sliding of the surfaces. Because oxygen is desirable in lubricants (to help maintain the oxide-film barrier between the sliding metals) these detached wear fragments can become oxidized to give hard oxide particles which *abrade* the surfaces in the way that sandpaper might.

Figure 28.9 shows how a hard material can "plough" wear fragments from a softer material, producing severe abrasive wear. Abrasive wear is not, of course, confined to indigenous wear fragments, but can be caused by dirt particles (e.g.,

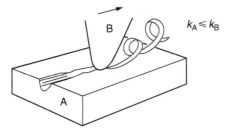

$k_A \leqslant k_B$

**FIGURE 28.9**
Abrasive wear.

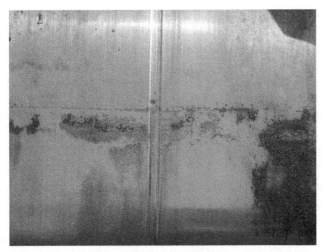

**FIGURE 28.10**
Abrasive wear track on crank-pin of marine diesel engine with a width of 8 mm and a depth of 0.2 mm.

sand) making their way into the system, or—in an engine—by combustion products: that is why it is important to filter the oil.

Obviously, the rate of abrasive wear can be reduced by reducing the load—just as in a hardness test. The particle will dig less deeply into the metal, and plough a smaller furrow. Increasing the hardness of the metal will have the same effect. Although abrasive wear is often bad (Figure 28.10) we would find it difficult to sharpen tools, or polish brass, or drill rock, without it.

## 28.6 SURFACE AND BULK PROPERTIES

Many considerations enter the choice of material for a bearing. It must have bulk properties that meet the need to support loads and transmit heat fluxes. It must be processable: that is, capable of being shaped, finished, and joined.

It must meet certain economic criteria: limits on cost, availability, and suchlike. If it can do all these things it must further have—or be given—necessary surface properties to minimize wear, and, when necessary, resist corrosion.

So, bearing materials are not chosen for their wear or friction properties (their "tribological" properties) alone; they have to be considered in the framework of the overall design. One way forward is to choose a material with good bulk properties, and then customize the surface with treatments or coatings. For the most part, it is the properties of the surface that determine tribological response, although the immediate subsurface region is obviously important because it supports the surface itself.

There are two general ways of tailoring surfaces. The aim of both is to increase the surface hardness, or to reduce friction, or all of these. The first is *surface treatment* involving only small changes to the chemistry of the surface. They exploit the increase in the hardness given by embedding foreign atoms in a thin surface layer: in carburizing (carbon), nitriding (nitrogen), or boriding (boron), the surface is hardened by diffusing these elements into it from a gas, liquid, or solid powder at high temperatures. Steels, which already contain carbon, can be surface-hardened by rapidly heating and then cooling their surfaces with a flame, an electron beam, or a laser. Elaborate though these processes sound, they are standard procedures, widely used, and to very good effect.

The second approach, that of *surface coating*, is more difficult, and that means more expensive. But it is often worth it. Hard, corrosion-resistant layers of alloys rich in tungsten, cobalt, chromium, or nickel can be sprayed onto surfaces, but a refinishing process is almost always necessary to restore the dimensional tolerances. Hard ceramic coatings such as $Al_2O_3$, $Cr_2O_3$, TiC, or TiN can be deposited by plasma methods and these not only give wear resistance but resistance to oxidation and other sorts of chemical attack as well. And—most exotic of all—it is now possible to deposit diamond (or something very like it) on to surfaces to protect them from almost anything.

## EXAMPLES

**28.1**  Explain the origins of friction between solid surfaces in contact.

**28.2**  The following diagram shows a compression joint for fixing copper water pipe to plumbing fittings. When assembling the joint the gland nut is first passed over the pipe followed by a circular olive made from soft copper. The nut is then screwed onto the end of the fitting and the backlash is taken up. Finally the nut is turned through a specified angle which compresses the olive on to the surface of the pipe. The angle is chosen so that it is just sufficient to make the cross-section of the pipe yield in compression over the length that is in contact with the olive. Show that the

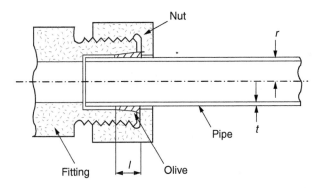

water pressure required to make the pipe shoot out of the fitting is given approximately by

$$p_w = 2\mu\sigma_y \left(\frac{t}{r}\right)\left(\frac{l}{r}\right)$$

where $\mu$ is the coefficient of friction between the olive and the outside of the pipe.

Calculate $p_w$ given the following information: $t = 0.65$ mm, $l = 7.5$ mm, $r = 7.5$ mm, $\mu = 0.15$, $\sigma_y = 120$ MN m$^{-2}$. Comment on your answer in relation to typical hydrostatic pressures in water systems. [*Hints:* The axial load on the joint is $p_w \pi r^2$; the radial pressure applied to the outside of the pipe by the olive is $P = \sigma_y t/r$.]

**Answer**

3.1 MN m$^{-2}$, or 31 bar

**28.3** Give examples, from your own experience, of situations where friction is (a) desirable, and (b) undesirable.

**28.4** Give examples, from your own experience, of situations where wear is (a) desirable, and (b) undesirable

**28.5** Bicycle chains often "stretch" during use. Referring to Example 12.4, this cannot be due to plastic extension or creep, because the material is hardened steel, with a factor of safety of 8.5 against yield. The stretch is in fact caused by wear between the links and the pins.

I (D.R. Jones) find that I need to shorten my bicycle chain by 25 mm (two chain pitches) every five years or so. Estimate the *radial* wear on the pins and holes, assuming all contacting surfaces wear at the same rate. The number of links in the chain is currently 110. Take other dimensions from Example 12.4.

**Answer**

0.06 mm (2.2 thou)

**28.6** Soft metal gaskets are often used for gas-tight or liquid-tight seals between steel surfaces. Examples are the use of soft copper washers under the heads of drain plugs in car engine sumps, or indium gaskets between bolted stainless steel flanges in high-vacuum equipment (indium melts at only 156°C, and is one of the softest metals known). Explain the principle involved in selecting a soft metal for such applications.

A washer for an oil drain plug has an outside diameter of 22 mm, an inside diameter of 15 mm, and is made from soft copper with a yield strength of 50 MN m$^{-2}$. The drain plug has a shank 15 mm in diameter, and is made from steel with a yield strength of 280 MN m$^{-2}$. The drain plug is torqued up until the copper yields. At this level of torque, estimate the ratio of the tensile stress in the shank to the yield stress of the steel. Comment on the practical significance of your answer.

I (DRHJ), being an engineer, change the oil and filter on my car myself, and always reuse the copper washer (it saves me money, and I can't be bothered to go to the garage and buy a new one). But before replacing it, I always soften it by heating it to dull red heat over the gas hob in the kitchen. Why is it important to do this?

**Answer**

Tensile stress in shank = 0.21 × yield stress of steel

**28.7** Electrical wiring circuits mainly use soft copper wires, and the ends are wired into plug sockets, switches, distribution boards, etc., using brass connectors fitted with brass grub screws to hold the wires in place. Why is it important to do the screws up tight? What can happen if the screws are not done up tight enough?

**28.8** The wheels of railway locomotives, carriages, and wagons are often pressed on to their axles, as shown in this diagram of a pressing operation.

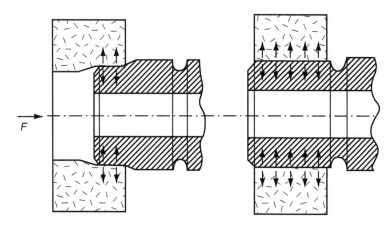

Experiments have been carried out in which steel wheels were repeatedly pressed on and off steel axles. Before the wheel was pressed on each time, the interference of fit was measured (the difference between the diameter of the axle and the diameter of the wheel bore). The results are given in the following table.

| Pressing Number | Interference before Each Pressing (μm) |
|---|---|
| 1 | 330 |
| 2 | 230 |
| 7 | 200 |
| 21 | 180 |

Which physical process was responsible for the observed decrease in interference?

In 1951, a new class of steam locomotives built by British Railways (the 70,000, or Britannia class) started failing in service owing to the driving and coupled wheels slipping on their axles. The axles were fitted with roller bearings, which meant that the wheelset (the finished assembly of two wheels, axle, and roller bearings) could not be put into the wheel balancing machine for adjusting the lead balance weights. So instead, the wheels were first pressed on to a dummy axle, the temporary assembly was balanced in the balancing machine, the wheels were then pressed off the dummy axle, and finally were pressed on to the real axle. Why do you think this unusual method of assembly might have made it more likely that the wheels would slip in service?

The YouTube clip shows No 70,000 Britannia herself, running in preservation in 1995. Serious mechanical engineering! *http://www.youtube.com/watch?v=YnXquACE9Ww*

This has the added benefit, at 1.30, of a loud squeal from the following locomotive (ex Great Western Railway No 7325), because of a "hot box", that is, a lubrication failure to one of the axle bearings! 7325 was taken off the train at Swansea as a result, before the axle got damaged.

28.9 The following diagram shows the arrangement of exhaust gas turbine and compressor wheel in a large turbocharger for a marine diesel engine. The compressor wheel is secured to the turbine shaft by an interference fit. The compressor wheel is aluminum alloy and the turbine shaft is steel. The shaft typically runs at 15,000 rpm.

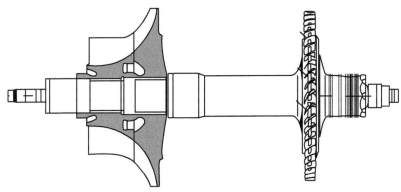

Given that the interference of fit is critical in this application, how would you assemble the compressor wheel to the turbine shaft without losing any interference? Justify your method with a simple calculation. The coefficient of thermal expansion of aluminum alloy is $22.5 \times 10^{-6}\,°C^{-1}$. The required interference of fit is 0.25 mm, on a shaft diameter of 110 mm.

28.10 This photograph shows a cattle grid on a bicycle path in Cambridge, England. There are many such grids in the city. The cross bars over which the cycles pass were originally round steel bars, galvanized so that they would not rust. In wet weather cyclists had to be very careful when crossing the grids, or they would skid sideways and fall off. Imagine that you are the City Engineer. Do you know what is causing this problem, and should you have foreseen it? How would you put this design error right—quickly and cheaply, of course? [A design feature, that they *did* get right, is the little ramp from the bottom of the pit to the upper edge of the cattle grid. It is to allow hedgehogs to escape if they have been unlucky enough to fall into the pit.]

– 52 11 41.85 N 0 07 00.22 E

## CHAPTER 29

# Case Studies in Friction and Wear

## CONTENTS

## 29.1 INTRODUCTION

In this chapter we examine three quite different problems involving friction and wear. The first involves most of the factors that appeared in Chapter 28: it is that of a round shaft or journal rotating in a cylindrical bearing. This type of *journal bearing* is common in all types of rotating or reciprocating machinery: the crankshaft bearings of an automobile are good examples. The second is quite different: it involves the frictional properties of ice in the design of skis and sledge runners. The third Case Study introduces us to some of the frictional properties of polymers: the selection of rubbers for anti-skid tires.

## 29.2 CASE STUDY 1: THE DESIGN OF JOURNAL BEARINGS

In the proper functioning of a well-lubricated journal bearing, the frictional and wear properties of the materials are, surprisingly, irrelevant. This is because the mating surfaces never touch: they are kept apart by a thin pressurized film of oil formed under conditions of *hydrodynamic lubrication*. Figure 29.1 shows a cross section of a bearing operating hydrodynamically. The load on the journal

Engineering Materials I: An Introduction to Properties, Applications, and Design, Fourth Edition

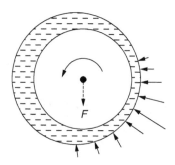

**FIGURE 29.1**
Hydrodynamic lubrication.

pushes the shaft to one side of the bearing, so that the working clearance is almost all concentrated on one side. Because oil is viscous, the revolving shaft drags oil around with it. The convergence of the oil stream toward the region of nearest approach of the mating surfaces causes an increase in the pressure of the oil film, and this pressure lifts the shaft away from the bearing surface.

Pressures of 10 to 100 atmospheres are common under such conditions. Provided the oil is sufficiently viscous, the film at its thinnest region is still thick enough to cause complete separation of the mating surfaces. Under *ideal* hydrodynamic conditions there is no asperity contact and no wear. Sliding of the mating surfaces takes place by shear in the liquid oil itself, giving coefficients of friction in the range 0.001 to 0.005.

Hydrodynamic lubrication is all very well when it functions properly. But when starting an engine up, or running slowly under high load, hydrodynamic lubrication is not effective, and we have to fall back on *boundary lubrication* (see Chapter 28).

Under these conditions some contact and wear of the mating surfaces will occur (this is why car engines last less well when used for short runs rather than long ones). Crankshafts are difficult and expensive to replace when worn, whereas bearings can be designed to be cheap and easy to replace as shown in Figure 29.2. It is thus good practice to concentrate as much of the wear as possible on the *bearing*—and, as we showed in our section on *adhesive wear* in the previous chapter, this is done by having a bearing material that is softer than the journal—or a journal that is harder than the bearing material.

Crankshaft journals can be "case-hardened" by special chemical and heat treatments (Chapter 28) to increase the surface hardness. (It is important not to harden the *whole* shaft because this will make it brittle and it might then break under shock loading.) Or bearing materials can be selected form the standard alloys which have been developed for this purpose (see Table 29.1).

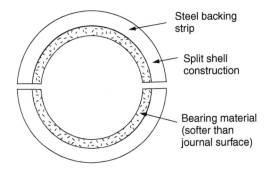

**FIGURE 29.2**
Replaceable bearing shells.

Table 29.1  Typical Bearing Alloys

| Alloy | $\sigma_y$ (MN m$^{-2}$) | Fatigue Strength |
|---|---|---|
| White metal (lead based) Pb – 10% Sb, 6% Sn | 30 | Low |
| White metal (tin based) Sn – 8% Sb, 4% Cu | 60 | Low |
| Aluminum based | | |
| Al – 40% Sn | 40 | Low |
| Al – 6% Sn, 1% Ni | 50 | Medium |
| Al – 11% Si, 1% Cu | 90 | Good |
| Copper based | | |
| Cu – 30% Pb | 60 | Medium |
| Cu – 20% Pb, 5% Sn (leaded bronze) | 125 | Good |
| Cu – 10% Sn, 1% P (phosphor bronze) | 230 | Good |

There are several other advantages to having a relatively soft bearing material; these are described next.

## Embeddability
Real bearings contain dirt, such as small hard particles of silica. If the particles are thicker than the oil film, they will generate abrasive wear. But if the bearing surface is soft enough, these particles will be pushed into the surface, and removed from circulation (Figure 29.3).

## Conformability
Slight misalignments of bearings can be self-correcting if plastic flow occurs easily in the bearing metal (Figure 29.4). But there is a compromise between load-carrying ability and conformability.

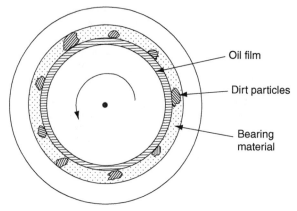

**FIGURE 29.3**
Dirt particles become embedded in a soft bearing material.

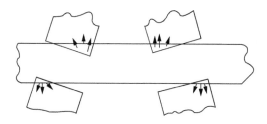

**FIGURE 29.4**
A conformable bearing material will flow to adjust to minor misalignments.

## Preventing seizure

If there is a breakdown in the oil supply, frictional heating will rapidly increase the temperature of the bearing, and this would normally lead to pronounced metal-to-metal contact, gross atomic bonding between journal and bearing, and seizure. A soft bearing material of low melting point will be able to shear in response to the applied forces, and may also melt locally. This helps to protect the journal from more severe damage, and also helps to avoid major component breakages which would result from the sudden locking-up of moving parts. Figures 29.5 and 29.6 show a typical example of overheating damage caused to the main crankshaft bearing of a marine diesel engine. The white metal bearing is badly scuffed, smeared, and oxidized, but there was no damage to the crankshaft journal.

When designing a bearing for a particular application, it is useful to think in terms of the pressure-velocity (or "PV") envelope inside which the bearing can function safely (Figure 29.7). We must ensure that the maximum pressure

**FIGURE 29.5**
Scuffed main crankshaft bearing of a marine diesel engine.

**FIGURE 29.6**
Close-up of the bearing surface in Figure 29.5.

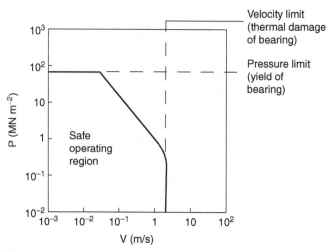

**FIGURE 29.7**

Pressure-velocity diagram for a sintered (porous) tin bronze bearing impregnated with PTFE (no additional lubrication *http://www.glycodur.de/en/lieferprogramm/gleitlagerwerkstoff/index.htm; http://www.glycodur.de/en/lieferprogramm/gleitlagertechnik/lager_pvbereich.htm*).

and sliding speed encountered in service will not make the material yield (see Table 29.1) or suffer heat damage. The PV envelope is not simply a property of the bearing material, but also the nature of the applied loading, the lubrication conditions, and the constructional details of the bearing. Although hydrodynamic lubrication should keep the surfaces apart, some frictional heating is generated even under ideal lubrication conditions. Much more heat can be generated if lubrication conditions revert to boundary lubrication or worse.

A classic example can be found in the railway locomotive Mallard which holds the world speed record for steam railway locomotives of 125.9 mph (*http://en.wikipedia.org/wiki/LNER_Class_A4_4468_Mallard*). During the record-breaking run on 3 July 1938 the crank-pin bearing for the inside cylinder overheated. Steam was shut off immediately after the measured distance had been completed, and the locomotive had to stop at the nearest station. No doubt the combination of high sliding speed and the harsh, reciprocating loading took the bearing outside the limits of the PV envelope.

Fluctuating loadings also occur, for example, in petrol and diesel engines, and are particularly pronounced in diesel engine crank-pin and crosshead bearings. Thus, in addition to the PV envelope, we also need to consider *fatigue* of the bearing material. This can lead to cracking in the bearing surface, and eventual detachment of pieces of bearing metal. Figure 29.8 shows a typical example of fatigue damage in a marine engine crank-pin bearing. As Table 29.1 shows, the fatigue strength depends strongly on the type of bearing material.

**FIGURE 29.8**
Marine diesel engine crank-pin bearing (upper shell). White metal missing because of fatigue cracking.

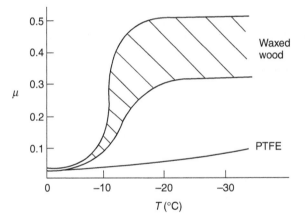

**FIGURE 29.9**
Friction of materials on ice at various temperatures.

## 29.3 CASE STUDY 2: MATERIALS FOR SKIS AND SLEDGE RUNNERS

Skis, both for people and for aircraft, used to be made of waxed wood. Down to about −10°C, the friction of waxed wood on snow is very low—$\mu$ is about 0.02—and if this were not so, planes equipped with skis could not take off from packed-snow runways, and the winter tourist traffic to Switzerland would drop sharply. Below −10°C, bad things start to happen (Figure 29.9): $\mu$ rises sharply to about 0.4. Polar explorers have observed this repeatedly. Wright, a member of the 1911–1913 Scott expedition, writes: "Below 0°F (–18°C) the friction (on the sledge runners) seemed to increase progressively as the temperature fell"; it caused the expedition considerable hardship. What determines the friction of skis on snow?

Ice differs from most materials in that its melting point *drops* if you compress it. It is widely held that pressure from the skis causes the snow beneath to melt, but this is nonsense: the pressure of a large person distributed over a ski lowers the melting point of ice by about 0.0001°C and, even if the weight is carried by asperities that touch the ski over only $10^{-3}$ of its nominal area, the depression of the melting point is still only 0.1°C. Pressure melting, then, cannot account for the low friction shown in Figure 29.9. But as the person starts to descend the ski slope, work is done against the frictional forces, heat is generated at the sliding surface, and the "velocity limit" of Figure 29.7 is exceeded. The heat melts a layer of ice, producing a thin film of water, at points where asperities touch the ski: the person hydroplanes along on a layer of water generated by friction. The principle is exactly like that of the lead–bronze bearing, in which local hot-spots melt the lead, producing a lubricating film of liquid which lowers $\mu$ and saves the bearing.

Below –10°C, heat is conducted away too quickly to allow this melting—and because their thermal conductivity is high, skis with exposed metal (aluminum or steel edges) are slower at low temperatures than those without. At these low temperatures, the mechanism of friction is the same as that of metals: ice asperities adhere to the ski and must be sheared when it slides. The value of $\mu$ (0.4) is close to that calculated from the shearing model in Chapter 28. This is a large value of the coefficient of friction—enough to make it very difficult for a plane to take off, and increasing by a factor of more than 10 the work required to pull a loaded sledge. What can be done to reduce it?

This is a standard friction problem. A glance at Figure 28.5 shows that, when polymers slide on metals and ceramics, $\mu$ can be as low as 0.04. Among the polymers with the lowest coefficients are PTFE ("Teflon") and polyethylene. By coating the ski or sledge runners with these materials, the coefficient of friction stays low, *even* when the temperature is so low that frictional heating is unable to produce a boundary layer of water. Aircraft and sports skis now have polyethylene or Teflon undersurfaces; the Olympic Committee has banned their use on bob-sleds, which already, some think, go fast enough.

## 29.4 CASE STUDY 3: HIGH-FRICTION RUBBER

So far we have talked of ways of reducing friction. But for many applications— brake pads, clutch linings, climbing boots, and above all, car tires—we want as much friction as we can get.

The frictional behavior of rubber is quite different from that of metals. In Chapter 28 we showed that when metallic surfaces were pressed together, the bulk of the deformation at the points of contact was plastic; and that the friction between the surfaces arose from the forces needed to shear the junctions at the areas of contact.

But rubber deforms *elastically* up to very large strains. When we bring rubber into contact with a surface, therefore, the deformation at the contact points is *elastic*. These elastic forces still squeeze the atoms together at the areas of contact, of course; adhesion will still take place there, and shearing will still be necessary if the surfaces are to slide. This is why car tires grip well in dry conditions. In *wet* conditions, the situation is different; a thin lubricating film of water and mud forms between rubber and road, and this will shear at a stress a good deal lower than previously, with dangerous consequences. Under these circumstances, another mechanism of friction operates to help prevent a skid.

It is illustrated in Figure 29.10. All roads have a fairly rough surface. The high spots push into the tire, causing a considerable local elastic deformation. As the tire skids, it slips forward over the rough spots. The region of rubber that was elastically deformed by the rough spot now relaxes, while the rubber just behind this region becomes compressed as it reaches the rough spot. Now, all rubbers exhibit some *anelasticity*; the stress–strain curve looks like Figure 29.11.

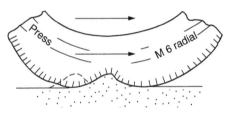

**FIGURE 29.10**
Skidding on a rough road surface deforms the tire material elastically.

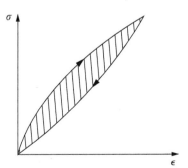

**FIGURE 29.11**
Work is needed to cycle rubber elastically.

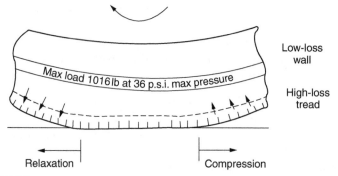

**FIGURE 29.12**

Anti-skid tires, with a high-loss tread (for maximum grip) and a low-loss wall (for minimum heating up).

As the rubber is compressed, work is done on it equal to the area under the upper curve; but if the stress is removed we do not get all this work back. Part of it is dissipated as heat—the part shown as the shaded area between the loading and the unloading curve. So to make the tire slide on a rough road we have to do work, even when the tire is well lubricated, and if we have to do work, there is friction. Special rubbers have been developed for high-loss characteristics (called "high-loss" or "high-hysteresis" rubbers) and these have excellent skid resistance even in wet conditions.

There is one obvious drawback of high-hysteresis rubber. In normal rolling operation, considerable elastic deformations still take place in the tire wall, and high-loss tires will consume fuel and generate considerable heat. The way out is to use a low-loss tire covered with a high-loss tread—another example of design using composite materials (Figure 29.12).

## EXAMPLES

**29.1** How does lubrication reduce friction?

**29.2** It is observed that snow lies stably on roofs with a slope of less than 24°, but that it slides off roofs with a greater slope. Skiers, on the other hand, slide on a snow-covered mountainside with a slope of only 2°. Why is this?

**29.3** A person of weight 100 kg standing on skis 2 m long and 0.10 m wide slides on the 2° mountain slope, at 0°C. Calculate the loss of potential energy when the ski slides a distance equal to its own length. Hence calculate the average thickness of the water film beneath each ski. (The latent heat of fusion of ice is 330 MJ m$^{-3}$.)

**Answers**

Work done 68 J; average film thickness = 0.5 μm

**29.4** How can friction between road and tire be maintained even under conditions of appreciable lubrication?

**29.5** In countries that have cold winters (e.g., Sweden) it is usual for road vehicles to be fitted with special tires in the winter months. These tires have hard metal studs set into the tread. Why is this necessary?

**29.6** Explain the principle behind hydrodynamic lubrication. Under what conditions is hydrodynamic lubrication likely to break down? What then saves the bearing?

**29.7** Why do plain bearings often consist of a hardened shaft running in a soft shell?

**29.8** Modern gas and diesel engines use multilayered bearing shells as shown in the following diagram. Tight machining and assembly tolerances mean that the bearing material can be made very thin, since conformability is no longer much of an issue. This also means that the bearing can stand a much greater pressure without the soft material yielding.

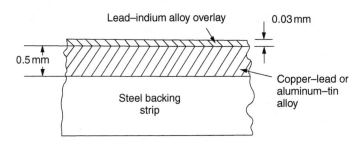

With reference to Example 11.2, modify the analysis for the situation where there is no friction between the bar and the upper anvil, but there is sticking friction between the bar and the lower anvil. This models the situation in the bearing (bar = soft bearing material, upper anvil = oil film, lower anvil = steel backing strip). Hence show that

$$F \leq 2.5wLk$$

Verify that the general formula now becomes

$$F \leq 2wLk\left(1 + \frac{w}{8d}\right)$$

For a bearing layer 0.5 mm thick, and a bearing shell 40 mm wide, find the ratio between the yield load of the bearing layer, and the yield load of an unconstrained sample of bearing material.

**Answer**

Ratio = 11; a huge gain in the effective strength of the bearing material

**29.9** The photographs at the top of the next page show the propeller shaft bearing from a large cargo vessel which was put into dry dock for repairs. The white metal lining of the bearing has suffered considerable damage in service. List the main features of the damage. What mechanism of failure would you suggest to explain these features?

Cargo vessel in dry dock for repairs. Note propeller and propeller shaft removed (just forward of the rudder). Lisnave shipyard, Sétubal, Portugal. – 38 28 40 N 8 47 21 W

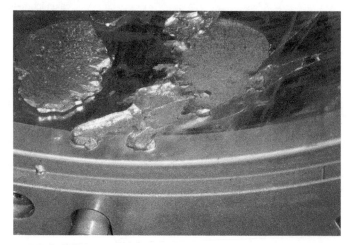

Propeller shaft white metal bearing

**29.10** The YouTube link shows a steam locomotive on the famous Darjeeling Himalaya 2 ft gauge line in India (New Jalpaiguri to Darjeeling, West Bengal – 27 01 00 N 88 14 51 E ); see *http://en.wikipedia.org/wiki/Darjeeling_Himalayan_Railway* and *http://www. youtube.com/watch?v=LwnmaB4OaIo&feature=related*

At the start of the clip, the driving wheels of the loco can be seen slipping on the rail because it is damp from the overnight dew. Shortly afterward, the two "sand boys" can be seen dropping sand on to the rails just in front of the loco to help the wheels get a grip. Why was the loco slipping in the first place, and why should sand stop the loco slipping?

## CHAPTER 30

# Final Case Study: Materials and Energy in Car Design

## CONTENTS

## 30.1 INTRODUCTION

The status of steel as the raw material of choice for the manufacture of car bodies rests principally on its *price*. It has always been the cheapest material that meets the necessary strength, stiffness, formability, and weldability requirements of large-scale car body production.

Until recently, the fact that steel has a density two-thirds that of lead was accorded little significance. But increasing sustainability awareness and legislative requirements concerning carbon emissions are now changing that view. Car makers are looking hard at alternative materials.

Engineering Materials I: An Introduction to Properties, Applications, and Design, Fourth Edition

## 30.2 ENERGY AND CARBON EMISSIONS

Energy is used to build a car, and energy is used to run it. High oil prices mean that the cost of the gas consumed during the life of a car is comparable with the cost of the car itself. Consumers now want more fuel-efficient cars, and more fuel-efficient cars have lower carbon emissions.

This trend is magnified by the taxation policies of some governments. Fuel duties have increased, so one way in which the motorist can offset the effects of additional fuel taxation is to reduce fuel consumption per mile. However, this strategy can be partly countered by imposing annual taxes on vehicles that relate to their carbon emissions. This acts as a driver to reduce engine power—effectively vehicle size. The extent to which governments really care about sustainability and carbon emissions—or are using these as a front for generating yet further taxation revenues—is open to question. But the way in which the economics are felt at consumer level is the same, and only likely to get worse in the longer term.

## 30.3 WAYS OF ACHIEVING ENERGY ECONOMY

It is clear from Table 30.1 that the energy content of the car itself—the steel, rubber, glass, and manufacturing process—is small: less than 10% that required to move the car. This means there is little point trying to save energy here; indeed (as we shall see) it may pay to use more energy to make the car (using, for instance, aluminum instead of steel) if this reduces fuel consumption.

We must focus, then, on reducing the energy used to move the car. The following are the two routes.

- *Improve engine efficiency.* Engines are already remarkably efficient: There is little more that can be done with internal combustion engines, although progress is being made with IC/electric hybrids and all-electric systems.
- *Reduce the weight of the car.* Figure 30.1 shows how the fuel consumption (g.p.m.) and the mileage (m.p.g.) vary with car weight. There is a linear correlation: halving the weight halves the g.p.m. This is why small cars are more economical than big ones: engine size and performance have some influence, but it is mainly the weight that determines the fuel consumption.

Table 30.1 Energy in the Manufacture and Use of Cars

| | |
|---|---|
| Energy to produce cars, per year | = 0.8% to 1.5% of total energy consumed by nation |
| Energy to move cars, per year | = 15% of total energy consumed by nation |
| Transportation of people and goods, total | = 24% of total energy consumed by nation |

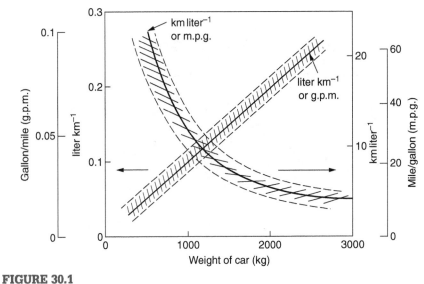

**FIGURE 30.1**
Fuel consumption of production cars.

We can reduce the size of cars, but the consumer does not like that. Or we can reduce the *weight* of the car by substituting lighter materials for those used now. Lighter cars not only use less fuel, but they also have lower carbon emissions— hence, the interest in producing lighter vehicles, reversing a consistent trend in the opposite direction.

## 30.4 MATERIAL CONTENT OF A CAR

As Figure 30.1 suggests, most cars weigh between 400 kg and 2500 kg. In a typical modern production car (Figure 30.2), this is made up as shown in Table 30.2.

## 30.5 ALTERNATIVE MATERIALS

### Primary mechanical properties

Candidate materials for substitutes must be lighter than steel, but structurally equivalent. For the engine block, the choice is obvious: aluminum (density 2.7 Mg m$^{-3}$) or possibly magnesium (density 1.8 Mg m$^{-3}$) replace an *equal volume* of cast iron (density 7.7 Mg m$^{-3}$) with an immediate weight reduction on this component of 2.8 to 4.3 times. The production methods remain almost unchanged. Most manufacturers have made this change; cars like the one shown in Figure 30.3 are a thing of the past.

**FIGURE 30.2**

The Volkswagen Passat—a typical modern pressed-steel body with no separate chassis. For a *given material* this "monocoque" construction gives a minimum weight-to-strength ratio. *(Courtesy of Volkswagen—photo credit to Brian Garland © 2004)*

| Table 30.2 Contributors to the Weight of Car | |
| --- | --- |
| **Material** | **Component** |
| 71% steel | Body shell; panels |
| 15% cast iron | Engine block; gear box; rear axle |
| 4% rubber | Tires; hoses |
| Balance | Glass, zinc, copper, aluminum, polymers |

**FIGURE 30.3**

The Morris Traveller—a classic of the 1950s—used wood as an integral part of the body shell.

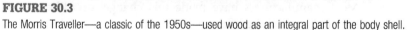

**Table 30.3** Properties of Candidate Materials for Car Bodies

| Material | Density $\rho$ (Mg m$^{-3}$) | Young's modulus, $E$ (GN m$^{-2}$) | Yield strength, $\sigma_y$ (MN m$^{-2}$) | $(\rho/E^{1/3})$ | $(\rho/\sigma_y^{1/2})$ |
|---|---|---|---|---|---|
| Mild steel⎫<br>High-strength steel⎭ | 7.8⎭ | 207 | 220⎫<br>up to 500⎭ | 1.32 | 0.53<br>0.35 |
| Aluminum alloy | 2.7 | 69 | 193 | 0.66 | 0.19 |
| GFRP (chopped fiber, molding grade) | 1.8 | 15 | 75 | 0.73 | 0.21 |

The biggest potential weight saving, however, is in the body panels, which make up 60% of the weight of the vehicle. Here the choice is more difficult. Candidate materials are given in Table 30.3.

How do we assess the possible weight saving? Just replacing a steel panel with an aluminum-alloy or fiberglass one of equal thickness (giving a weight saving that scales as the density) is unrealistic: both substitutes have much lower moduli, and thus will deflect (under given loads) far more; and one of them has a much lower yield strength, and might undergo plastic flow.

If (as with body panels) *elastic* deflection is what counts, the logical comparison is for a panel of equal *stiffness*. And if resistance to *plastic* flow counts (as with fenders) then the proper thing to do is compare sections with equal resistance to plastic flow.

The *elastic* deflection $\delta$ of a panel under a force $F$ (Figure 30.4) is given by

$$\delta = \frac{Cl^3 F}{Ebt^3} \tag{30.1}$$

where $t$ is the sheet thickness, $l$ the length, and $b$ the width of the panel. The constant $C$ depends on how the panel is held: at two edges, at four edges, and so on—it does not affect our choice. The mass of the panel is

$$M = \rho btl \tag{30.2}$$

The quantities $b$ and $l$ are determined by the design (they are the dimensions of the door, trunk lid, etc.). The only variable in controlling the stiffness is $t$. Now from Equation (30.1)

$$t = \left(\frac{Cl^3 F}{\delta Eb}\right)^{1/3} \tag{30.3}$$

Substituting this expression for $t$ in Equation (30.2) gives the mass of the panel shown in Equation (30.4).

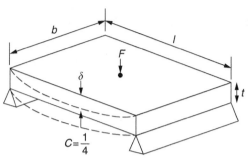

**FIGURE 30.4**
Elastic deflection of a car-body panel.

$$M = \left\{ C\left(\frac{F}{\delta}\right) l^6 b^2 \right\}^{1/3} \left(\frac{\rho}{E^{1/3}}\right) \tag{30.4}$$

For a given *stiffness* $(F/\delta)$, *panel dimensions* $(l, b)$, and *edge-constraints* $(C)$, the lightest panel is the one with the smallest value of $\rho/E^{1/3}$.

We can do a similar analysis for *plastic* yielding. A panel with the section shown in Figure 30.5 yields at a load

$$F = \left(\frac{Cbt^2}{l}\right) \sigma_y \tag{30.5}$$

The panel mass is given by Equation (30.2). The only variable is $t$,

$$t = \left(\frac{Fl}{Cb\sigma_y}\right)^{1/2} \tag{30.6}$$

from Equation (30.5). Substituting for $t$ in Equation (30.2) gives

$$M = \left(\frac{Fbl^3}{C}\right)^{1/2} \left(\frac{\rho}{\sigma_y^{1/2}}\right) \tag{30.7}$$

The panel with the smallest value of $\rho/\sigma_y^{1/2}$ is the one we want.

We can now assess candidate materials more sensibly on the basis of the data given in Table 30.3. The values of the property groups are shown in the last two columns. For most body panels (*elastic* deflection determines design) high-strength steel offers no advantage over mild steel: it has the same value of $\rho/E^{1/3}$. GFRP is better (a lower $\rho/E^{1/3}$ and weight), and aluminum alloy is better still—that is one of the reasons aircraft are made of aluminum. But the weight saving in going from steel to aluminum is not a factor of 3 (the ratio of the densities) but only 2 (the ratio of $\rho/E^{1/3}$) because the aluminum panel has to be thicker to compensate for its lower $E$.

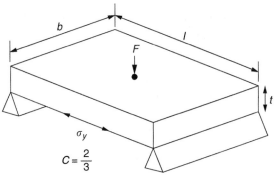
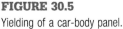

**FIGURE 30.5**
Yielding of a car-body panel.

High-strength steel *does* offer a weight saving for strength-limited components: fenders, front and rear header panels, engine mounts, bulkheads, and so forth—the weight saving $\left(\rho/\sigma_y^{1/2}\right)$ is a factor of 1.5. Both aluminum alloy and fiberglass offer potential weight savings of up to 3 times $\left(\rho/\sigma_y^{1/2}\right)$ on these components. This makes possible a saving of at least 30% on the weight of the vehicle—if an aluminum engine block is used, the overall weight saving is larger still. These are very substantial savings—sufficient to achieve a 50% increase in mileage per gallon without any decrease in size of car or increase in engine efficiency. So they are obviously worth examining more closely. What, then, of the *other* properties required of the substitute materials?

## Secondary properties

Although resistance to deflection and plastic yielding are obviously of first importance in choosing alternative materials, other properties enter into the selection. Table 30.4 lists the conditions imposed by the service environment.

*Elastic* and *plastic* deflection we have dealt with already. The toughness of steel is so high that *fracture* of a steel panel is seldom a problem. But what about the other materials? The data for toughness are given in Table 30.5.

But what is the proper way to use toughness values? Suppose the panel is loaded up to its yield load (above this load we *know* it will begin to fail by plastic flow so it does not matter whether other failure mechanisms also appear)—what is the maximum crack size that is still stable? If this is large enough that it should not appear in service, we are satisfied; if not, we must increase the section. This crack size is given conservatively by

$$\sigma_y\sqrt{\pi a} = K_c = \sqrt{EG_c}$$

**Table 30.4 The Service Environment of the Average Car**

| Environment | Condition | |
|---|---|---|
| Loading | Static → Elastic or plastic deflection | |
| | Impact → Elastic or plastic deflection | |
| | Impact → Fracture | |
| | Fatigue → Fatigue fracture | |
| | Long-term static → Creep | |
| Physical | $-40°C < T < 60°C$ | |
| | 10% < relative humidity < 100% | |
| Chemical | UV light | |
| | Dust particles | |
| | Water | Gas |
| | Oil | Antifreeze |
| | Brake fluid | Salt |
| | Transmission fluid | Air |

**Table 30.5 Properties of Body-Panel Materials: Toughness, Fatigue, and Creep**

| Material | Toughness $G_c$ (kJ m$^{-2}$) | Tolerable Crack Length (mm) | Fatigue | Creep |
|---|---|---|---|---|
| Mild steel | ≈100 | ≈140 ⎫ | OK | OK |
| High-strength steel | ≈100 | ≈26 ⎭ | | |
| Aluminum alloy | ≈20 | ≈12 | OK | OK |
| GFRP (chopped fiber, molding grade) | ≈37 | ≈30 | OK | Creep above 60°C |

from which

$$a_{max} = \frac{EG_c}{\pi\sigma_y^2}$$

The resulting crack lengths are given in Table 30.5. A panel with a longer crack will fail by "tearing"; one with a short crack will fail by general yield—it will bend permanently. Although the crack lengths are shorter in replacement materials than in steel, they are still large enough to permit the replacement materials to be used.

*Fatigue* (Chapter 17) is always a potential problem with any structure subject to varying loads: anything from the loading due to closing the door to that caused by engine vibration can, potentially, lead to failure. The fatigue strength of all these materials is adequate.

*Creep* (Chapter 20) is not normally a problem a designer considers when designing a car body with metals: the maximum service temperature reached is 60°C, and neither steel nor aluminum alloys creep significantly at these temperatures. But GFRP does, so extra reinforcement or heavier sections may be necessary where temperatures exceed this value.

More important than creep or fatigue in current car design is the *effect of environment.* A significant part of the cost of a car is contributed by the manufacturing processes designed to prevent rusting; and these processes only partly work—it is body rust that ultimately kills a car, since the mechanical parts (brake discs, etc.) can be replaced relatively easily.

Steel is particularly bad in this regard. Under ordinary circumstances, aluminum is much better. Although the effect of salt on aluminum is bad, heavy anodizing will slow down even that form of attack to tolerable levels (yacht masts are made of anodized aluminum alloy).

So aluminum alloy is good: it resists all the fluids likely to come in contact with it. What about GFRP? The strength of GFRP is reduced by up to 20% by continuous immersion in most of the fluids—even salt-water—with which it is likely to come into contact; but (as we know from fiberglass boats) this drop in strength is not critical, and it occurs without visible corrosion, or loss of section. In fact, GFRP is much more corrosion-resistant, in the normal sense of "loss-of-section," than steel

## 30.6 PRODUCTION METHODS

The biggest penalty one has to pay in switching materials is likely to be the higher material and production costs. *High-strength steel*, of course, presents almost no problem. The yield strength is higher, but the section is thinner, so that only slight changes in punches, dies, and presses are necessary, and once these are paid for, the extra cost is just that of the material.

At first sight, the same is true of *aluminum alloys.* But because they are heavily alloyed (to give a high yield strength) their *ductility* is low. If expense is unimportant, this does not matter; some early Rolls-Royce cars (Figure 30.6) had aluminum bodies which were formed into intricate shapes by laborious hand-beating methods, with frequent annealing of the aluminum to restore its ductility. But in mass production we should like to deep draw body panels in one operation—and then low ductility is much more serious. The result is a loss of design flexibility: there are more constraints on the use of aluminum alloys than on steel; and it is this, rather than the cost, that is the greatest obstacle to the wholesale use of aluminum in cars.

**FIGURE 30.6**

A 1932 Rolls-Royce. Mounted on a separate steel chassis is an all-aluminum hand-beaten body by the famous coach building firm of James Mulliner. Any weight advantage due to the use of aluminum is totally outweighed by the poor weight-to-strength ratio of separate-chassis construction; but the bodywork remains immaculate after 80 years of continuous use!

**FIGURE 30.7**

A Lotus Esprit, with a GFRP body (but still mounted on a steel chassis—which does not give anything like the weight saving expected with an all-GFRP monocoque structure). *(Courtesy of Group Lotus Ltd.)*

GFRP looks as if it would present production problems: you may be familiar with the tedious hand layup process required to make a fiberglass boat or canoe. But mass-production methods have now been developed to handle GFRP. Most modern cars have GFRP components (fenders, facia panels, internal panels) and a few have GFRP bodies (Figure 30.7), usually mounted on a steel chassis; the full weight savings will only be realized if the *whole* load-bearing

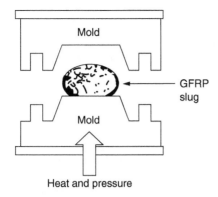

**FIGURE 30.8**
Compression molding of car-body components.

structure is made from GFRP. In producing GFRP car panels, a slug of polyester resin, with *chopped* glass fibers mixed in with it, is dropped into a heated split mold (Figure 30.8).

As the polyester used is a *thermoset* it will "go off" in the hot mold, after which the solid molding can be ejected. Modern methods allow a press like this to produce one molding per minute—still slower than steel pressing, but practical. Molding brings certain advantages. It offers design flexibility—particularly in change of section, and sharp detail—which are not easily achieved with steel. And GFRP moldings often result in consolidation of components, reducing assembly costs.

## 30.7 CONCLUSIONS

| For | Against |
|---|---|
| **A. High-strength steel** | |
| Retains all existing technology | Weight saving only appreciable in designing against *plastic* flow |
| *Use in high-stress areas—but fatigue may then become a problem.* | |
| **B. Aluminum alloy** | |
| Large weight saving in both body shell and engine block | Unit cost higher |
| Retains much existing technology | Deep drawing properties poor—loss in design flexibility |

*Continued*

| For | Against |
|---|---|
| Corrosion resistance good | |
| *Aluminum alloy offers saving of up to 40% in total car weight. The increased unit cost is offset by the lower running cost of the lighter vehicle, and the greater recycling potential of the aluminum.* | |
| **C. GFRP** | |
| Large weight saving in body shell | Unit cost higher |
| Corrosion resistance good | Massive changes in manufacturing technology |
| Gain in design flexibility and parts consolidation | Designer must cope with some creep |
| *GFRP offers savings of up to 30% in total car weight, at some increase in unit cost and considerable capital investment in new equipment. Recycling problems still have to be overcome.* | |

# Symbols and Formulae

## CONTENTS

## LIST OF PRINCIPAL SYMBOLS

| Symbol | Meaning (units) |
|--------|-----------------|
| *Note:* | Multiples or submultiples of basic units indicate the unit suffixes typically used with materials data. |
| $a$ | side of cubic unit cell (nm) |
| $a$ | crack length (mm) |
| $A$ | constant in fatigue crack-growth law |
| $A$ | constant in creep law $\dot{\varepsilon}_{ss} = A\sigma^n e^{-Q/\bar{R}T}$ |
| $b$ | Burgers vector (nm) |
| $c$ | concentration ($m^{-3}$) |
| $D$ | diffusion coefficient ($m^2\ s^{-1}$) |
| $D_0$ | pre-exponential constant in diffusion coefficient ($m^2\ s^{-1}$) |
| $E$ | Young's modulus of elasticity ($GN\ m^{-2}$) |
| $f$ | force acting on unit length of dislocation line ($N\ m^{-1}$) |
| $F$ | force (N) |
| $g$ | acceleration due to gravity on the Earth's surface ($m\ s^{-2}$) |
| $G$ | shear modulus ($GN\ m^{-2}$) |
| $G_c$ | toughness (or critical strain energy release rate) ($kJ\ m^{-2}$) |

*Continued*

Engineering Materials I: An Introduction to Properties, Applications, and Design, Fourth Edition

| Symbol | Meaning (units) |
| --- | --- |
| $H$ | hardness (MN m$^{-2}$ or kg mm$^{-2}$) |
| $J$ | diffusion flux (m$^{-2}$ s$^{-1}$) |
| $k$ | shear yield strength (MN m$^{-2}$) |
| $k$ | Boltzmann's constant $\bar{R}/N_A$ (J K$^{-1}$) |
| $K$ | bulk modulus (GN m$^{-2}$) |
| $K$ | stress intensity factor (MN m$^{-3/2}$) |
| $K_c$ | fracture toughness (critical stress intensity factor) (MN m$^{-3/2}$) |
| $\Delta K$ | $K$ range in fatigue cycle (MN m$^{-3/2}$) |
| $m$ | constant in fatigue crack growth law (dimensionless) |
| $n$ | creep exponent in $\dot{\varepsilon}_{ss} = A\sigma^n e^{-Q/\bar{R}T}$ |
| $N$ | number of fatigue cycles |
| $N_A$ | Avogadro's number (mol$^{-1}$) |
| $N_f$ | number of fatigue cycles leading to failure (dimensionless) |
| $Q$ | activation energy per mole (kJ mol$^{-1}$) |
| $r_0$ | equilibrium interatomic distance (nm) |
| $\bar{R}$ | universal gas constant (J K$^{-1}$ mol$^{-1}$) |
| $S_0$ | bond stiffness (N m$^{-1}$) |
| $t_f$ | time-to-failure (s) |
| $T$ | line tension of dislocation (N) |
| $T$ | absolute temperature (K) |
| $T_M$ | absolute melting temperature (K) |
| $U^{el}$ | elastic strain energy (J) |
| $\gamma$ | (true) engineering shear strain (dimensionless) |
| $\Delta$ | dilatation (dimensionless) |
| $\varepsilon$ | true (logarithmic) strain (dimensionless) |
| $\varepsilon_f$ | (nominal) strain after fracture; tensile ductility (dimensionless) |
| $\varepsilon_n$ | nominal (linear) strain (dimensionless) |
| $\varepsilon_0$ | permittivity of free space (F m$^{-1}$) |
| $\dot{\varepsilon}_{ss}$ | steady-state tensile strain-rate in creep (s$^{-1}$) |
| $\mu_k$ | coefficient of kinetic friction (dimensionless) |
| $\mu_s$ | coefficient of static friction (dimensionless) |
| $\nu$ | Poisson's ratio (dimensionless) |
| $\rho$ | density (Mg m$^{-3}$) |
| $\sigma$ | true stress (MN m$^{-2}$) |
| $\sigma_n$ | nominal stress (MN m$^{-2}$) |
| $\sigma_{TS}$ | (nominal) tensile strength (MN m$^{-2}$) |
| $\sigma_y$ | (nominal) yield strength (MN m$^{-2}$) |
| $\tilde{\sigma}$ | ideal strength (GN m$^{-2}$) |
| $\Delta\sigma$ | stress range in fatigue (MN m$^{-2}$) |
| $\tau$ | shear stress (MN m$^{-2}$) |

# SUMMARY OF PRINCIPAL FORMULAE

## Chapter 2

Exponential growth

$$\frac{dC}{dt} = \frac{r}{100} C, \; t_D = \frac{100}{r} \log_e 2$$

$C$ = consumption rate (ton per year)
$r$ = fractional growth rate (percent per year)
$t$ = time
$t_D$= doubling time

## Chapter 3

Stress, strain, Poisson's ratio, elastic moduli

$$\sigma = \frac{F}{A}, \; \tau = \frac{F_s}{A}, \; p = -\frac{F}{A}, \; v = -\frac{\text{lateral strain}}{\text{tensile strain}}$$

$$\varepsilon_n = \frac{u}{l}, \; \gamma = \frac{w}{l}, \; \Delta = \frac{\Delta V}{V}$$

$$\sigma = E\varepsilon_n, \; \tau = G\gamma, \; p = -K\Delta$$

when $v = 1/3$, $K = E$, and $G = (3/8)E$

Stresses and strains in three dimensions

$$\varepsilon_1 = \frac{\sigma_1}{E} - v\frac{\sigma_2}{E} - v\frac{\sigma_3}{E}$$

$$\varepsilon_2 = \frac{\sigma_2}{E} - v\frac{\sigma_1}{E} - v\frac{\sigma_3}{E}$$

$$\varepsilon_3 = \frac{\sigma_3}{E} - v\frac{\sigma_1}{E} - v\frac{\sigma_2}{E}$$

$$\Delta = \varepsilon_1 + \varepsilon_2 + \varepsilon_3$$

Stresses in thin-walled tube under internal pressure

Hoop stress

$$\sigma = \frac{pr}{t}$$

Axial stress (when tube carries longitudinal pressure loading)

$$\sigma = \frac{pr}{2t}$$

Bending stress in beam

$$\sigma = \frac{Mz}{I}$$

Bending deflection of beams

$$\delta = C_1 \left( \frac{Fl^3}{EI} \right)$$

Vibration frequency of beams

$$f = C_2 \sqrt{\frac{EI}{Ml^3}}$$

Buckling load of beams

$$F_{cr} = C_3 \left( \frac{EI}{l^2} \right)$$

$F(F_s)$ = normal (shear) component of force
$A$ = area
$u(w)$ = normal (shear) component of displacement
$\sigma(\varepsilon_n)$ = true tensile stress (nominal tensile strain)
$\tau(\gamma)$ = true shear stress (true engineering shear strain)
$p(\Delta)$ = external pressure (dilatation)
$v$ = Poisson's ratio
$E$ = Young's modulus
$G$ = shear modulus
$K$ = bulk modulus.
$\varepsilon_1, \varepsilon_2, \varepsilon_3 \ (\sigma_1, \sigma_2, \sigma_3)$ = principal strain (stress) components in 1, 2, and 3 directions
$r$ = tube radius
$t$ = wall thickness
$M$ = bending moment
$z$ = distance from neutral axis
$I$ = second moment of area of beam cross section
$C_1$ = constant depending on support and loading geometry
$l$ = length of beam
$C_2$ = constant depending on mass distribution and loading geometry
$M$ = vibrating mass
$C_3$ = constant depending on support and loading geometry

## Chapter 6

Young's modulus (longitudinal) of unidirectional composite

$$E = V_f E_f + (1 - V_f) E_m$$

Young's modulus (transverse) of unidirectional composite

$$E = 1 / \left\{ \frac{V_f}{E_f} + \frac{(1 - V_f)}{E_m} \right\}$$

Subscripts f and m = reinforcement and matrix, respectively
V = volume fraction

## Chapter 8

Nominal/true stress–strain

$$\sigma_n = \frac{F}{A_0}, \ \sigma = \frac{F}{A}, \ \varepsilon_n = \frac{u}{l_0} = \frac{l - l_0}{l_0}, \ \varepsilon = \int_{l_0}^{l} \frac{dl}{l} = \ln\left(\frac{l}{l_0}\right)$$

$A_0 l_0 = Al$ for *plastic* deformation; or for elastic or elastic/plastic deformation when $v = 0.5$. Thus,

$$\sigma = \sigma_n(1 + \varepsilon_n)$$

Also

$$\varepsilon = \ln(1 + \varepsilon_n)$$

Work of deformation, per unit volume

$$U = \int_{\varepsilon_{n,1}}^{\varepsilon_{n,2}} \sigma_n d\varepsilon_n = \int_{\varepsilon_1}^{\varepsilon_2} \sigma d\varepsilon$$

For linear-elastic deformation *only*

$$U = \frac{\sigma_n^2}{2E}$$

Hardness

$$H = F/A$$

$\sigma_n$ = nominal stress
$A_0(l_0)$ = initial area (length)
$A(l)$ = current area (length)
$\varepsilon$ = true strain

## Chapters 9 and 10

The dislocation yield-strength

$$\tau_y = \frac{2T}{bL}$$

$$\sigma_y = 3\tau_y$$

Grain-size effect

$$\tau_y = \beta d^{-1/2}$$

$T = $ line tension (about $Gb^2/2$)
$b = $ Burgers vector
$L = $ obstacle spacing
$\sigma_y = $ yield strength
$d = $ grain size
$\beta = $ constant

## Chapter 11

Shear yield stress

$$k = \sigma_y/2$$

Hardness

$$H \approx 3\sigma_y$$

Necking starts when

$$\frac{d\sigma}{d\varepsilon} = \sigma$$

Fully plastic bending moments of cross-sections

$M_p = \sigma_y a^3/4$ (square, side $a$)

$M_p = \sigma_y bd^2/4$ (rectangle, breadth $b$, depth $d$)

$M_p = 4\sigma_y r^3/3$ (round bar, radius $r$)

$M_p = 4\sigma_y (r_1^3 - r_2^3)/3$ (tube, external and internal radii $r_1$ and $r_2$)

$M_p = 4\sigma_y tr^2$ (thin-walled tube)

Shear-yielding torques

$\Gamma = 2\pi k r^3/3$ (round bar)

$\Gamma = 2\pi k(r_1^3 - r_2^3)/3$ (rube)

$\Gamma = 2\pi k r^2 t$ (thin-walled tube)

## Chapters 13 and 14

The stress intensity

$$K = Y\sigma\sqrt{\pi a}$$

Fast fracture occurs when

$$K = K_c = \sqrt{EG_c}$$

$a$ = crack length
$Y$ = dimensionless constant
$K_c$ = critical stress intensity or fracture toughness
$G_c$ = critical strain energy release rate or toughness

## Chapter 15

Tensile strength of brittle material

$$\sigma_{TS} = \frac{K_c}{\sqrt{\pi a_m}}$$

Modulus of rupture

$$\sigma_r = \frac{6M_r}{bd^2} = \frac{3Fl}{2bd^2}$$

$$\sigma_{TS} = \frac{\sigma_r}{\left\{2(m+1)^2\right\}^{1/m}}$$

Weibull equation

$$P_s(V) = \exp\left\{-\frac{V}{V_0}\left(\frac{\sigma}{\sigma_0}\right)^m\right\} \quad \text{(constant stress)}$$

$$P_s(V) = \exp\left\{-\frac{1}{\sigma_0^m V_0}\int_V \sigma^m dV\right\} \quad \text{(varying stress)}$$

$$P_s = 1 - P_f = 1 - (j - 0.375)/(n + 0.25)$$

$K_c$ = fracture toughness
$a_m$ = size of longest microcrack (crack depth for surface crack, crack half-length for buried crack)
$M_r$ = bending moment to cause fracture
$b, d$ = width and depth of beam
$F, l$ = fracture load and span of beam
$P_s$ = survival probability of component
$P_f$ = failure probability of component

$V$ = volume of component
$V_0$ = volume of test specimen
$\sigma$ = tensile stress in component
$\sigma_0$ = normalizing stress ($P_s = 1/e = 0.37$)
$m$ = Weibull modulus
$n$ = number of test results
$j$ = rank

## Chapters 17 and 18

Relation between total strain amplitude and number of reversals to failure, $2N_f$

$$\frac{\Delta\varepsilon^{tot}}{2} \approx \frac{\sigma_f'(2N_f)^b}{E} + \varepsilon_f' \, (2N_f)^c$$

Basquin's law (high cycle)

$$\frac{\Delta\sigma}{2} \approx \sigma_f' \, (2N_f)^b$$

Effect of tensile mean stress (strain approach)

$$\frac{\Delta\varepsilon^{tot}}{2} \approx \frac{(\sigma_f' - \sigma_m)(2N_f)^b}{E} + \varepsilon_f' \, (2N_f)^c$$

Effect of tensile mean stress (stress approach)

$$\frac{\Delta\sigma_{\sigma m}}{2} \approx \frac{\Delta\sigma(\sigma_f' - \sigma_m)}{2\sigma_f'}$$

Crack growth law

$$\Delta K = Y\Delta\sigma\sqrt{\pi a}$$

$$\frac{da}{dN} = A(\Delta K)^m$$

Failure by crack growth

$$N_f = \int_{a_0}^{a_f} \frac{da}{A(\Delta K)^m}$$

$$SCF_{eff} = S(SCF - 1) + 1$$

$\sigma_f' \, (\varepsilon_f')$ = true fracture stress (strain)
$b,c$ = constants ($b \approx -0.05$ to $-0.12$, $c \approx -0.5$ to $-0.7$)
$\Delta\sigma$ = stress range (tensile for crack growth law)
$\sigma_m$ = tensile mean stress
$\Delta K$ = stress intensity range
$A, m$ = constants

$SCF_{eff}$ = effective stress concentration factor
$SCF$ = stress concentration factor
$S$ = notch sensitivity factor

# Chapter 20
Creep rate

$$\dot{\varepsilon}_{ss} = A\sigma^n e^{-Q/\bar{R}T}$$

$\dot{\varepsilon}_{ss}$ = steady-state tensile strain-rate
$Q$ = activation energy
$\bar{R}$ = universal gas constant
$T$ = absolute temperature
$A, n$ = constants

# Chapter 21
Fick's law

$$J = -D\frac{dc}{dx}$$

Arrhenius's law

$$\text{Rate} \propto e^{-Q/\bar{R}T}$$

Diffusion coefficient

$$D = D_0 e^{-Q/\bar{R}T}$$

Diffusion distance

$$x \approx \sqrt{Dt}$$

$J$ = diffusive flux
$D$ = diffusion coefficient
$c$ = concentration
$x$ = distance
$D_0$ = pre-exponential factor
$t$ = time

# Chapter 22
Rate of diffusion creep

$$\dot{\varepsilon}_{ss} = \frac{C'\sigma e^{-Q/\bar{R}T}}{d^2}$$

$C'$ = constant; $d$ = grain size

## Chapter 24

Linear growth law for oxidation

$$\Delta m = k_L t; \quad k_L = A_L e^{-Q_L/\bar{R}T}$$

Parabolic growth law for oxidation

$$(\Delta m)^2 = k_p t; \quad k_p = A_p e^{-Q_p/\bar{R}T}$$

$\Delta m$ = mass gain per unit area

$k_L, k_P, A_L, A_P$ = constants

## Chapter 26

Uniform loss rate in $\mu m$/year

$$87,600 \, (w/\rho A t) = 3.27 \, (ai/zp)$$

$w$ = weight metal lost in mg

$\rho$ = metal density in Mg m$^{-3}$ or g cm$^{-3}$

$A$ = corroding area in cm$^2$

$t$ = exposure time in hours

$a$ = atomic weight of metal

$i$ = current density at corroding surface in $\mu A$ cm$^{-2}$

$z$ = number electrons lost per metal atom

## Chapter 28

True contact area $a \approx P/\sigma_y$

$P$ = contact force

## MAGNITUDES OF PROPERTIES

The listed properties for most structural materials are in the ranges shown in this table.

| | |
|---|---|
| Moduli of elasticity, $E$ | 2–200 GN m$^{-2}$ |
| Densities, $\rho$ | 1–10 Mg m$^{-3}$ |
| Yield strengths, $\sigma_y$ | 20–200 MN m$^{-2}$ |
| Toughnesses, $G_c$ | 0.2–200 kJ m$^{-2}$ |
| Fracture toughnesses, $K_c$ | 0.2–200 MN m$^{-3/2}$ |

# Physical Constants and Conversion of Units

## PHYSICAL CONSTANTS (SI UNITS)

| | |
|---|---|
| Absolute zero of temperature | $-273.2°C$ |
| Avogadro's number, $N_A$ | $6.022 \times 10^{23}\,mol^{-1}$ |
| Base of natural logarithms, e | 2.718 |
| Boltzmann's constant, $k$ | $1.381 \times 10^{-23}\,J\,K^{-1}$ |
| Charge on electron | $1.602 \times 10^{-19}\,C$ |
| Faraday's constant, $F$ | $9.649 \times 10^4\,C\,mol^{-1}$ |
| Gas constant, $R$ | $8.314\,J\,mol^{-1}\,K^{-1}$ |
| Magnetic moment of electron | $9.274 \times 10^{-24}\,A\,m^{-2}$ |
| Permittivity of vacuum, $\varepsilon_0$ | $8.854 \times 10^{-12}\,F\,m^{-1}$ |
| Planck's constant, $h$ | $6.626 \times 10^{-34}\,J\,s$ |
| Rest mass of electron | $9.110 \times 10^{-31}\,kg$ |
| Rest mass of neutron | $1.675 \times 10^{-27}\,kg$ |
| Rest mass of proton | $1.673 \times 10^{-27}\,kg$ |
| Unified atomic mass constant | $1.661 \times 10^{-27}\,kg$ |
| Velocity of light in vacuum | $2.998 \times 10^8\,m\,s^{-1}$ |
| Volume of perfect gas at STP (i.e., at 0°C; and under one atmosphere pressure, or $1.013 \times 10^5\,N\,m^{-2}$) | $22.4 \times 10^{-3}\,m^3\,mol^{-1}$ (approximate) |

## CONVERSION OF UNITS – GENERAL

| | | |
|---|---|---|
| Angle | 1 rad | $= 57.30°$ |
| Density | $1\,g\,cm^{-3}$ | $= 1\,Mg\,m^{-3}$ |
| Diffusion Coefficient | $1\,cm^2\,s^{-1}$ | $= 10^{-4}\,m^2\,s^{-1}$ |
| Dynamic Viscosity | 1 P | $= 0.1\,N\,m^{-2}\,s$ |
| Force | 1 kgf | $= 9.807\,N$ |
| | 1 lbf | $= 4.448\,N$ |
| | 1 dyn | $= 10^{-5}\,N$ |
| Length | 1 mile | $= 1.609\,km$ |
| | 1 foot | $= 304.8\,mm$ |
| | 1 inch | $= 25.4\,mm$ |
| | 1 Å | $= 0.1\,nm$ |
| Mass | 1 ton (long) | $= 1.017\,Mg$ |
| | 1 tonne | $= 1.000\,Mg$ |
| | 1 ton (short) | $= 0.908\,Mg$ |
| | 1 flask (Hg) | $= 34.50\,kg$ |
| | 1 lb mass | $= 0.454\,kg$ |
| | 1 troy oz | $= 31.103\,g$ |
| | 1 avoirdupois oz | $= 28.35\,g$ |
| | 1 carat | $= 0.20\,g$ |
| Stress Intensity | $1\,ksi\,\sqrt{in}$ | $= 1.10\,MN\,m^{-3/2}$ |
| Surface Energy | $1\,erg\,cm^{-2}$ | $= 1\,mJ\,m^{-2}$ |
| Temperature | 1°F | $= 0.556\,K$ |
| | 32°F corresponds to 0°C | |
| Volume | 1 UK gallon | $= 4.546 \times 10^{-3}\,m^3$ |
| | 1 US gallon | $= 3.785 \times 10^{-3}\,m^3$ |
| | 1 liter | $= 10^{-3}\,m^3$ |

## CONVERSION OF UNITS – STRESS AND PRESSURE[*]

| | MN m$^{-2}$ | dyn cm$^{-2}$ | lb in$^{-2}$ | kgf mm$^{-2}$ | bar | long ton in$^{-2}$ |
|---|---|---|---|---|---|---|
| MN m$^{-2}$ | 1 | $10^7$ | $1.45 \times 10^2$ | 0.102 | 10 | $6.48 \times 10^{-2}$ |
| dyn cm$^{-2}$ | $10^{-7}$ | 1 | $1.45 \times 10^{-5}$ | $1.02 \times 10^{-8}$ | $10^{-6}$ | $6.48 \times 10^{-9}$ |
| lb in$^{-2}$ | $6.89 \times 10^{-3}$ | $6.89 \times 10^4$ | 1 | $7.03 \times 10^{-4}$ | $6.89 \times 10^{-2}$ | $4.46 \times 10^{-4}$ |
| kgf mm$^{-2}$ | 9.81 | $9.81 \times 10^7$ | $1.42 \times 10^3$ | 1 | 98.1 | $63.5 \times 10^{-2}$ |
| bar | 0.10 | $10^6$ | 14.48 | $1.02 \times 10^{-2}$ | 1 | $6.48 \times 10^{-3}$ |
| long ton in$^{-2}$ | 15.44 | $1.54 \times 10^8$ | $2.24 \times 10^3$ | 1.54 | $1.54 \times 10^2$ | 1 |

## CONVERSION OF UNITS – ENERGY[*]

| | J | erg | cal | eV | BThU | ft lbf |
|---|---|---|---|---|---|---|
| J | 1 | $10^7$ | 0.239 | $6.24 \times 10^{18}$ | $9.48 \times 10^{-4}$ | 0.738 |
| erg | $10^{-7}$ | 1 | $2.39 \times 10^{-8}$ | $6.24 \times 10^{11}$ | $9.48 \times 10^{-11}$ | $7.38 \times 10^{-8}$ |
| cal | 4.19 | $4.19 \times 10^7$ | 1 | $2.61 \times 10^{19}$ | $3.97 \times 10^{-3}$ | 3.09 |
| eV | $1.60 \times 10^{-19}$ | $1.60 \times 10^{-12}$ | $3.83 \times 10^{-20}$ | 1 | $1.52 \times 10^{-22}$ | $1.18 \times 10^{-19}$ |
| BThU | $1.06 \times 10^3$ | $1.06 \times 10^{10}$ | $2.52 \times 10^2$ | $6.59 \times 10^{21}$ | 1 | $7.78 \times 10^2$ |
| ft lbf | 1.36 | $1.36 \times 10^7$ | 0.324 | $8.46 \times 10^{18}$ | $1.29 \times 10^{-3}$ | 1 |

1 kWh = 3.600 MJ. 1 ton coal = 8000 kWh (thermal) = 2500 kWh (electric)
1 barrel oil (0.133 ton) = 0.23 tce (ton coal equivalent). 1 ton oil = 1.7 tce

## CONVERSION OF UNITS – POWER[*]

| | kW | erg s$^{-1}$ | hp | ft lbf s$^{-1}$ |
|---|---|---|---|---|
| kW (kJ s$^{-1}$) | 1 | $10^{10}$ | 1.34 | $7.38 \times 10^2$ |
| erg s$^{-1}$ | $10^{-10}$ | 1 | $1.34 \times 10^{-10}$ | $7.38 \times 10^{-8}$ |
| hp | $7.46 \times 10^{-1}$ | $7.46 \times 10^9$ | 1 | $5.50 \times 10^2$ |
| ft lbf s$^{-1}$ | $1.36 \times 10^{-3}$ | $1.36 \times 10^7$ | $1.82 \times 10^{-3}$ | 1 |

[*]To convert row unit to column unit, multiply it by the number at the row–column intersection: for example,
1 MN m$^{-2}$ = 10 bar

# References

Ashby, M.F., 2005. Materials selection in mechanical design, third ed. Elsevier.

Ashby, M.F., Cebon, D., 1996. Case studies in materials selection. Granta Design.

Ashby, M.F., Johnson, K., 2002. Materials and design—the art and science of material selection in product design. Elsevier.

ASM, 1999. Metals handbook, 2nd desktop edition. ASM.

Bowden, F.P., Tabor, D., 1950. The friction and lubrication of solids, Part 1. Oxford University Press.

Bowden, F.P., Tabor, D., 1965. The friction and lubrication of solids, Part 2. Oxford University Press.

British Standards Institution, 1993. BS 7608: Code of practice for fatigue design and assessment of steel structures. BSI.

Broek, D., 1989. The practical use of fracture mechanics. Kluwer.

Calladine, C.R., 1985. Plasticity for engineers. Ellis Horwood.

Chapman, P.F., Roberts, F., 1983. Metal resources and energy. Butterworths.

Charles, J.A., Crane, F.A.A., Furness, J.A.G., 1997. Selection and use of engineering materials, third ed. Butterworth-Heinemann.

Cottrell, A.H., 1964. The mechanical properties of matter. Wiley.

Cottrell, A.H., 1977. Environmental economics. Edward Arnold.

Crawford, R.J., 1998. Plastics engineering, third ed. Butterworth-Heinemann.

Davies, G., 2004. Materials for automotive bodies. Elsevier.

Fontana, M.G., 1987. Corrosion engineering, third ed. McGraw-Hill.

Frost, H.J., Ashby, M.F., 1982. Deformation mechanism maps. Pergamon.

Gale, W., Totemeier, T., 2003. Smithells metals reference book, eighth ed. Elsevier.

Gordon, J.E., 1988. The new science of strong materials, or why you don't fall through the floor, second ed. Princeton University Press.

Gordon, J.E., 2003. Structures: Or why things don't fall down. Da Capo Press.

Hertzberg, R.W., 1996. Deformation and fracture of engineering materials, fourth ed. Wiley.

Hull, D., 2001. Introduction to dislocations, fourth ed. Butterworth-Heinemann.

Hull, D., Clyne, T.W., 1996. An introduction to composite materials, second ed. Cambridge University Press.

Hutchings, I.M., 1992. Tribology: Functions and wear of engineering materials. Butterworth-Heinemann.

**467**

Kittel, C., 1996. Introduction to solid state physics, seventh ed. Wiley.

Kubaschewski, O., Hopkins, B.E., 1962. Oxidation of metals and alloys, second ed. Butterworths.

Lawn, B.R., 1993. Fracture of brittle solids, second ed. Cambridge University Press.

Lewis, P.R., Reynolds, K., Gagg, C., 2003. Forensic materials engineering: Case studies. CRC Press.

Llewellyn, D.T., Hudd, R.C., 1998. Steels: Metallurgy and applications, third ed. Butterworth-Heinemann.

McEvily, A.J., 2002. Metal failures. Wiley.

Murakami, Y., 1987. Stress intensity factors handbook. Pergamon.

Polmear, I.J., 1995. Light alloys, third ed. Butterworth-Heinemann.

Powell, P.C., Ingen Housz, A.J., 1998. Engineering with polymers, second ed. Stanley Thornes.

Schijve, J., 2001. Fatigue of structures and materials. Kluwer.

Seymour, R.B., 1987. Polymers for engineering applications. ASM International.

Shewmon, P.G., 1989. Diffusion in solids, second ed. TMS Publishers.

Ward, I.M., 1983. Mechanical properties of solid polymers, second ed. Wiley.

Young, W.C., Budynas, R.G., 2001. Roark's formulas for stress and strain, seventh ed. McGraw-Hill.

# Index

*Note*: Page numbers followed by *f* indicate figures and *t* indicate tables.

**469**

CPI Antony Rowe
Eastbourne, UK
June 22, 2015